Evolutionary Games in Natural, Social, and Virtual Worlds

Evolutionary Games in Natural, Social, and Virtual Worlds

DANIEL FRIEDMAN AND BARRY SINERVO

OXFORD
UNIVERSITY PRESS

Oxford University Press is a department of the University of Oxford. It furthers the University's objective of excellence in research, scholarship, and education by publishing worldwide. Oxford is a registered trade mark of Oxford University Press in the UK and certain other countries.

Published in the United States of America by Oxford University Press
198 Madison Avenue, New York, NY 10016, United States of America.

Library of Congress Cataloging-in-Publication Data
Names: Friedman, Daniel, 1947– author. | Sinervo, Barry, author.
Title: Evolutionary games in natural, social, and virtual worlds / Daniel Friedman and Barry Sinervo.
Description: Oxford ; New York : Oxford University Press, [2016] | Includes bibliographical references and index.
Identifiers: LCCN 2015034942 | ISBN 9780199981151 (alk. paper)
Subjects: LCSH: Game theory. | Evolution.
Classification: LCC HB144 .F746 2016 | DDC 519.3–dc23 LC record available at http://lccn.loc.gov/2015034942

ISBN 978–0–19–998115–1

9 8 7 6 5 4 3 2 1
Typeset in Arno
Printed on acid-free paper
Printed in the United States of America

CONTENTS

PREFACE

Classic evolutionary theory goes back to biologists Charles Darwin (1859) and Alfred Wallace (1858), and to precursors such as economist Thomas Malthus (1798). In the 20th century that theory became the unifying principle of biology, and is sometimes referred to as the neo-Darwinian Synthesis or modern synthesis. **Classic game theory**, founded in 1944 by mathematician John von Neumann and economist Oskar Morgenstern, studies strategic interaction among agents with differing objectives. As developed in the second half of the 20th century by numerous researchers (including twelve Nobel laureates), game theory focuses on equilibrium. It provides a common language across many of the social sciences.

Evolutionary game theory (EGT) is a cross-fertilization of the two classic theories. EGT focuses on how fitness emerges from strategic interaction, and on the resulting evolutionary dynamics. Launched by biologist John Maynard Smith and mathematician George Price in the 1970s, EGT has been developed in the last few decades by theorists in biology, social science, mathematics, and, most recently, in computer science and engineering. In the last two decades, EGT has been trumpeted in numerous surveys and special issues of academic journals and in several textbooks.

The excitement comes in part from the elegance and sweep of EGT, but perhaps in larger part from the promise that applications of EGT will change the intellectual trajectory of life sciences, social sciences, and engineering. However, that promise is not yet realized—applications so far are scattered, and most applied researchers find EGT inaccessible.

This book is intended to help rectify the situation—our aim is to introduce EGT to applied researchers in all fields. Part I assembles many of the most useful techniques and presents them in a manner accessible to graduate students (and advanced undergraduates) in biology, economics, engineering, and allied disciplines. In particular, it shows how to take EGT models to the data. Part II collects exemplary applications from many fields, and exposits them in a consistent manner. These chapters are intended to serve as templates for applied work everywhere.

The book grew out of a class at University of California, Santa Cruz that mixed PhD students from economics, computer science, and biology (plus a few strays from other disciplines) with advanced undergraduates. Students formed teams, usually led by graduate students, to pursue original research. They applied the techniques they learned to problems in biology and ecology, to questions in economics and politics and other social sciences, and even to issues arising in virtual worlds created by computer scientists. Some of these projects eventually led to solid academic publications.

In that same spirit, we hope that individual researchers and research teams across many disciplines will read this book as a monograph that inspires new projects. The book also can be used as a primary or supplementary text for classes in game theory and applications.

Each chapter begins with an overview, written in conversational style, that highlights the issues and techniques to be covered. There are already many valuable books and innumerable articles on most of our topics, so we close each chapter with a general discussion and notes that point out connections with existing literature. A set of exercises (many recycled from homework assignments, exams, and student projects) and a bibliography round out most chapters.

Our students, and our intended readers, vary considerably in their mathematical preparation. We assume familiarity with basic calculus, linear algebra, and probability theory, and try to build on that gradually throughout Part I. Several chapters have mathematical appendices to satisfy the curiosity of well-prepared readers; other readers can ignore them on first pass and return to them later if that seems worthwhile.

Throughout the book we develop a series of numerical tools for simulating games and drawing figures. These include Excel spreadsheets of discrete games of asexual and sexual creatures (Chapters 2–4), code written in [R] for ODE solvers of continuous games (Chapters 1–3), and a cellular automata package written in R with a graphical user interface (Chapter 6). These packages are available on the Friedman, Daniel, and Sinervo, Barry (2015), *Evolutionary Game Theory: Simulations*. UC Santa Cruz. Software. doi:10.7291/d1mw2x. Readers who have access to the commercial package Mathematica may also want to access the solver called Dynamo, created by Sandholm and coauthors available on their website.

Chapter 1 begins Part I, Basics. The chapter introduces evolution's basic dynamic elements: fitness, shares, state space geometry, the replicator equation, and stability concepts. The chapter briefly notes applications to bacteria, memory chips, internal organization of firms, taboos and conventions, and stock market strategies in settings where there is no strategic interaction. It also distinguishes memes from genes, and reviews simple haploid and diploid models of gene transmission. An appendix connects the replicator equation to relative entropy.

Chapter 2 takes a first look at strategic interdependence, known to biologists as frequency dependent selection. It distinguishes positive interaction effects (increasing returns or synergy) from negative effects (e.g., congestion). The chapter introduces Maynard Smith's famous Hawk-Dove game, and goes on to show that there are only two other generic types of 2×2 games: Dominant strategy (DS), illustrated with data on RNA viruses, and Coordination (CO). The remaining sections show how these 2×2 games represent edges of the state space in games with three alternative strategies. We include data on male side-blotched lizards, and show that they play a variant of rock-paper-scissors.

Chapter 3 analyzes more complex forms of interaction. It introduces helpful geometric techniques for analyzing 3×3 and higher-dimensional games and shows how to extend them to nonlinear games. It then considers two-population games, both the usual asymmetric analysis and own-population effects. The chapter also introduces more general dynamics than replicator, including several forms of monotone dynamics. Finally, it explains how to estimate the payoff matrix from panel data on *Uta stansburiana*, using maximum likelihood techniques and state-of-the-art multinomial logit and Dirichlet error models.

Chapter 4 connects dynamic steady states analyzed in previous chapters to the classic static concepts of Nash equilibrium and ESS. It applies these ideas to established models of Fisherian runaway sexual selection. A technical appendix makes accessible three main techniques of stability analysis: eigenvalues of (projected) Jacobian matrices, Lyapunov functions, and Liouville's theorem.

Chapter 5 shows how social structures can modulate strategic interaction. It covers Bergstrom's model of assortative matching and the Price equation. It also introduces Hadamard products of matrices that provide a unified description of transmission across generations via memes and genes, with or without crossover and recombination, including an example of haploid sexual selection drawn from Fisherian runaway.

Chapter 6 formalizes spatial strategic interaction as games on grids. It develops cellular automata examples ranging from Conway's classic Game of Life to complex cooperation, and revisits several earlier examples of assortative and dissassortative interactions in an explicit spatial framework. The chapter features simulations of viscous game dynamics with emergent features such as cooperative groups in public goods games.

Part II applies the techniques to a variety of phenomena in the natural, social, and virtual worlds. Chapter 7 documents general RPS interactions in mating systems for isopods, damselflies, fish, and birds. It applies Hadamard products to show how mate preferences can change one sort of game into another. A new co-evolutionary model describes predator/prey interactions with learning.

Chapter 8 opens with some general perspectives on mechanisms behind share dynamics in human society, ranging from diploid genetics to individual learning.

It reports laboratory data of profit-motivated human subjects choosing among alternative strategies in several simple games, develops alternative models of learning rules and decision rules, and demonstrates how to fit these models to the data.

Chapter 9 considers "plastic" strategies that respond to contingencies, and also life-cycle strategies. We classify the various usages of the term plasticity in game theoretic literature. We apply the framework to data collected on elephant seal mating behavior and life history at Año Nuevo beaches 1970–2012.

Chapter 10, contributed by computer science professor Manfred K. Warmuth, shows how replicator dynamics (in a setting with no frequency dependence) correspond to multiplicative updates studied in the context of online learning. The updates learn very quickly (a "blessing") but they also wipe out potentially valuable variety that may be important when the environment changes (the "curse"). The chapter presents several different techniques developed in machine learning to lift the curse, and also different techniques seen in Nature. The chapter asks what machine learning can learn from Nature and vice versa.

Chapter 11, contributed by engineering professor John Musacchio, analyzes congestion effects in road and packet-switched communication networks. It presents the Braess paradox and obtains results bounding the degree of inefficiency in equilibrium ("the price of anarchy"). The chapter includes a treatment of Pigovian taxes, of competition amongst profit-maximizing link-owners, and of analogies to electrical circuits.

Chapter 12, contributed by economics professor Matthew McGinty, applies evolutionary games to environmental economics and international trade. It shows how evolutionary game theory can be used for policy analysis.

Chapter 13 opens with a simple diagram capturing the tension that all social creatures face between self interest and the social good. It then reviews two famous devices for alleviating the tension, Hamilton's kin selection and Trivers' bilateral reciprocity, and demonstrates their isomorphism. Next it reviews the "kaleidoscope" niches that our ancestors inhabited in Pleistocene times, emphasizing the intense selection for flexible forms of cooperation within stable groups of individuals. It then presents humans' novel coping device, the moral system, which builds on but goes well beyond the earlier devices. It shows how the content of the moral code evolves on short timescales to adapt to novel environments, and concludes with a brief discussion of new adaptations (e.g., egalitarian to hierarchical moral codes) to accommodate pastoral life styles ten thousand years ago.

Chapter 14 meditates on the classic theme of speciation. It first presents the idea of evolutionary branching in adaptive dynamics. After discussing morph loss, it notes cellular automata models that create boundaries between proto-species. It concludes with some standard bifurcation models and a repurposed Hotelling model as useful ingredients for further work.

Besides an index, the last part of the book includes a jargon glossary, and a summary of mathematical notation. As always, the idea is to make the material accessible to readers from different disciplines.

We are pleased to acknowledge the many debts we incurred in writing this book.

Most directly, we are indebted to our colleagues Matthew McGinty, John Musacchio, and Manfred Warmuth for contributing (respectively) Chapters 12, 11, and 10. Their chapters provide much-needed disciplinary diversity, and they worked hard with us to make their ideas accessible to all our readers. Shlomo Moran, Shay Moran, and Michal Derezinski helped debug Chapter 10. We are also indebted to Marek Czachor and Maciej Kuna who clarified our thinking on the connections (developed in the appendix of Chapter 1) between relative entropy and replicator dynamics.

Drafts of all chapters were reviewed by outside experts. We received very useful suggestions from Ted Bergstrom, Bill Sandholm, Franjo Weissing, and Stu West, as well as others whose names were not disclosed to us.

Generations of UCSC students helped shape this book, and the final versions of the chapters owe much to them. In particular, Jacopo Magnani developed the ML estimator for Chapter 3, Morgan Maddren wrote software for Chapter 6, Carla Sette, James Lamb, and Dustin Polgar analyzed the RoPaScLiSp game in Chapter 7, Sameh Habib helped analyze the e-seal data in Chapter 9, Sean Laney wrote the index, and Christina Louie wrote the glossaries and checked equations.

Of course, we own all remaining errors and obscurities, whose number would be far greater without their help.

The authors are grateful to many people at Oxford University Press. By his enthusiasm and persistence, acquisitions editor Terry Vaughn convinced us to sign with Oxford. After Terry's retirement, Scott Parris exhibited great skill and patience (major virtues in an acquisitions editor!) in extending deadlines, helping us find chapter reviewers, and working with us to improve the book. Cathryn Vaulman made preparing the manuscript almost painless. Sunoj Sankaran at Newgen took charge of the production process, and Jenifer Rod at OUP managed the marketing. To all, much thanks!

Finally, Dan's greatest debt is to his wife, Penny Hargrove, for her patience and forbearance as the book neared completion. Barry's greatest debts are to his wife, Jeanie Vogelzang, for similar support and patience during the entire process of writing, and to his son Ari, whose enthusiasm for biology continuously introduces new material on evolutionary games to his Isi (Finnish for dad).

PART ONE

Basics

1

Population Dynamics

Evolutionary game theory employs techniques originally developed in very different branches of mathematics and social sciences, as well as biology. To get started, we will put to one side ideas from classic game theory—they will begin to appear in Chapter 2—and introduce the basic ideas of evolutionary dynamics.

We begin by discussing fitness, a notoriously slippery concept. Drawing on standard analytic geometry and probability theory, we introduce state spaces that capture the current state of an evolving population. Our interest is in the distribution of traits or behaviors within each population. The traits may be biological and governed by genes, or may instead have a social or a virtual basis that can be described by "memes."

The next idea is a dynamical process that specifies how the state evolves (i.e., how the population distribution changes over time). Here we draw on basic calculus and introduce difference equations and differential equations. Our focus is on replicator dynamics, a particular but widely used evolutionary process. Alternatives to replicator dynamics will be considered in later chapters, but we introduce basic ideas of diploid population genetics at the end of this chapter.

This last major topic is more specialized, but sexier. For humans and most other creatures of interest, an individual's genes come from two parents (sexual tranmission), and the genes are deployed in paired copies (diploid). We discuss the necessary modifications to replicator dynamics that these complications entail, and cover some other basic ideas from genetics. The first appendix spells out the mathematical details; it is accessible to readers with patience and a taste for algebra.

The second appendix takes a deeper look at the usual ("haploid") replicator equation. It uses techniques from calculus, differential equations, and probability theory to point out connections to entropy and to explore the continuous time limit of discrete replicator dynamics. To our knowledge, these connections have not previously appeared in published literature. Of course, this appendix can be skipped with no loss of continuity by readers not interested in such subtleties.

As with most chapters of this book, the final two sections are called exercises and notes. All readers should look over the exercises, since they often point to useful extensions of the basic ideas. Working out the exercises will, we predict, actually help build readers' strength and skill in applying the techniques.

The notes provide, section by section, additional details, tangents, and guides to the literature.

1.1 FITNESS

Evolution is about how populations change over time. To begin, suppose that there are two or more alternative strategies or traits $i = 1,2,\ldots$ in a single population. Let s_i denote the fraction of the population (or *share*) using strategy i. The basic evolutionary question is how each s_i changes over time.

Fitness is the key. Denoted by w_i, it refers to the ability of a strategy (or trait) to increase its share. In some situations fitness can be measured directly, for example, via tracking gene frequencies from one cohort to the next. In other cases, we can't observe fitness directly, but there might be a good observable proxy, such as profit or foraging success.

The idea of shares and fitness is best understood via examples. In a population of *Escherichia coli* bacteria, some fraction $s_1 > 0$ is resistant to penicillin, and the remaining fraction, $s_2 = 1 - s_1 > 0$ is not resistant. If there is no penicillin around, then the non-resistant trait has higher fitness, so $w_1 < w_2$, since that trait allows a more efficient metabolism. Over time, s_1 decreases and s_2 increases. On the other hand, when there is a high concentration of penicillin in the neighborhood then $w_1 >> w_2$, and s_1 quickly rises towards 1.0.

Suppose that there are only three types of memory chips that can be used in a computer. In some environments, their fitnesses w_1, w_2, w_3 depend only on the relative speeds. In other environments, price or capacity or some combination might determine fitness. The chip with highest fitness will gain share, and the share will dwindle for the chip with lowest fitness. Of course, according to Moore's famous law, after a few months a new chip will be introduced that has yet higher speed and greater capacity at no higher price. It will gain share at the expense of existing chips, until a still fitter chip appears.

The idea of fitness also applies to economic questions. For example, an organization can use an old- fashioned hierarchy ($i = 1$: workers report to foremen who report to floor managers who report to assistant vice presidents who report to VPs who report to EVPs who report to the CEO) or a flexible governance structure ($i = 2$: teams form for specific tasks and dissolve when they are completed). In a new industry like smartphone software, the flexible stucture is likely to generate greater profits, so $w_1 < w_2$ and s_2 will increase. In a mature industry like coal mining, it is more likely that $w_1 > w_2$, and s_1 approaches 1.0.

Social customs can be analyzed from the same perspective. A taboo against eating pork products ($i = 1$, say) kept its adherents healthier in historical times and places when trichonosis and other diseases were prevalent. Thus it achieved high fitness and in some places became universally adopted, so there $s_1 = 1.0$.

The custom of primogeniture ($i = 1$: the oldest surviving son inherits the entire family estate) was rare in many times and places compared to "fair" division among all surviving sons ($i = 2$). However, in feudal Europe, a single large estate could better protect itself than a collection of smaller estates, so once again $w_1 > w_2$, and s_1 rose towards 1.0.

In modern times, stock market investors choose between "fundamentalist" strategies (e.g., you buy a stock whose price is below your estimate of its fundamental value, the present value of its future earnings) and "chartist" strategies (e.g., you buy a stock whose price recently exceeded 1.3 times its average over the last 6 months). Warren Buffett and other fundamentalists gained a larger share of invested wealth in most of the late 20th century, but one can find other times and places in which the chartists do better and increase their share.

One last example. Since around 1800, economies oriented towards market exchange have typically generated increased gross domestic product (GDP) faster than centrally planned economies or feudal economies. GDP is a good proxy for fitness, since greater GDP gives a country important military and diplomatic advantages. Hence the share of the world's economy dominated by the market has increased over time and now is, by most definitions, at least $s_1 = 0.8$. On the other hand, when Russia attempted a quick transition to a market economy in the early 1990s, the results were disastrous, arguably because Russia lacked the necessary legal and political infrastructure. Here as in previous examples, the fitness of a strategy is not absolute; it depends on the environment.

1.2 TRADEOFFS AND FITNESS DEPENDENCE

We defined the fitness of a trait as the ability to increase its share over time. But where does that ability come from, and how does it depend on environment and other factors? Traits often bring some advantages and some disadvantages relative to alternative traits, and then assessing fitness comes down to analyzing the tradeoffs.

A financial tradeoff

Investors trade off risk against average return. According to standard theory (backed by a fair amount of historical evidence), some financial assets (such as U.S. Treasury bills) reliably pay a steady income, while other assets (such as newlylisted stocks) on average pay a higher return, but are riskier in the sense that the return sometimes is low or even negative.

The tradeoff between risk and return depends on the investor's age. A young professional should target high returns and accept a fair amount of risk, because things should average out over the decades remaining before retirement. Her fitness or payoff (economists might try to measure this as the expected utility of consumption over her lifetime) is highest with a fairly risky portfolio. On the other hand, a retiree whose income mainly comes from his investments is much less able to tolerate risk, so he should accept a lower average return. His fitness is maximized with a much lower-risk portfolio.

A biological tradeoff

The fitness of any biological trait has two components, survival and fecundity. Individuals with a given trait have to survive to maturity if they are to pass the trait on to later generations (i.e., to their progeny). The more survivors, the greater the fitness. Second and equally important, the surviving individuals have to be fecund, and actually produce viable progeny.

Often there is a tradeoff between the two components, and different strategies may tilt the tradeoff one way or the other. Thus what biologists call an r-strategy (say $i = 1$) is to produce large numbers of progeny by provisioning them relatively poorly. For example, certain plant species produce vast numbers of tiny seeds that don't germinate well except on open ground. By contrast, what biologists call a K-strategy (which we will label $i=2$) produces relatively few but well-provisioned progeny. For example, other sorts of plants produce relatively few seeds, but with devices like shells or hooks that help them germinate and get established even with a thick ground cover. It is not hard to think of similar examples for birds (a few large well-tended eggs vs. lots of small quick-hatching eggs), and a host of other creatures.

Some mathematics can shed light on the tradeoff. Population growth is governed by the logistic equation

$$\dot{N} = rN\left(1 - \frac{N}{K}\right). \tag{1.1}$$

The instantaneous change $\dot{N} = \frac{dN}{dt}$ in the population size N is a function of the maximal growth rate r and carrying capacity K, which are fixed parameters in equation (1.1). Consider r-strategists with a high maximal growth rate r_1 but low carrying capacity K_1 versus K-strategists with a higher carrying capacity $K_2 > K_1$ but lower maximal growth rate $r_2 < r_1$. Compare the logistic equations for the two populations, N_1 and N_2,

$$\dot{N}_1 = r_1 N_1 \left(1 - \frac{N_1}{K_1}\right) \tag{1.2}$$

$$\dot{N}_2 = r_2 N_2 \left(1 - \frac{N_2}{K_2}\right). \tag{1.3}$$

The dynamics are simple if each type resides in a separate population, initially small. The r-strategists will increase quickly at first and soon approach the carrying capacity, K_1. The K-strategists will increase more slowly and but eventually will approach a larger population size $K_2 > K_1$. Figure 1.1 shows a case where the K-strategy does better most of the time.

Variations on equations (1.2) and (1.3) were used by Sinervo et al. (2000) to capture the frequency and density dependent game played out by lizards that lay small clutches of large eggs (K-strategist) versus large clutches of small eggs (r-strategist) in the same population. See the notes for more comments on such density-dependent games.

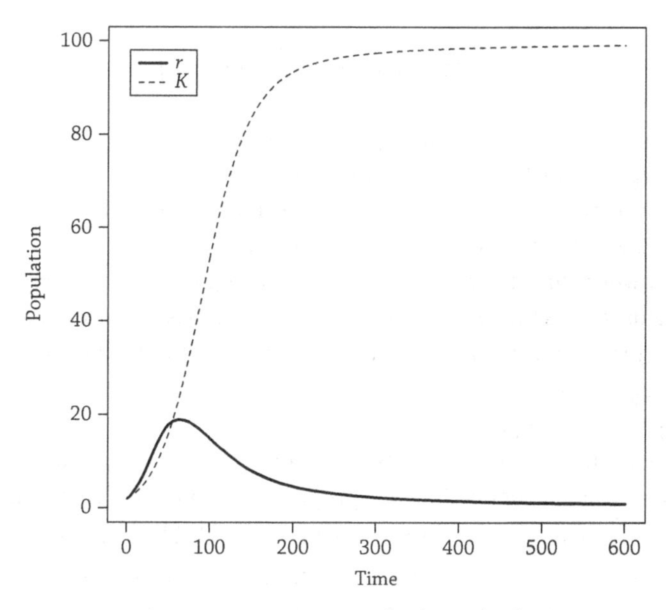

Figure 1.1 The rK-game in equations (1.4) and (1.5) solved for parameters ($r_1 = 0.08, r_2 = .05, K = 50, c_{11} = 2, c_{12} = c_{22} = c_{21} = 0.5$) that allow the K-strategy to prevail.

1.3 DEPENDENCE ON ENVIRONMENT, DENSITY, AND FREQUENCY

The reader doubtless can think of many other tradeoffs of a similar nature—speed versus accuracy, functionality versus elegance, income versus personal satisfaction, to name a few. Of course, any dominated strategy (one that does worse in both respects) is always less fit, but what determines fitness among strategies that are not dominated?

We've noted all along that the environment or context matters, beginning with the impact of antibiotics on bacterial fitness and more recently we mentioned how the of stage of life impacts the fitness of investment strategies. However, fitness is not determined entirely by such external factors. There are two crucial internal factors as well. Biologists refer to them as density dependence and frequency dependence.

We can begin to understand these factors by considering a variation on the equations used to describe r- versus K-strategists, when both types are mixed in a single population. Here we must track the raw numbers in a population, and also look at the shares $s_1 = \frac{N_1}{N_1+N_2}$ and $s_2 = 1 - s_1 = \frac{N_2}{N_1+N_2}$. The system is described by the coupled differential equations

$$\dot{N}_1 = r_1 N_1 \left(1 - c_{11}\frac{N_1}{K} - c_{12}\frac{N_2}{K}\right) \tag{1.4}$$

$$\dot{N}_2 = r_2 N_2 \left(1 - c_{22}\frac{N_2}{K} - c_{21}\frac{N_1}{K}\right) \tag{1.5}$$

where K is the overall carrying capacity. The parameters c_{11} and c_{22} specify the own-type competitive effects, which are often assumed to be 1.0 in the typical formulation of Lotka-Volterra competition, while c_{ij} for $i \neq j$ specify the competitive effect of the other type j on type i. In these equations, a K-strategist type has lower values for c, since lowering c is equivalent to raising K in the denominator. For example, in Figure 1.1, strategy 2 has a lower c than strategy 1's value $c_{11} = 2$, so it is the K-strategy (shown as a dotted line). Even though the r-strategy 1 increases share faster at first, it is eventually overtaken by the K strategy in this example.

Even the basic r-K model already incorporates density dependence. Fitness of strategy i can be defined as its proportional rate of change, that is, its growth rate $\dot{N}_i/N_i = \ln \dot{N}_i$. Thus in equations (1.2)–(1.3), we see that fitness is given by the expression $w_i = r_i \left(1 - \frac{N_i}{K_i}\right)$. Since this expression is decreasing in the population size (or density) N_i, we have what is called negative density dependence. Interested readers can easily check that, in the more complicated r-K model, fitness again depends negatively on density.

Frequency dependence occurs when fitness w_i depends on the current distribution of strategies within the population. Such dependence is exemplified in the coupled r-K model (1.4)–(1.5) when the cross-coefficients $c_{12} \neq 0$ and/or $c_{21} \neq 0$. Frequency dependence will be central to all later chapters of this book, but we put it aside for the rest of this chapter.

1.4 STATE SPACE GEOMETRY

An evolving system is modeled by listing the alternative strategies $i = 1,2,\ldots,n$, and keeping track of the number of individuals N_i employing them in each population, as well as the total population $N_T = \sum_{j=1}^{n} N_j$.

Most work in evolutionary game theory assumes away density dependence. The justification is either that the total population size hovers near a steady state value N_T, or that the important features of the population are insensitive to N_T within the relevant range. For the most part, we will follow tradition, but occasionally we will note cases where density dependence is important, and expand our analysis accordingly.

When we ignore density dependence and treat population size as constant, the analysis simplifies and allows us to focus our attention on the shares $s_i = \frac{N_i}{N_T}$. Why? Because then fitness w_i can be defined as the growth rate of shares, as a chapter-end exercise invites you to verify. For standard applications, then, the *state* of the system is the vector of shares $\mathbf{s} = (s_1, s_2, \ldots, s_n)$, and an evolutionary model seeks to describe how the state $\mathbf{s}$ changes over time.

The smallest number of individuals choosing a strategy is zero so $N_i \geq 0$, and any relevant population contains some individuals so $\sum_{i=1}^{n} N_i > 0$. It follows that shares are non-negative ($s_i \geq 0$) and that they sum to $\sum_{i=1}^{n} s_i = \frac{\sum_{i=1}^{n} N_i}{\sum_{j=1}^{n} N_j} = 1$. Therefore any possible state of a given population is represented by a point in the *n-simplex*, the set of vectors whose components are non-negative and sum to 1. The evolutionary process is visualized as the state moving over time within its simplex.

Simplexes have a very nice geometry. To begin with the simplest case, consider two alternative strategies for individuals in a single population, as in Figure 1.2. The raw population vector is $\mathbf{N} = (N_1, N_2)$, a point in the positive quadrant of the 2-dimensional plane. Multiplying or dividing the components of $\mathbf{N}$, any positive number can be visualized as a move along the ray from the origin (0,0) through $\mathbf{N}$. To get the state s corresponding to N, simply divide the components of N by the sum $N_1 + N_2$. This picks out the point where the ray picks out the point where that ray intersects the line segment $s_1 + s_2 = 1$. Hence the simplex is the 1-dimensional line segment that connects the point $\mathbf{s} = (s_1, s_2) = (1,0)$, where

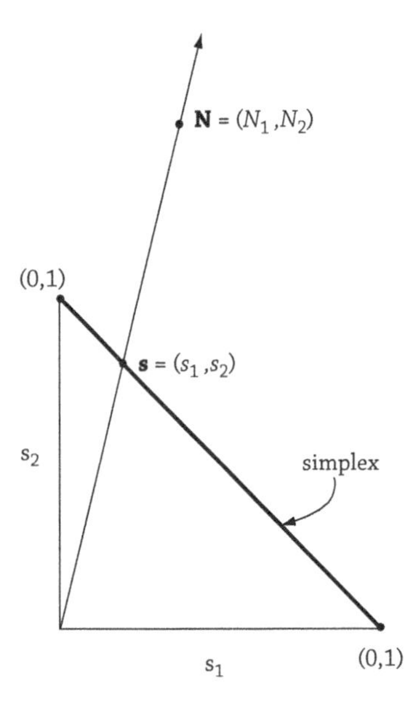

Figure 1.2 The geometry of the two-strategy simplex.

everyone uses the first strategy, to the point $(s_1, s_2) = (0, 1)$, where everyone uses the second strategy. Alternatively, we can visualize this simplex by looking only at its first component s_1, with s_2 understood to be $1 - s_1$, as in Panel A of Figure 1.3.

Often we will consider three possible strategies in a single population. The raw population vector $\mathbf{N} = (N_1, N_2, N_3)$ now is a point in the positive orthant of 3-dimensional space, and the state $\mathbf{s} = (s_1, s_2, s_3)$ lies in a 2-dimensional subspace satisfying the constraint $s_1 + s_2 + s_3 = 1$. The simplex now is the equilateral triangle whose corners are the points $\mathbf{e}_1 = (1, 0, 0)$, $\mathbf{e}_2 = (0, 1, 0)$, and $\mathbf{e}_3 = (0, 0, 1)$. In Panel B of of Figure 1.3, this triangle is laid out flat on the page. Each corner represents a state in which everyone uses the same strategy i, and the opposite edge represents all states in which only the other two strategies are used. The share of strategy i increases steadily as we move back towards that corner via line segments parallel to the opposite edge.

The coordinates of any state $\mathbf{s}$ describe it as a weighted average of the corners of the triangle, $\mathbf{s} = (s_1, s_2, s_3) = s_1\mathbf{e}_1 + s_2\mathbf{e}_2 + s_3\mathbf{e}_3$. These sort of coordinates are known as Barycentric, which perhaps explains why they are so loved by one of the authors. Occasionally we will represent the same simplex in a different way. The projection of the simplex into the 2-dimensional $s_1 - s_2$ plane is the right triangle $\{(s_1, s_2) : s_i \geq 0, s_1 + s_2 \leq 1\}$. It also represents the state because, by setting $s_3 = 1 - s_1 - s_2$, we recover the omitted coordinate.

What if there are four alternative strategies for a single population? It is hard to visualize the 4-dimensional space in which the N_i's reside, but the simplex

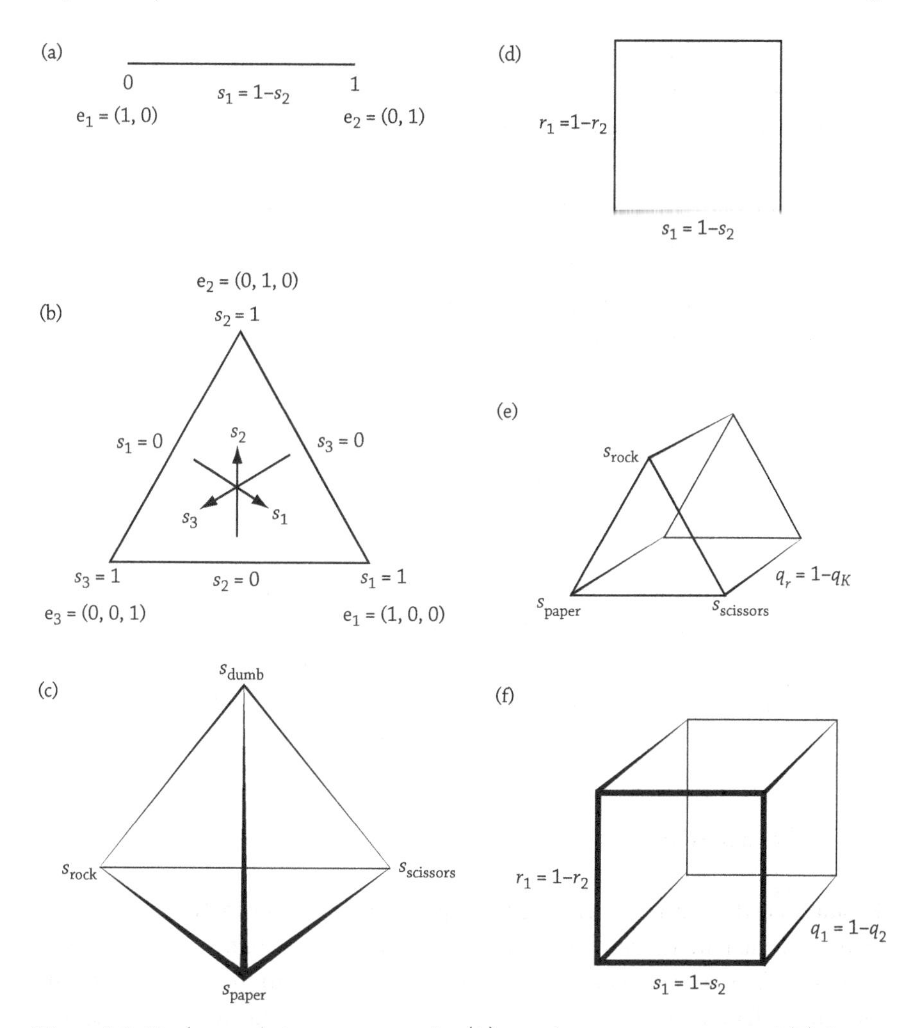

Figure 1.3 Single-population state spaces for (A) two alternative strategies, and (B) three alternative strategies. Panel C is for a game with four strategies, called rock, paper, scissors, and dumb, discussed in Chapter 3. State spaces for simple two-population games are shown in D and E, while F is the state space for a game where each of three populations has two alternative strategies.

is 3-dimensional because it satisfies the linear constraint $s_1 + s_2 + s_3 + s_4 = 1$. The corners of this simplex are the points $\mathbf{e}_1 = (1,0,0,0)$, $\mathbf{e}_2 = (0,1,0,0)$, $\mathbf{e}_3 = (0,0,1,0)$, and $\mathbf{e}_4 = (0,0,0,1)$, and they can be arranged in 3-dimensional space to form the equilateral tetrahedron shown in Panel C of Figure 1.3. Again, each corner represents a pure state where everyone uses the indicated strategy, and the opposite face (a triangle) represents a mixed state where only the other three strategies are used. Edges represent states in which two strategies are used and the other two are not. Barycentric coordinates are natural here as well.

It is hard to draw on the page the $(n-1)$ dimensional simplex corresponding to $n=5$ or more alternative strategies in a single population. Some readers may be able to visualize the corners, edges, and other aspects of these higher dimensional simplexes, but if you can't, don't worry! We will seldom need them in this book.

Later chapters sometimes will consider evolutionary games involving two or more distinct populations. For example, in the biological realm, a predator population may co-evolve with a prey population, or a bumblebee population with a flower population. In a digital domain, the population of content providers may co-evolve with the population of end-users (consumers). Panel D illustrates the simplest such state space—one population has two alternative strategies with shares s_1 and $s_2 = 1 - s_1$, and it interacts with a second population also with two alternative strategies with shares denoted r_1 and $r_2 = 1 - r_1$. The state space is the cartesian product of the state spaces for each population. In the present case, we have the cartesian product of two line segments, which is the unit square.

Panels E and F show two other possibilities. One population has three alternative strategies and the other has two in Panel E, so the result is a triangular prism, the cartesian product of a triangle and a line segment. In Panel F there are three distinct populations, each with two alternative strategies. The state space is the cube, the cartesian product of three line segments.

1.5 MEMES AND GENES

Evolution is the central organizing principle of biology. Its genetic mechanisms, some of which are quickly summarized in Section 1.10, are now are fairly well understood. However, many of the examples already mentioned—Internet content providers and users, stock market investors, and food taboos, for example—are not governed by genetic mechanisms. There is no gene for markets or for stock market fundamentalism. Rather, individuals chose among alternative strategies, consciously or unconsciously, and over time the strategy shares change within the population.

This book examines regularities that apply to any sort of evolution, whether or not the alternative strategies are coded in genes. When genes are not directly involved, sometimes it is helpful to have an analogous word to describe the carriers of alternative strategies. Then, following Dawkins (1976), we will refer to *memes*.

Like genes, the fitness of memes has two components: persistence within individuals (survival) and transmission across individuals (fecundity). But these components work differently for memes than for genes.

Genes typically persist in individuals from birth to death. By contrast, an individual may switch memes many times within a single lifetime. Perhaps even more important, genes generally are transmitted vertically, from parents to their progeny. Memes can transmit vertically, but they can also transmit horizontally, from peer to peer (especially among adolescents!), or diagonally. They can even transmit reverse vertically: his kids persuaded one of the authors to buy and use a smartphone.

You might think that memes are more complicated to analyze because of all the possible transmission routes and individual switches. Yet we will soon see that genes can also be quite complicated, especially when two parents are involved in transmission. In any application, whether it involves genes or memes, you have to think through the components of fitness, and to identify the factors that seem most important.

Finding a useful simplification is often challenging, and it takes scientific insight and creativity. Things become more routine once you have built a simplified model, because then you can apply techniques explained in textbooks like the one in your hands.

1.6 FINITE POPULATIONS AND RANDOMNESS

For a finite population $N_T = \sum_{i=1}^{n} N_i < \infty$, the possible states are lattice points in the simplex; that is, each share in each population belongs to the set $\{0, \frac{1}{N_T}, \frac{2}{N_T}, \ldots, \frac{N_T-1}{N_T}, 1\}$. Even in continuous time, dynamics consist of *jumps* from one lattice point to another, and will generally be *stochastic*, that is, the jumps are random to some degree. To describe such dynamics, mathematicians begin with Markov processes.

For many purposes, models that assume an infinite population are very good approximations of dynamic behavior in large finite populations. In continuous time, such models feature changes in shares that are *smooth*, with no jumps, and are *deterministic*, with no randomness. This simplifies the mathematics considerably.

In this book we focus on smooth deterministic dynamic models. Changes in shares in continuous time are described by systems of ordinary differential equations (ODEs), and in discrete time by systems of difference equations. We occasionally will deal with applications in which randomness is important (e.g., genetic drift), and then we mention what happens when we model finite populations in more realistic detail. Also, we supplement a deterministic model with stochastic elements when we use data to estimate the model's parameters.

1.7 REPLICATOR DYNAMICS IN DISCRETE TIME

Consider now a simple model with two alternative strategies, $s = (s_1, s_2)$. Of course, shares are non-negative $(s_i \geq 0)$ and sum to unity $(s_1 + s_2 = 1)$. The initial state $s(0)=(s_1(0),s_2(0))$ is given. Suppose that the population is observed periodically (e.g., every summer) and the fitnesses W_1, W_2 are constant.

The discrete time replicator equation then describes the population over time. At times $t=0,1,2,\ldots$ the state is given by

$$s_i(t+1)=\frac{W_i}{\overline{W}}s_i(t), \quad i=1,2, \tag{1.6}$$

where $\overline{W}=s_1W_1+s_2W_2$ is the average fitness in the population.

The equation says that the share increases or decreases according to relative fitness. That is, the current share $s_i(t)$ is multiplied by its relative fitness $\frac{W_i}{\overline{W}}$. If relative fitness is greater than 1.0—(i.e., if fitness W_i is greater than average fitness $\overline{W}$)—then i's share will increase. It will decrease if relative fitness is less than 1.0.

For example, suppose $W_1 = 4$ and $W_2 = 3$ and the initial state is $s(0) = (s_1(0),s_2(0)) = (0.6,0.4)$. Then $\overline{W}=s_1W_1+s_2W_2=4(0.6)+3(0.4)=3.6$ and the next state has shares $s_1(1) = \frac{W_i}{\overline{W}}s_1(0) = \frac{4}{3.6}0.6 = 2/3 > 0.6$, and $s_2(1) = \frac{3}{3.6}0.4=1/3<0.4$.

Equation (1.6) is written for the case with only two alternative strategies. But the replicator equation looks exactly the same for any finite number $n > 2$ of alternatives: just replace $i=1,2$ by $i=1,2,\ldots,n$, and set $\overline{W}=\sum_{i=1}^{n}s_iW_i$.

Do replicator dynamics keep the state in the simplex? That would mean that at all times the shares are non-negative and sum to 1.0. If we have such a state at time t and apply (1.6), then we are asking: (a) are the shares still positive at time $t+1$, and (b) do they still sum to 1.0? Both answers are "Yes" in the last numerical example, but what assurance do we have more generally?

Recall that fitness W_i is defined as the expected number of descendents in the next generation. This can't be a negative number. Unless the entire species immediately vanishes, at least one trait (or strategy) must have positive fitness. Therefore, average fitness is positive and the ratio $\frac{W_i}{\overline{W}}$ is guaranteed to be non-negative. Thus we have an affirmative answer to (a): share i at time $t+1$ is the product of a nonnegative share at time t and a non-negative ratio, and so it must be non-negative.

Question (b) also has an affirmative answer, as shown by the following algebra:

$$\sum_{i=1}^{n}s_i(t+1)=\sum_{i=1}^{n}\frac{W_i}{\overline{W}}s_i(t)=\frac{1}{\overline{W}}\sum_{i=1}^{n}W_is_i(t)=\frac{1}{\overline{W}}\overline{W}=1.0. \tag{1.7}$$

Replicator dynamics are very straightforward, and can be taken as the definition of fitness. Yet there are situations where the natural dynamics are different, and then alternative dynamics must be considered. We will begin to do so in Chapter 3.

1.8 REPLICATOR DYNAMICS IN CONTINUOUS TIME

One limitation of (1.6) is that it only gives the shares at discrete time intervals. In some cases that is all you would want to know. For example, adult mayflies live only for a few weeks each year, and their progeny are dormant most of the year. So it is reasonable to focus on shares of alternative mayfly traits only once each year.

However, in many other applications you might want to know how shares change in continuous time. For example, advertisers track brand shares every month (or week or day), not just once a year, and you might want to know how the fundamentalist share in the stock market changes minute by minute when you are about to make a big trade.

To deal with shorter time intervals, we generalize (1.6) by writing

$$s_i(t+\Delta t) = \left[\frac{W_i}{\overline{W}}\right]^{\Delta t} s_i(t), \tag{1.8}$$

where Δt is the time interval of interest. This equation agrees with (1.6) when $\Delta t = 0$ or 1, but it also makes sense for intermediate values of Δt. Now take the natural logarithm of both sides to obtain

$$\ln s_i(t+\Delta t) = \Delta t\left[\ln W_i - \ln \overline{W}\right] + \ln s_i(t). \tag{1.9}$$

Subtract $\ln s_i(t)$ from both sides and divide both sides by Δt to obtain

$$\frac{\ln s_i(t+\Delta t) - \ln s_i(t)}{\Delta t} = \left[\ln W_i - \ln \overline{W}\right]. \tag{1.10}$$

To get the continuous time replicator equation, take the limit $\Delta t \to 0$ and the left-hand side of (1.10) becomes $\frac{d\ln s_i(t)}{dt} = \frac{1}{s_i}\frac{ds_i}{dt} = \frac{\dot{s}_i}{s_i}$. Using the notation $w_i = \ln W_i$ and $\bar{w} = \ln \overline{W}$, the right-hand side of (1.10) becomes $[w_i - \bar{w}]$. Multiply through by s_i and we have the elegant system of ordinary differential equations,

$$\dot{s}_i(t) = (w_i - \bar{w})s_i(t), \ i = 1, \ldots, n, \tag{1.11}$$

known as **continuous replicator dynamics.**

The derivation shows that the growth rate $\frac{\dot{s}_i}{s_i}$ of each share i is equal to its fitness w_i in excess of the population average $\bar{w}$. Here fitnesses are in logs, hence can be negative as well as positive, which often turns out to be an advantage in empirical work. Note that average fitness of the w's corresponds to the geometric mean fitness of the W's, since $\bar{w} = \sum_{i=1}^{n} s_i w_i$ is the natural log of $W_1^{s_1} W_2^{s_2} \cdots W_n^{s_n}$. See the first part of the Appendix B for further explanation of why $\bar{w}$ in (1.11) is the geometric rather than the arithmetic average.

It is easy to check that (1.11) keeps the state in the simplex. All s_i remain non-negative because no growth rate $\frac{\dot{s}_i}{s_i}$, whether positive or negative, can turn a positive number into a negative number. Summing (1.11) over all strategies, we get $\sum_{i=1}^{n} \dot{s}_i(t) = \sum_{i=1}^{n} (w_i - \overline{w}) s_i(t) = \sum_{i=1}^{n} w_i s_i(t) - \overline{w} \sum_{i=1}^{n} s_i(t) = \overline{w} - \overline{w} \cdot 1 = 0$. Thus there is no change over time in the sum of the shares, so it remains at 1.0.

Consider again the simple example $W_1 = 4$, $W_2 = 3$, and $s(0) = (0.6, 0.4)$. Then $w_1 = \ln 4 \approx 1.4$ and $w_2 = \ln 3 \approx 1.1$. We get $\bar{w} = 0.6 w_1 + 0.4 w_2 \approx 1.27$. Then equation (1.11) says that $\dot{s}_1(0) = (w_i - \bar{w}) s_i(t) \approx (1.4 - 1.27) 0.6 = 0.078$. Thus $s_1(1) \approx s_1(0) + 1 \cdot \dot{s}_1(0) \approx 0.678$, which is fairly close to the answer using discrete replicator, $s_1(1) = 2/3 \approx 0.667$.

Much more will be said later about replicator dynamics. The second part of Appendix B discusses connections with entropy. Later chapters will show how replicator dynamics work when fitnesses are not constant.

1.9 STEADY STATES AND STABILITY

Of special interest are situations where the system has settled down. These situations are easier to observe, and often tell us much of what we want to know about the system. This section will introduce the main ideas from the dynamic perspective: steady states, stability, and basins of attraction. Chapter 4 will link this dynamic perspective to the traditional static perspective of game theory, and will develop the ideas in much more detail.

A system is in **steady state** if the shares of all strategies are constant over time. That is, $\Delta s_i(t) = 0$ or $\dot{s}_i(t) = 0$ at all times $t \geq 0$. Inspection of (1.6) or (1.11) reveals that, under replicator dynamics, a state $s^* = (s_1^*, \ldots, s_n^*)$ is steady iff (shorthand for "if and only if") for each strategy i, either $s_i^* = 0$, or $W_i = \overline{W}$ (or both). That is, all surviving strategies achieve the same average fitness.

In the cases considered so far in this chapter, fitness is constant for each strategy. Except for the coincidence that two different strategies have exactly the same fitness, the steady states must be in pure strategies, at a corner (i.e., vertex) of the simplex. Later we will see that non-constant fitnesses (or even just two-parent genetics) routinely allow steady states in the interior of the simplex.

Not all steady states are equally important. In the simple example with $W_1 = 4$ and $W_2 = 3$, we saw that from the initial state $s(0) = (0.6, 0.4)$ the share s_1 increased over time and s_2 decreased. It turns out (as you will see from the chapter-end exercises) that over time the shares converge to the steady state $s^* = (s_1^*, s_2^*) = (1, 0)$, and the same is true from almost all other initial states. The system never approaches the other steady state $s^* = (0, 1)$ unless it started there.

In some sense, then, $s^* = (1,0)$ is a stable steady state and $s^* = (0,1)$ is unstable. Dynamical systems researchers have settled on the following definitions to formalize these ideas. They apply to any dynamical process (replicator or otherwise) and any number of alternative strategies.

A steady state s^* is **Lyapunov stable** if, for any initial state $s(0)$ sufficiently near s^*, the entire trajectory $\{s(t) : t \geq 0\}$ remains near s^*. The full definition involves epsilons and deltas to pin down what is meant by "near." It will not enlighten many readers so we relegate it to the notes. A steady state s^* is **locally asymptotically stable** if, for any initial state $s(0)$ sufficiently near s^*, the trajectory $\{s(t) : t > 0\}$ actually converges to s^*, denoted $s(t) \to s^*$ as $t \to \infty$. The open set of initial conditions from which the trajectory converges to a particular steady state s^* is called its **basin of attraction**. Thus an unstable steady state has an empty basin of attraction. If the entire interior of the simplex is included in the basin of attraction, then the steady state is **globally asymptotically stable**.

Figure 1.4 illustrates the simple numerical example. You can see that the steady state $s^* = (1,0)$ is Lyapunov stable, locally asymptotically stable and, indeed, globally asymptotically stable, while the other steady state $s^* = (0,1)$ is unstable in all three senses.

A warning is in order. Like many other authors, when we casually say "stable" we usually mean locally Lyapunov stable *and* asymptotically stable. Such steady states are also called "sinks" and are marked in our diagrams by a filled-in black dot. We usually say "neutrally stable" when we mean Lyapunov stable but not asymptotically stable, and mark such steady states by a circle with a dot inside ($\odot$). We will use the jargon more carefully when it is important to do so.

In state spaces of dimension 2 or more, an unstable steady state can have an open set of trajectories that diverge from it, but still have a lower-dimensional set of trajectories (called the "saddle path" if it is 1-dimensional or more generally called the "stable manifold") that converge to it. Such steady states are called "saddles" and are marked by a circled x ($\otimes$). If all trajectories diverge from it, the steady state is called a "source" and marked with an open circle.

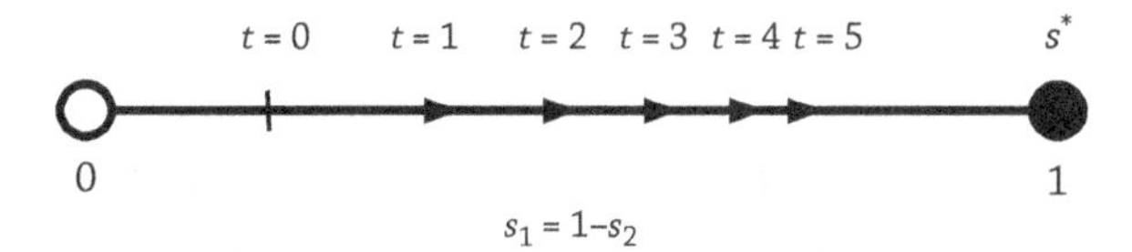

Figure 1.4 Phase portrait for continuous replicator (1.11). Fitnesses are $w_1 = 4$ and $w_2 = 3$. The length and the direction of arrows indicate the absolute value and the sign of the time derivative $\dot{s}$. Stable steady states are denoted by solid dots and unstable steady states by open circles.

Some intuition may be helpful on how stability in one dimension relates to stability in higher dimensions. In n dimensions, consider n different lines that go through a steady state and that span the entire space. If the steady state is stable on each of the lines separately, then it is a sink (i.e., is stable overall). If it is stable on some lines and unstable on others, then it is a saddle. (Think of a pass in the mountains: you look down both ways along a through route but look up in both directions along the ridge line.) If it is unstable on all lines then it is a source. Chapter 4 will formalize this intuition, including some necessary qualifications.

1.10 SEXUAL DYNAMICS

We now turn to biology, where trait transmission takes three distinct forms. The replicator equations discussed so far describe the first and simplest form, haploid asexual transmission. This form describes gene transmission quite precisely in some cases, including many bacterial and viral systems, as well as transmission via the Y chromosome or via maternal mitochondria. A second form, haploid sexual transmission, is specified below. It describes transmission in some algae and other single-celled organisms, and also is a reasonable description of meme transmission in some cases.

The third form, diploid sexual transmission, is similar to the second form but a bit more complex. It applies to most familiar biological creatures, including humans. Unfortunately, complete diploid models often have no closed-form solution and are not very tractable analytically. Numerical solutions exist, but may not be very robust or insightful.

When is it worthwhile to use more complex and realistic models of transmission? Fortunately, in many cases you don't need a complete solution. Decent approximations may be all you really want, and then a haploid simplification often can capture the most important dynamic features (e.g., when a novel strategy can increase its share). However, in some biological applications we really do need to account for diploidy, and in others we need to confirm that a haploid approximation is good enough.

Basic genetics

This rest of this section explains a classic model of sexual transmission. It can be skimmed quickly on first reading, since only occasionally do we later need to refer to it.

Some background facts and terminology will get us started. You, like most plants and animals and textbook authors, are a diploid creature (i.e., you carry double sets of genes). How many genes are we talking about? Humans are thought to carry approximately 23,000 individual genes. A fruitfly has half the number of genes (13,500), but some "simple" organisms like the water flea *Daphnia* carry 31,000 genes, far more than humans. Genes reside on what are called autosomal chromosomes. Humans have 22 pairs of autosomal chromosomes (the pairs make us diploid) and we also carry one pair of sex chromosomes, the X and Y. Each chromosome is a very long double helical coil of the now-famous molecule called DNA.

Locus (plural loci) is the physical location of a gene on the chromosome. Each gene can come in alternative versions, called **alleles**. The **genotype** of an individual specifies which allele it carries in each locus under consideration.

The basic idea in haploid sexual selection is to combine propagules from two parents, each supplying one copy of each gene. These two propagules fuse to form a zygote carrying two copies of the genes that then divides twice to form four daughter cells, each with recombined versions of the parents' genes. Diploid sexual reproduction adds a few additional complications that we will ignore for the present.

Figure 1.5 spells this out, and shows how haploid sexual reproduction leads to new combinations via a structured random process called **segregation**. The figure tracks two loci labeled p and t, and each of them has two alternative alleles, labeled with the suffixes 1 and 2. The female parent has genotype p1t2—that is, both pairs of one chromosome have allele 1 at locus p and both pairs of another chromosome have allele 2 at locus t. The male parent has genotype p2t1, and the figure begins with both parents contributing gametes, one of each of the chromosome pairs. Figure 1.5 concludes that progeny are equally likely to have any one of four genotypes, including the original parental genoytpes p1t2 and p2t1, as well as the newly recombined genotypes p1t1 and p2t2. Thus sex serves a powerful role in recombining the genotypes of two parents. For the number of genes (and alleles at each locus) seen in humans, the number of recombined genotypes, if we include the process of recombination along the length of chromosomes, is an astronomically large number (yes, more than the number of stars in the known universe).

Before proceeding, a comment on memes may be in order. Often children receive multiple memes from two parents, and these may be recombined. For example, one child may be chatty like her mother and engage in social causes like her dad, while her brother may be laconic like his dad and shun causes like his mom. Haploid inheritance and diploid inheritance are not the only possibilities. The chapter notes mention exceptions and mixed cases.

In general, haploid and diploid dynamics are complicated by finite population size effects, migration, and population structure. What do we mean by population structure? Take this book's two authors, for example. Author DF's genes are drawn from a pool shared by eastern European Jews, while author BS's come from a Finnish gene pool. Historically these gene pools had little contact, but

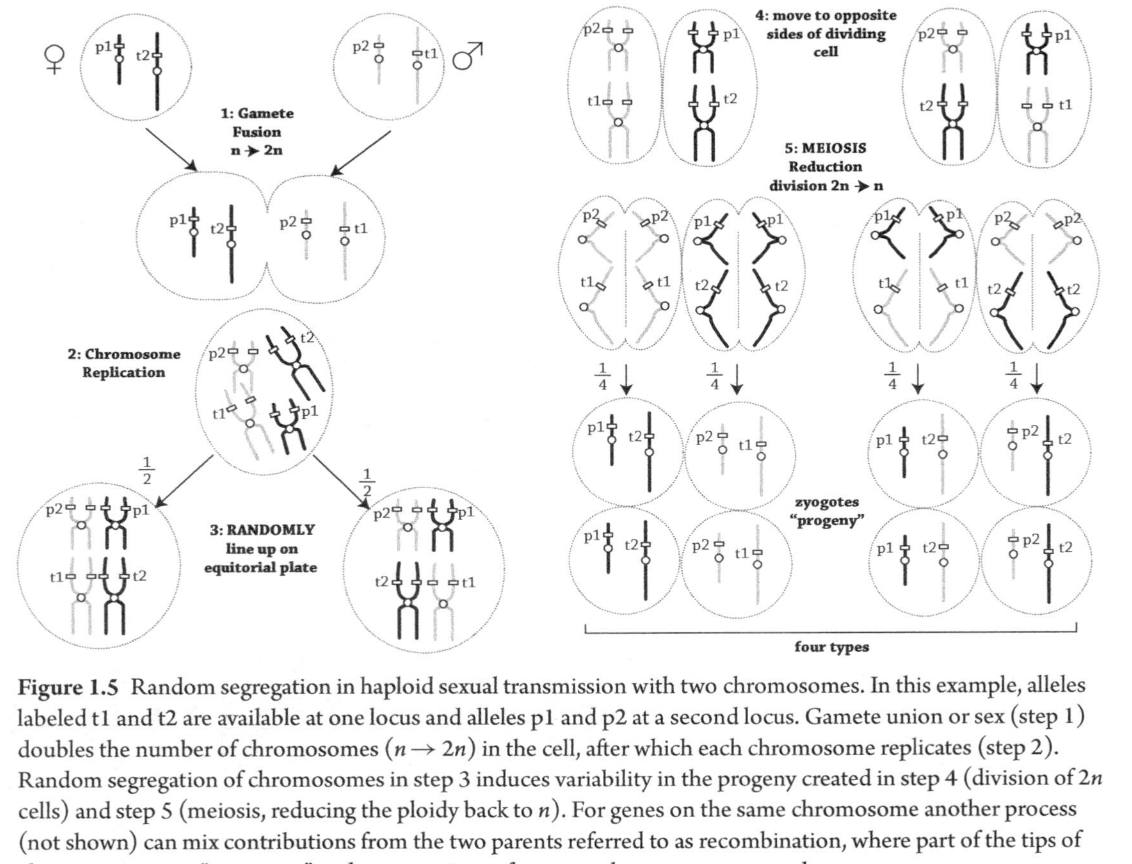

Figure 1.5 Random segregation in haploid sexual transmission with two chromosomes. In this example, alleles labeled t1 and t2 are available at one locus and alleles p1 and p2 at a second locus. Gamete union or sex (step 1) doubles the number of chromosomes ($n \rightarrow 2n$) in the cell, after which each chromosome replicates (step 2). Random segregation of chromosomes in step 3 induces variability in the progeny created in step 4 (division of $2n$ cells) and step 5 (meiosis, reducing the ploidy back to n). For genes on the same chromosome another process (not shown) can mix contributions from the two parents referred to as recombination, where part of the tips of chromosomes can “cross over” and swap positions from one chromosome to another.

now both authors find themselves in the fluid population structure of California's central coast. DF's progeny also have genes from a variety of northern and western European pools, and they have migrated to interact with yet different human gene pools in China and in the southern United States, while BS's progeny still resides in Santa Cruz.

The point is that human population structure is very hard to actually pin down. Chapters 5 and 6 will show how evolutionary games in populations with fairly simple spatial structures can give rise to amazing emergent patterns. Even without spatial structure, we will see in Chapter 4 that mate selection can dramatically affect the dynamics.

The Hardy-Weinberg model

For now, however, we put aside spatial structure and other complications, and consider completely random gene combinations. The classic model, developed independently about 100 years ago by mathematician G. H. Hardy and geneticist Wilhelm Weinberg, studies a large diploid population with a well-mixed ("panmictic") gene pool and no migration. That is, at the reproduction phase of the life cycle, all surviving individuals release gametes into a common pool of haploid egg and haploid sperm. Sperm and egg then fuse randomly to form the next generation of diploid progeny.

The law of large numbers enables us to easily compute the shares or gene frequencies for a single locus with two alleles labeled A and a. The three genotypes are AA, Aa, and aa, whose numbers are denoted N_{AA}, N_{Aa}, and N_{aa}. The total count of A's in the gene pool is $N_A = 2N_{AA} + N_{Aa}$, and is $N_a = 2N_{aa} + N_{Aa}$ for a alleles, with the overall total count $N_T = N_A + N_a = 2[N_{AA} + N_{Aa} + N_{aa}]$. The allele shares are $s_A = \frac{N_A}{N_T} = \frac{2N_{AA}+N_{Aa}}{2[N_{AA}+N_{Aa}+N_{aa}]}$ and $s_a = \frac{N_a}{N_T} = \frac{2N_{aa}+N_{Aa}}{2[N_{AA}+N_{Aa}+N_{aa}]}$. One can confirm that the homozygotes' diploid shares are given simply by multiplying:

$$s_{AA} = s_A s_A = s_A^2, \text{ and } s_{aa} = s_a s_a = s_a^2. \tag{1.12}$$

while heterozygotes' share, taking into account that each allele can come via the egg or via the sperm, is

$$s_{Aa} = s_A s_a + s_a s_A = 2 s_A s_a. \tag{1.13}$$

Fisher's fundamental theorem

We'll now see how selection works in this simple Hardy-Weinberg model. To begin, note that each genotype has its own fitness. Here fitness differences are due entirely to different survival rates, since fecundity is the same for all genotypes in this panmictic model.

Table 1.1 summarizes equations (1.12)–(1.13) and includes general expressions for fitness of the three genotypes. The parameters m_{Aa} and m_{aa} capture fitness shortfalls of the Aa and aa genotypes relative to the AA genotype. Figure 1.6 shows the possible cases. The generic case is drawn for $0 < m_{Aa} < m_{aa}$, the neutral case is $0 = m_{Aa} = m_{aa}$, while the last panel shows overdominance, with $m_{Aa} < 0 < m_{aa}$. Of course, the negative shortfall for Aa means that the heterozygote genotype has greater fitness than the homozygote AA.

Table 1.1. Diploid Shares and Fitness

Genotype:	**AA**	**Aa**	**aa**
shares in diploid progeny genotypes after mating	s_A^2	$2s_As_a$	s_a^2
absolute fitness during survival phase	W_{AA}	W_{Aa}	W_{aa}
relative fitness during survival phase	1	$1-m_{Aa}$	$1-m_{aa}$

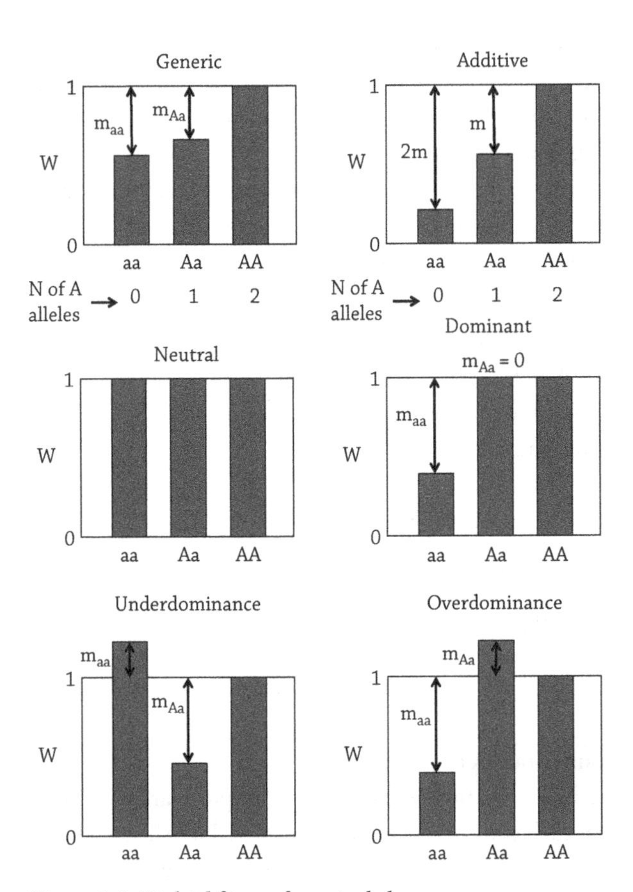

Figure 1.6 Diploid fitness for a single locus.

As usual, average fitness is $\overline{W} = \Sigma_i W_i s_i$. After normalizing as in Table 1.1, we have

$$\overline{W} = s_A^2 + 2s_A s_a(1 - m_{Aa}) + s_a^2(1 - m_{aa}) = s_A^2 + 2s_A s_a + s_a^2 - 2s_A s_a m_{Aa} - s_a^2 m_{aa}.$$

Since $s_A^2 + 2s_A s_a + s_a^2 = (s_A + s_a)^2 = 1^2 = 1$, this simplifies (after substituting out $s_A = 1 - s_a$) to

$$\overline{W} = 1 - 2m_{Aa}s_a + 2m_{Aa}s_a^2 - m_{aa}s_a^2. \tag{1.14}$$

Beginning with the expressions in Table 1.1, Appendix A derives expressions for Δs_a, the change in allele share in this model from one generation to the next. Using many steps of straightforward algebra, and taking the derivative of (1.14) with respect to s_a, it obtains the following elegant conclusion:

$$\Delta s_a = \frac{s_A s_a}{2\overline{W}} \frac{d\overline{W}}{ds_a}. \tag{1.15}$$

This last expression is a version of Fisher's fundamental theorem, a pillar of evolutionary biology. The factor $s_A s_a$ is the binomial variance for the haploid shares s_A, $s_a = 1 - s_A$. To interpret the remaining factor, recall the following math fact: for any variable V that depends on a quantity x, the logarithmic derivative $\frac{d\ln V}{dx} = \frac{1}{V}\frac{dV}{dx}$ represents the proportional rate of change in V as x increases. Hence the factor $\frac{1}{2\overline{W}}\frac{d\overline{W}}{ds_a}$ is the proportional rate of change in average fitness as the allele share s_a increases, that is, it is the strength of selection.

Thus equation (1.15) simply says that the rate of change in allele frequency (or share) is half the variance in allele frequency times the strength of selection. Figure 1.7 spells this out in a diagram. The variance factor captures the idea that

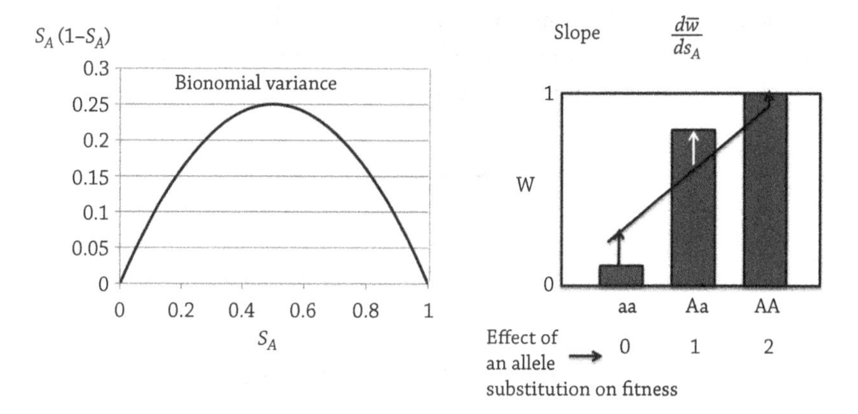

Figure 1.7 The two factors that change haploid shares.

a given strength of selection changes allele frequency more rapidly when both alleles have equal shares than when one of them is rare.

Overdominance and interior steady states

An example will illustrate some of the implications of Fisher's theorem. Sickle cell anemia in humans is caused by a single point mutation, call it a, to the normal gene, call it A, for hemoglobin. Normal red blood cells are disk-shaped, but the alternative allele causes them to collapse into a sickle shape. In environments where malaria is prevalent, a moderate amount of sickling increases fitness because the body recycles its red blood cells faster when they sickle, and the malaria parasite (which lives on red blood cells) doesn't have as much time to grow. However, lots of sickling prevents the red blood cells from transporting much oxygen, greatly reducing fitness.

In a malarial environment, then, the fitness shortfall m_{aa} is positive, reflecting the greatly reduced fitness, but the shortfall m_{Aa} for the mixed genotype is negative, indicating enhanced fitness or "overdominance." Field studies by the World Health Organization produced the estimates $m_{Aa} = -0.6$ and $m_{aa} = 0.8$.

Suppose that the initial prevalence of sickle cell allele is small, say $s_a(0) = 0.001$. Then $\overline{W} \approx W_{AA} = 1$ and $\frac{d\overline{W}}{ds_a} \approx -2m_{Aa} = -2(-0.6) = 1.2$. Fisher's equation then says $\Delta s_a = \frac{s_A s_a}{2\overline{W}} \frac{d\overline{W}}{ds_a} \approx \frac{1 \cdot s_a}{2 \cdot 1} 1.2 = 0.6 s_a$ (i.e., at first, the sickle cell gene's share grows exponentially, by almost 60% per period.)

To find steady states, we solve $0 = \Delta s_a$. Of course, as usual, we have the steady states $s_a = 0$ and $s_A = 0$. More interestingly, we have an interior steady state if $\frac{d\overline{W}}{ds_a}$ has a root between 0 and 1. Using equation (1.14), the condition is $0 = \frac{d\overline{W}}{ds_a} = -2m_{Aa} + [4m_{Aa} - 2m_{aa}]s_a$, or $s_a^* = m_{Aa}/[2m_{Aa} - m_{aa}] = -0.6/[2(-0.6) - 0.8] = 0.3$. Thus, in malarial environments, we get an interior steady state due to overdominance, with about a 30% share for the sickle cell gene. Simulations suggest, and analytical methods introduced in Chapter 3 show, that the steady state is globally stable, as shown in Figure 1.8.

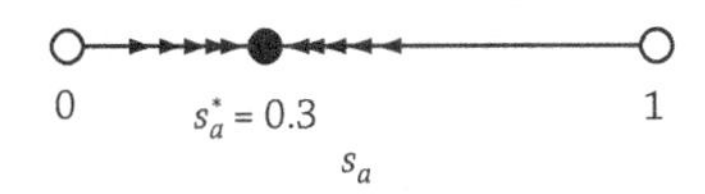

Figure 1.8 Phase portrait for the sickle cell allele, a. The stable interior steady state at $s_a^* = 0.3$ implies genotype shares approximately $s_{aa}^* = 0.3^2 = 0.09 = 9\%, s_{AA}^* = 0.7^2 = 49\%$, and $s_{aA}^* = 2(0.7)(0.3) = 42\%$ in a malarial environment.

1.11 DISCUSSION

We have not yet begun to present evolutionary games, but the stage has been set. You now have in mind the basic ideas—that populations of individuals may carry alternative versions of memes and/or genes, and we intend to track how the shares of each alternative grow or shrink over time.

Fitness encapsulates the factors that enable the shares to increase, and the basic dynamics can be captured in replicator equations in discrete time ($t \in \{0, 1, 2, \ldots\}$) or in continuous time ($t \in [0, \infty)$). The equations describe how the share vector moves as a point within a simplex. The share vector may ultimately settle down into a corner point (so that only one strategy survives), or to a point representing a mix of several strategies that each have maximal fitness.

We also summarized some of the basic biology of genes and how they are transmitted. In the most interesting case, called overdominance, a mix of genes can coexist.

In the rest of the book, we will see how strategic considerations (or frequency dependence) enter the picture, and will discover many sorts of evolutionary games that then are played.

APPENDIX A: DERIVATION OF THE FISHER EQUATION

The Fisher equation has intrinsic importance, and its derivation illustrates mathematical tricks (attributed to the evolutionary biologist J. B. S. Haldane) that we will use over and over in later chapters.

The first task is to find an expression for s'_a, the share of allele a in the next generation. Table 1.1 shows that the genotype aa will have relative frequency $s_a^2(1 - m_{aa})$, and the genotype Aa (which has half as many a alleles) will have relative frequency $2s_A s_a(1 - m_{Aa})$ in the next gene pool. To obtain shares, we normalize by mean fitness $\overline{W}$ and find that

$$s'_a = \frac{s_a^2 - m_{aa}s_a^2 + \frac{1}{2}(2s_A s_a - m_{Aa}2s_A s_a)}{\overline{W}}. \tag{1.16}$$

Recall from equation (1.14) that

$$\overline{W} = 1 - 2m_{Aa}s_a + 2m_{Aa}s_a^2 - m_{aa}s_a^2, \tag{1.17}$$

which (because $s_A = 1 - s_a$) can also be written

$$\overline{W} = 1 - 2s_A s_a m_{Aa} - s_a^2 m_{aa}. \tag{1.18}$$

The desired equation is for the change in a's share $\Delta s_a = s_a' - s_a$, so we subtract $s_a = s_a \frac{\overline{W}}{\overline{W}} = s_a \frac{1-2s_A s_a m_{Aa} - s_a^2 m_{aa}}{\overline{W}}$ from both sides of equation (1.16). This gives us

$$\Delta s_a = s_a' - s_a = \frac{s_a^2 - m_{aa}s_a^2 + (s_A s_a - m_{Aa}s_A s_a)}{\overline{W}} - \frac{s_a(1 - 2s_A s_a m_{Aa} - s_a^2 m_{aa})}{\overline{W}} \tag{1.19}$$

Noting the common factor $\frac{s_a}{\overline{W}}$, write

$$\Delta s_a = s_a \frac{s_a - m_{aa}s_a + s_A - m_{Aa}s_A - 1 + 2s_A s_a m_{Aa} + s_a^2 m_{aa})}{\overline{W}} \tag{1.20}$$

and use $s_a + s_A - 1 = 0$ to simplify to

$$\Delta s_a = s_a \frac{(-m_{aa}s_a - m_{Aa}s_A + 2s_A s_a m_{Aa} + s_a^2 m_{aa})}{\overline{W}}. \tag{1.21}$$

Now collect the terms in m_{aa} and factor out $s_a m_{aa}$ so

$$\Delta s_a = s_a \frac{(-m_{Aa}s_A + 2s_A s_a m_{Aa} + s_a m_{aa}(-1 + s_a))}{\overline{W}}, \tag{1.22}$$

substitute $-s_A = -1 + s_a$

$$\Delta s_a = s_a \frac{(-m_{Aa}s_A + 2s_A s_a m_{Aa} - s_a s_A m_{aa})}{\overline{W}} \tag{1.23}$$

and factor out s_A to obtain

$$\Delta s_a = s_a s_A \frac{(-m_{Aa} + 2s_a m_{Aa} - s_a m_{aa})}{\overline{W}}. \tag{1.24}$$

Finally, differentiate (1.18) to obtain $\frac{1}{2}\frac{d\overline{W}}{ds_a} = -m_{Aa} + 2s_a m_{Aa} - s_a m_{aa}$, and substitute this into (1.24) to obtain the desired result

$$\Delta s_a = \frac{s_a s_A}{2\overline{W}} \frac{d\overline{W}}{ds_a}. \tag{1.25}$$

One of the Exercises below invites you to practice these algebraic skills on the simplex by solving for Δs_A.

APPENDIX B: REPLICATOR DYNAMICS, MEAN FITNESS, AND ENTROPY

Mean fitness in the continuous replicator equation

A careful reader may note that equation (1.10) suggests that the appropriate mean fitness is $\ln \overline{W} = \ln(\sum_{i=1}^{n} s_i W_i)$. We now explain why the continuous replicator equation (1.11) uses the slightly different expression $\bar{w} = \overline{\ln W} = \sum_{i=1}^{n} s_i \ln W_i = \sum_{i=1}^{n} s_i w_i$. We are indebted to Marek Czachor and Maciej Kuna for clarifying our thinking on this point and for showing us a version of the computation below.

A very careful reader will note that (1.8), the equation from which (1.10) was derived, requires that $\overline{W}^{\Delta t}$ be the s-weighted sum of the $W_i{}^{\Delta t}$'s. Therefore $\ln \overline{W} = \ln(\sum_{i=1}^{n} s_i W_i)$ is actually not the correct expression in (1.10) when $\Delta t < 1$; instead we should use $\ln \overline{W} = \ln(\sum_{i=1}^{n} s_i W_i^{\Delta t})^{1/\Delta t}$. To get the continuous replicator equation, then, we need to take the limit as $h = \Delta t \to 0$. The third line of the following computation applies L'Hospital's rule and the fifth line applies the calculus fact $\frac{d}{dx} c^x = c^x \ln c$.

$$\begin{aligned}
\lim_{h\to 0} \ln \overline{W} &= \lim_{h\to 0} \ln \left(\sum_i s_i W_i^h \right)^{1/h} \\
&= \lim_{h\to 0} \frac{\ln\left(\sum_i s_i W_i^h\right)}{h} \\
&= \frac{d}{dh} \ln \sum_i s_i W_i^h \Big|_{h=0} \\
&= \frac{\sum_i s_i \frac{dW_i^h}{dh}}{\sum_i s_i W_i^h} \Big|_{h=0} \\
&= \frac{\sum_i s_i W_i^h \ln W_i |_{h=0}}{\sum_i s_i} \\
&= \sum_i s_i \ln W_i = \sum_i s_i w_i. \qquad (1.26)
\end{aligned}$$

Thus, in the limit, the proper weighted average is indeed $\sum_i s_i w_i = \bar{w}$. Economists may see parallels to the demonstration that Cobb-Douglas is a special case of CES.

Entropy and the replicator equation

Warmuth (2012) points out that the discrete replicator equation can be thought of as Nature's way of maximizing fitness given inertia in the form of relative entropy. To explain, first note that the *entropy of state* $s = (s_1, \ldots, s_n)$ in the simplex *relative to a given state* $b = (b_1, \ldots, b_n)$ in the simplex is

$$D(s,b) = \sum_{i=1}^{n} s_i \ln \frac{s_i}{b_i}. \qquad (1.27)$$

Thus $D(s,b)$, which is also known as *Kullback-Leibler divergence*, is the s-weighted average logarithmic distance of state s from state b. It is a natural way to say proportionately how far apart one point is from another point in the simplex.[1]

Given the current state $b = s(t)$, Nature maximizes fitness $w \cdot s = \sum_{i=1}^{n} s_i w_i$ of the next state s, net of the entropy cost $D(s,b)$ of the move, subject to the constraint that the next state $s = s(t+1)$ remains in the simplex. From an interior state b it is prohibitively costly to move to (much less past) the boundary, so to ensure that s remains in the simplex we need only impose the constraint $\sum_{i=1}^{n} s_i = 1$. Assigning multiplier λ to that constraint, we have the Lagrangian

$$L(s) = w \cdot s - D(s,b) + \lambda[\sum_{i=1}^{n} s_i - 1]. \tag{1.28}$$

The first-order conditions for Nature's optimum are, for $i = 1,\ldots,n$

$$0 = \frac{\partial L}{\partial s_i} = w_i - \frac{\partial}{\partial s_i}[s_i \ln \frac{s_i}{b_i}] + \lambda = w_i - \ln s_i + \ln b_i - 1 + \lambda, \tag{1.29}$$

so $\ln s_i = \ln b_i + w_i - 1 + \lambda$ or, exponentiating,

$$s_i = b_i e^{w_i} e^{-1+\lambda}. \tag{1.30}$$

Taking the first order condition with respect to λ and thus imposing the condition that the shares s_i in (1.30) sum to 1, we obtain

$$e^{1-\lambda} = \sum_{i=1}^{n} b_i e^{w_i} = \sum_{i=1}^{n} b_i W_i = \sum_{i=1}^{n} s_i(t) W_i = \overline{W}. \tag{1.31}$$

That is, the normalization factor in (1.30) is the mean fitness $\overline{W}$ of the current state $b = s(t)$ using the usual discrete time fitnesses $W_i = e^{w_i}$. Since s_i in (1.30) is shorthand for next period's share $s_i(t+1)$, we see that (1.30) is precisely the discrete time replicator equation

$$s_i(t+1) = \frac{W_i}{\overline{W}} s_i(t), \quad i = 1,\ldots,n. \tag{1.32}$$

Thus we can interpret simple (haploid, asexual, constant fitness) replicator dynamics as maximizing net fitness, step by entropy-penalized step within the simplex.

1. Two caveats may be worth mentioning. First, one can extend the definition of D to points s on the boundary of the simplex by the convention (consistent with L'Hospital's rule) that $0 \ln 0 = 0$. Second, D is not symmetric and doesn't obey the triangle inequality, so it is not a true metric. On the other hand, by Gibbs' inequality (which follows from the inequality $\ln x \leq x - 1$ valid for $x > 0$), we have $D(s,b) \geq 0$ with equality only at $s = b$.

For simplicity, the derivation took time step $\Delta t = 1$ and assigned a price of 1 to the relative entropy of the move. More generally, the price should lower when there is more time to adjust or, equivalently, keep the price at 1 but put weight Δt on next period's fitness. This yields the Lagrangian

$$L(s) = \Delta t\, w \cdot s - D(s,b) + \lambda \left[\sum_{i=1}^{n} s_i - 1 \right], \tag{1.33}$$

and the previous derivation yields the replicator equation

$$s_i(t+\Delta t) = \frac{e^{w_i \Delta t}}{\overline{W}} s_i(t), \quad i = 1,\ldots,n, \tag{1.34}$$

where now $\overline{W} = \sum_{j=1}^{n} s_j(t) e^{w_j \Delta t}$. Equation (1.34) is useful when data might be sampled at irregular times.

Recalling that continuous time mean fitness is $\bar{w} = \sum_{j=1}^{n} w_j s_j(t)$, we obtain

$$\ln \dot{s}_i = \frac{\dot{s}_i}{s_i} = w_i - \bar{w}, \tag{1.35}$$

which is indeed equivalent to (1.11).

Finally, it may be worth noting that the solution to (1.11) can be written in closed form when fitnesses w_i are all constant. Given an arbitrary initial condition $s(0)$ in the simplex, integrate (1.35) from time 0 to time $t > 0$ to obtain $\ln s_i(t) - \ln s_i(0) = tw_i - t\bar{w}$. Exponentiating we get

$$s_i(t) = \frac{s_i(0) e^{tw_i}}{e^{t\bar{w}}} = \frac{s_i(0) e^{tw_i}}{\sum_{j=1}^{n} s_j(0) e^{tw_j}}, \quad i = 1,\ldots,n, \tag{1.36}$$

where the last equality comes from imposing the constraint (already verified for (1.11)), that the shares sum to 1.

EXERCISES

1. Construct a spreadsheet (such as Excel) to keep track of shares of three alternative strategies under discrete replicator dynamics. For now, assume that the payoffs (or fitnesses) of each strategy are constant. *Suggestion:* Near the top of the spreadsheet, enter fitnesses W_1, W_2, W_2 and initial state $s_1(0)$, $s_2(0)$, $s_3(0)$. Below lay out columns with headings t, $\bar{W}$, $s_1(t)$, $s_2(t)$, $s_3(t)$. Then, in successive rows, use the replicator equation to compute the entries under these columns for $t = 1,2,\ldots,30$ or so.

2. Try a few examples with your spreadsheet, varying the W_i's and the $s_i(0)$'s, and then answer the following questions:
   (a) Does the population ever oscillate?
   (b) When does it seem to converge to a steady state?
   (c) When (if ever) does the steady state lie on a corner, or in a face or in the interior of the simplex?

   Extra credit: prove that your answers are true and complete.
3. Can memes be diploid? Can memes undergo recombination? Explain briefly.
4. (a) The argument at the end of Section 1.10 showed that there is a unique interior steady state in the overdominance case ($m_{Aa} < 0 < m_{aa}$) of the Hardy-Weinberg single locus model. Show that the same is true in the underdominance case ($m_{Aa} > 0 > m_{aa}$).
   (b) Adapt the spreadsheet in exercise (1) above to confirm that the interior steady state is stable in the overdominance case but that it is unstable (and the corner steady states are stable) in the underdominance case.
5. Derive an expression for Δs_A using steps parallel to those used to derive Fisher's equation for Δs_a. This will allow you to practice a variety of substitutions of shares on a simplex.
6. Demonstrate that $w_i = \frac{\dot{N}_i}{N_i}$ and $w_i = \frac{\dot{s}_i}{s_i}$ are equivalent definitions of fitness when the total population $N_T = \sum_j N_j$ is constant.
   Hint: Differentiate with respect to t the definition $s_i = \frac{N_i}{N_T}$ of share i (recalling that total population is constant) to obtain $\dot{s}_i = \frac{\dot{N}_i}{N_T}$. Then divide both sides by the definition of share i to obtain the desired result.

NOTES

Section 1.1 did not include a formal definition of fitness. In general, fitness refers to the instantaneous expected growth rate of a trait or strategy, but matters quickly become complicated when one must consider traits that affect future generations (such as parental care or sex ratios) or where fitness is density or frequency dependent. For example, see Metz et al. (1992, 2001) for thoughts on these matters. In later chapters of this book we will consider inclusive fitness, which stretches the concept to deal with social interactions.

For a readable overview of Russia's transition in the 1990s, see Friedman (2008, Ch. 5). For more discussion of investors' tradeoffs, see any personal financial advisor or finance textbook.

Equations (1.4) and (1.5) can also describe quite rich dynamics in situations where a K-strategist does well against the other type (e.g., c_{ij} for $i \neq j$) but relatively poorly against self types c_{ij} for $i = j$. Sinervo et al. (2000) discuss r-K tradeoffs in more depth in the color genotypes of female side-blotched lizards. In particular they include explicit frequency dependent selection on the r_i parameters, rather than just on the parameter K_i, though they also adopted the use of a different K_i for each strategy type. The Sinervo et al. (2000) version was empirically motivated by the frequency dependence observed in the density cycles of female side-blotched lizards. As an historical side note, Dennis Chitty (1958) first developed the idea of polymorphisms and natural selection as the driver behind lemming and other small rodent cycles, but never made the direct link to genotypes.

Economists use the word *homothetic* to refer to the assumption that (over the relevant range of population sizes) the process is not density dependent. For example, for homothetic preferences, the Marginal Rate of Substitution (MRS) at the point **s** in Figure 1.2 would define the MRS of any point along the ray through that point.

Readers should realize that the word "meme" is controversial. Some writers are quite enthusiastic about the concept and the word (e.g., Blackmore 2000). Others endorse the concept, but think the word misleads by emphasizing imitation (which comes from the Greek word "memesis") and close analogies to genes; they prefer "social evolution" or "cultural selection," or similar terms (e.g., Boyd and Richerson 1985). Some biologists (though seemingly fewer each year) resist the concept altogether, and regard evolution as entirely (or almost entirely) genetic (e.g., Sober and Wilson 1999). We prefer to use the word because it helps keep our sentences shorter, but must remind the reader that serious applied work requires the researcher to think through how evolutionary dynamics should be specified for that application, be it biological, social, or virtual.

To be thorough, we should also mention here that genes are not *always* transmitted vertically. Exceptions include so-called jumping genes, first reported in McClintock (1950).

Here is the formal definition of stability. A steady state s^* is (Lyapunov) *stable* if $\forall \epsilon > 0 \; \exists \delta > 0$ s.t. $|s(0) - s^*| < \delta \Rightarrow |s(t) - s^*| < \epsilon \; \forall t > 0$. There are bizarre examples of states that are asymptotically stable but not stable. You are very unlikely to encounter them in applications.

The structure of all population genetic models necessarily requires diploid genetics for diploid organisms like animals and plants. Classic evolutionary game theory that originates with Maynard Smith (1982) typically ignores this detail. Maynard Smith (1982) was however, very clear that he only considered the broad class of phenotypic models, which ignores this diploid structure to genes. The theoretical development of the field of population genetics (Provine 1971) includes an incredibly rich set of tools for analyzing the dynamics and stability of frequency dependent and general games of selection. Even though frequency dependent selection received extensive treatment in the field of population genetics (e.g., Wright 1968 and many others) the main focus was on the equilibrium solutions. Here, we shift the focus to include not only equilibrium solutions (e.g., NE and ESS) but also the dynamics of evolutionary change under frequency dependent selection. We also supply links to population genetics throughout the book. As pointed out by Weissing (1991), many population genetic phenomena classified as gene interaction or epistasis can be better understood as frequency dependent phenomena.

BIBLIOGRAPHY

Blackmore, Susan. *The meme machine*. Oxford University Press, 2000.

Boyd, Robert, and Peter J. Richerson. *Culture and the evolutionary process*. University of Chicago Press, 1985.

Chitty, Dennis. "Regulation in numbers through changes in viability." *Canadian Journal of Zoology* 38 (1958): 99–113.

Dawkins, Richard. *The selfish gene*. Oxford University Press, 1976.

Friedman, Daniel. *Morals and markets: An evolutionary account of the modern world*. Macmillan, 2008.

Maynard Smith, John. *Evolution and the theory of games*. Cambridge University Press, 1982.

McClintock, Barbara. "The origin and behavior of mutable loci in maize." *Proceedings of the National Academy of Sciences* 36, no. 6 (1950): 344–355.

Metz, Johan A. J., and Mats Gyllenberg. "How should we define fitness in structured metapopulation models? Including an application to the calculation of evolutionarily stable dispersal strategies." *Proceedings of the Royal Society of London B: Biological Sciences* 268, no. 1466 (2001): 499–508.

Metz, Johan A. J., Roger M. Nisbet, and Stefan A. H. Geritz. "How should we define 'fitness' for general ecological scenarios?" *Trends in Ecology and Evolution* 7, no. 6 (1992): 198–202.

Provine, William B. *The origins of theoretical population genetics*. University of Chicago Press, 1971.

Sinervo, Barry, Erik Svensson, and Tosha Comendant. "Density cycles and an offspring quantity and quality game driven by natural selection." *Nature* 406, no. 6799 (2000): 985–988.

Sober, Elliott, and David Sloan Wilson, eds. *Unto others: The evolution and psychology of unselfish behavior*. Harvard University Press, 1999.

Warmuth, Manfred. *Computer Science 272 Lecture Notes*. 2012.

Weissing, Franz J. "Evolutionary stability and dynamic stability in a class of evolutionary normal form games." In *Game equilibrium models I* I. Eshel and R. Selten, eds., pp. 29–97. Springer Berlin Heidelberg, 1991.

Wright, Sewall. *Evolution and the genetics of populations*, 4 vols. University of Chicago Press, 1968.

2

Simple Frequency Dependence

Some sorts of strategies suffer from congestion—the larger the share, the lower the payoff. For example, the impact of a political ad is probably lower when everybody has already seen dozens that day. Other strategies gain fitness when they are more prevalent. For example, infrastructure investment (e.g., in social media networks) to improve voter turnout probably has little impact until it reaches some critical mass.

Such effects are known to biologists as negative and positive frequency dependence. They are known to economists as decreasing or increasing returns, and management gurus call them synergy. We ignored such influences on fitness in Chapter 1, but they are central to evolutionary game theory. Indeed, game theory is all about strategic interaction, that is, how the distribution of other players' strategies affects the payoffs of your own strategy.

This chapter takes a first look at such strategic interdependence, and shows that it can take many forms beyond simple negative and positive frequency dependence. We begin with the simplest case, linear frequency dependence between two alternative strategies in a single unified population.

The first example is the famous Hawk-Dove game, as formulated by Maynard Smith (1982). After looking at a numerical example, we analyze Maynard Smith's two-parameter version of the game, and apply discrete time replicator dynamics. As in the Chapter 1 example of sickle cell anemia, we find an interior steady state in HD games, but here it is due to strategic interaction rather than diploid genetics. It turns out that the forces of congestion prevail in HD games. We will also identify a key boundary condition ignored in previous analyses, and will see how continuous time replicator dynamics simplify the situation.

We then consider other types of games with two alternative strategies. In coordination (CO) type games, the forces of increasing returns prevail. There is an interior steady state, but it is unstable. As in the underdominance case of diploid dynamics, this steady state separates the states that evolve towards fixation on the first pure strategy from the states that evolve towards fixation on the second pure strategy. Some literature refers to this case as bi-stable.

There is only one other type of game in our setting with two alternative strategies. We refer to as DS, because it has a dominant strategy (DS) whose share is forced to 100% (fixation, as biologists call it) under replicator dynamics. As one example, we show how data for certain RNA viruses shows that they play the most notorious DS game, known as prisoner's dilemma (PD). There are, we note, also much more benign games of DS type.

The last three sections show how to use these simplest sorts of frequency dependence as building blocks for more complex games. One important form of complexity is nonlinear frequency dependence between two strategies. Another is linear frequency dependence among three alternative strategies, as illustrated in the famous rock-paper-scissors game. The data show that male side-blotched lizards play a generalized version of this game.

2.1 THE HAWK-DOVE GAME

In the parlance of the Vietnam War era, hawk refers to an aggressive but costly strategy, while dove refers to a low-cost submissive strategy. Maynard Smith's Hawk-Dove game involves two animals (not necessarily birds) that meet at a resource. Seizing the resource adds to an individual's fitness. Each animal can adopt one of two strategies. Strategy H, or hawk, is to behave aggressively until victory or serious injury (perhaps death), while strategy D, or dove, is to display threateningly but to retreat if the opponent is aggressive.

The animals then reproduce as haploids, and in the next generation progeny follow the parental strategy. For now we assume that generations are discrete and non-overlapping; the adult population of hawks and doves dies before the progeny generation matures. Thus parents never play the game with their progeny.

Numerical example

Consider the following payoff (or fitness) table.

	Player 2:	
Player 1:	H	D
H	$(-1,-1)$	$(4,0)$
D	$(0,4)$	$(2,2)$

The first element of each vector is the payoff to (or fitness increment of) the row player (player 1) while the second element is the payoff to the column player (player 2). This game is symmetric in that the off-diagonal elements are mirror images of each other, and the vectors on the diagonal have the same elements. Consequently the payoffs to each player depend on her choice and her opponent's choice, but not on whether she is a column player or row player. If I pick H and my opponent picks D, for example, then I get 4 and my opponent gets 0, whether I am the row or the column player. This payoff symmetry arises from the fact that we are dealing with a single population. In the next chapter we will study games where row players and column players belong to separate populations, and there the payoff bimatrix need not be symmetric.

For single-population games, the symmetry of payoffs makes redundant the double-entry layout. We can simply show the row player's payoffs in an single-entry table as below, and not lose any information.

	Encounter rate:	
	s_H	$1-s_H$
Choice:	H	D
H	-1	4
D	0	2

The top line, labeled "Encounter rate," now gives the population shares of H (namely, s_H) and D (s_D, written as $1-s_H$) in the current adult cohort.

To see the decreasing returns (or congestion or negative frequency dependent) aspect of this game, notice that if everyone else is playing H, then I get payoff -1 if I also play H, but I get payoff $0 > -1$ if I instead play D. Likewise, I'm better off going against the crowd if they are all playing D: in this case I get 4 if I play H, but only $2 < 4$ if I also play D.

Can some mix of H and D persist from one generation to the next? To find out, we begin by writing out the fitnesses of the two strategies, W_H and W_D. Each strategy faces a population that is composed of hawks with share s_H and doves with share $s_D = 1-s_H$. Therefore, fitness (or expected payoff) to the row player is given by those shares times the corresponding payoffs listed in the table:

$$W_H = -1s_H + 4(1-s_H) = 4-5s_H \quad \textbf{(2.1)}$$

because any H player receives payoff -1 in the fraction s_H of encounters and payoff 4 in the remaining encounters. Likewise, the fitness of any D player is

$$W_D = 0s_H + 2(1-s_H) = 2-2s_H. \quad \textbf{(2.2)}$$

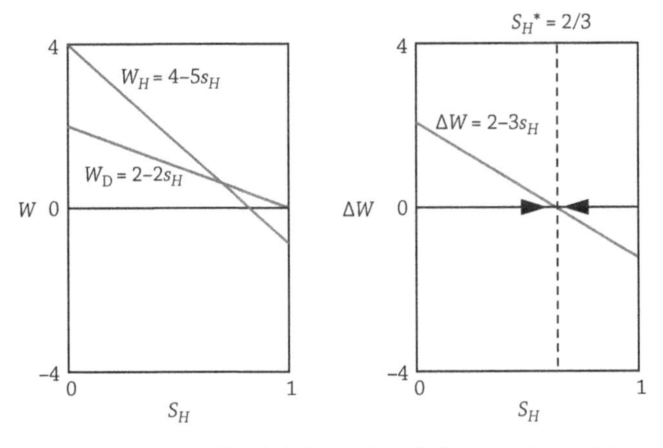

Figure 2.1 Fitness of hawk (H) and dove (D) in Panel A, and their difference ΔW in Panel B, as a functions of the share s_H of hawks.

It turns out that what really matters is the payoff difference, or relative fitness of the first strategy, denoted $\Delta W = W_H - W_D$. Thus

$$\Delta W = -s_H + 2(1 - s_H) = 2 - 3s_H. \tag{2.3}$$

See Figure 2.1 for graphs of W_H, W_D, and ΔW as functions of s_H.

Now consider the mix s^* that gives equal payoff, $\Delta W = 0$. From (2.3) we see that $s_H^* = \frac{2}{3}$, so $s_D^* = \frac{1}{3}$. Is this mix stable as we move from one generation to the next? To find out, first note from (2.3) that $\Delta W > 0$ when $s_H < \frac{2}{3}$. Thus H has the higher fitness and its share s_H should increase when it is below $\frac{2}{3}$. On the other hand, if $s_H > \frac{2}{3}$ then $\Delta W < 0$, so in this case the share of H should decrease back towards $\frac{2}{3}$. It seems that, as the arrows indicate in Figure 2.7 Panel B, the $s_H^* = \frac{2}{3}$ mix is stable. In Chapter 4 we will see that it is indeed a stable equilibrium known as an evolutionarily stable state (ESS).

2.2 H-D PARAMETERS AND DYNAMICS

While numerical examples help illustrate the issues, in most applications the models are written in terms of parameters. This chapter will show how the the dynamics (and the steady states) change as we vary the parameters, and how to estimate realistic parameter values from the data.

For the HD game, there are two important parameters: the value of the resource $v > 0$ and the cost $c > 0$ of losing a battle over the resource. When two animals play H, they each (as far as one can say from the outside) have a $\frac{1}{2}$ probability of losing, so the average cost is $\frac{c}{2}$. Likewise, each of them gains an average resource value of

$\frac{v}{2}$. The expected payoff for each of the two H strategists thus is $\frac{v-c}{2}$. When two animals play D, neither incurs the cost and each gains on average half the resource value, so both get payoff $\frac{v}{2}$. When a H strategist meets a D, the H always seizes the resource at no cost, so the H payoff is v and D gets 0. This all is summarized in the following parametrized payoff table.

	Encounter rate:	
	s_H	$1-s_H$
Choice:	H	D
H	$\frac{v-c}{2}$	v
D	0	$\frac{v}{2}$

You can see that we used $v=4$ and $c=6$ in the numerical example.

Row elements of the payoff matrix reflect payoffs of any individual player against each strategy available to other players. The shares shown above the columns (here s_H and $1-s_H$) aggregate the strategies used by all other players. Usually we assume that there are so many other players that the fractions are not affected by the choice of player 1, but this is not necessary to write in the table. Occasionally we will comment on what happens when the number of players is not so large.

The other implicit assumption is that the row player is equally likely to be matched with any other player, and so the payoffs from each possible strategy by other (column) players is proportional to its share. In Chapters 4, 5, and 6, we will relax this assumption in various ways. For example, players might be able to seek out and exploit other strategies at rates greater than random, that is, the fraction of D encounters might be greater than s_D. But for now we ignore such potential complications.

With these simplifying assumptions, the fitness of any strategy W_i for $i=H,D$ is obtained from the rate of encountering each strategy $j=H,D$ multiplied by the payoff W_{ij} from the encounter. Thus

$$\begin{aligned} W_H &= s_H W_{HH} + (1-s_H)W_{HD}, \\ &= s_H \frac{v-c}{2} + (1-s_H)v. \end{aligned} \tag{2.4}$$

Likewise,

$$\begin{aligned} W_D &= s_H W_{DH} + (1-s_H)W_{DD} \\ &= s_H \cdot 0 + (1-s_H)\frac{v}{2}. \end{aligned} \tag{2.5}$$

The payoff difference now is

$$\Delta W = W_H - W_D \tag{2.6}$$
$$= s_H \frac{v-c}{2} + (1-s_H)\frac{v}{2} = \frac{1}{2}(v - s_H c).$$

Can hawks invade a population of all doves? The question reduces to checking whether whether $\Delta W > 0$ when $s_H = 0$. From (2.6), or directly from the payoff matrix, we see that $\Delta W(0) = W_{HD} - W_{DD} = \frac{v}{2}$. This is positive since $v > 0$, by definition of resource value. So the answer to the first question is affirmative.

Can doves invade a population of all hawks? The answer is yes, as long as $\Delta W < 0$ (i.e., D is fitter) when $s_H = 1$. Clearly $\Delta W(1) = W_{HH} - W_{DH} = \frac{v-c}{2}$, which is positive iff $v > c$. In this case, D cannot invade and H can persist as a pure strategy.

The case $v < c$ is more interesting. In this case, pure H cannot persist, and (as in the numerical example) a stable mixture s_H^* can be found by solving $\Delta W = 0$. Inspection of the last expression in (2.6) reveals

$$s_H^* = \frac{v}{c}. \tag{2.7}$$

To check dynamics, recall the the discrete time replicator equation from Chapter 1. The shares of $i = H, D$ at time $t+1$ are given by:

$$s_i(t+1) = \frac{W_i}{\overline{W}} s_i(t), \quad i = H, D, \tag{2.8}$$

where

$$\overline{W} = s_H W_H + (1 - s_H) W_D \tag{2.9}$$

is average fitness in the population. It is easy to see that s_H increases whenever $\frac{W_H}{\overline{W}} > 1$ (i.e., whenever H yields a higher payoff than D, i.e., whenever $\Delta W > 0$ in equation (2.6), i.e., whenever $s_H < s_H^*$). Thus in the general case, as in the numerical example, s_H increases when it is below the steady state s_H^* and decreases when it is above the steady state.

Zone of total destruction

There is a problem with the discrete replicator equation (2.8) applied to Maynard Smith's (1982) formulation of the Hawk-Dove game. We can get a negative value of W_H, and hence a negative value of $s_H(t+1)$, contrary to the definition of shares and state space.

What could a negative value actually mean? In a biological system, W_H is a composite of reproduction and survival. One can perhaps imagine overkill, or negative survival numbers. But if no H survive, then reproduction is zero, not negative, so each W_i must be bounded below by zero. As noted in Chapter 1, it follows that $\overline{W}$ always is greater than zero.

So we have to modify the definition of W_H when equation (2.4) goes negative. The critical value of s_H, call it s_c, is where W_H is zero:

$$W_H = v - s_c\frac{v}{2} - s_c\frac{c}{2} = 0, \tag{2.10}$$

with solution

$$s_c = \frac{2v}{v+c}. \tag{2.11}$$

Thus, if $s_H(t) > s_c$ then the unmodified replicator equation says that $s_H(t+1) < 0$. The interpretation is that hawks in this cohort keep killing one another off and have no progeny. So $s_H \in [\frac{2v}{v+c}, 1]$ is a zone of total destruction. If we ever reach that zone, then the meek (D) shall inherit the earth, that is, $s_D = 1$ and $s_H = 0$ at time $t+1$ and forever after. In principle the HD game will not enter this zone during normal dynamics, but you might start out the game within this zone (e.g., starting conditions for $s_H(t=0)$), and in this case you would observe abnormal dynamics. See the problem at the end of this chapter, and an Excel spreadsheet derivation that reproduces the dynamics of the HD game.

When we use the replicator equations we have to guard against this. Negative values for W are not permissible. We either exclude that zone from the domain of the replicator equation, or else treat it as the total destruction endgame with W_H set to 0 instead of a negative number.

Continuous replicator

The total destruction problem does not arise for replicator dynamics in continuous time. Values of $w = ln(W)$ can go negative, but this simply describes negative growth rates or declining frequencies. Actual shares can never go negative, as can be seen from a little algebra. Since $\bar{w} = s_H w_H + (1-s_H)w_D$, we have

$$\begin{aligned}\dot{s}_H &= (w_H - \bar{w})s_H = (w_H - s_H w_H - (1-s_H)w_D)s_H = (1-s_H)(w_H - w_D)s_H \\ &= s_H(1-s_H)\Delta w,\end{aligned} \tag{2.12}$$

where $\Delta w = w_H - w_D$. When $\Delta w < 0$, the hawk growth rate $\dot{s}_H/s_H$ is negative but this means that s_H shrinks towards zero while remaining positive. Likewise for doves: for any parameter values, the growth rate $\dot{s}_D/s_D = -\dot{s}_H/(1-s_H)$ can be negative but s_D remains positive even if it shrinks towards zero.

Equation (2.12) is interesting in its own right. It shows that the direction of change in the state ($\dot{s}_H$ positive or negative) is the same as the sign of the fitness advantage (Δw positive or negative). Indeed, the rate of change $\dot{s}_H = -\dot{s}_D$ in the shares is equal to the fitness advantage times the factor $s_H(1-s_H)$, the binomial variance associated with the current shares. This factor goes to 0, and adjustment slows to a crawl, as we approach either endpoint. Thus the continuous replicator equation is in some ways parallel to the version of Fisher's fundamental equation derived near the end of Chapter 1.

2.3 THE THREE KINDS OF 2×2 GAMES

Besides convergence to an interior steady state as in the HD game, what other sorts of behavior can occur with 2×2 payoff tables? Let's give the two alternative strategies the labels $i,j = A,B$, and explore the possibilities.

Begin with a fully parametrized version of the 2×2 payoff table, suitable for continuous time analysis:

Table 2.1. Payoffs in a General 2×2 Game

	Encounter rate:	
	s_A	$1-s_A$
Choice:	A	B
A	w_{AA}	w_{AB}
B	w_{BA}	w_{BB}

Following the analysis of earlier sections, the fitness of strategy A is

$$w_A = w_{AA}s_A + w_{AB}(1-s_A), \tag{2.13}$$

and the fitness of the alternative strategy is

$$w_B = w_{BA}s_A + w_{BB}(1-s_A). \tag{2.14}$$

The payoff advantage of A over B then is

$$\Delta w = (w_{AA} - w_{BA})s_A + (w_{AB} - w_{BB})(1-s_A). \tag{2.15}$$

Equation (2.15) shows that the payoff advantage depends entirely on certain combinations of parameters, and these combinations turn out to have helpful interpretations. A biologist would describe the combination $w_1 = w_{AB} - w_{BB}$ as the fitness advantage of rare mutant strategy A when strategy B is common, while a game theorist would describe w_1 as the payoff advantage of A over B when playing against B. The analogous combination for the other strategy is $w_2 = w_{BA} - w_{AA}$, the advantage of rare mutant strategy B over common strategy A.

Using these parameter combinations, we can rewrite the payoff advantage in (2.15) as

$$\Delta w = -w_2 s_A + w_1(1 - s_A) = w_1 - (w_1 + w_2)s_A. \qquad \textbf{(2.16)}$$

The last expression shows that the payoff advantage line has vertical intercept w_1 and slope $-(w_1 + w_2)$. By setting $\Delta w = 0$ and solving for s_A, we see that $s_A^* = w_1/(w_1 + w_2)$ is the equal fitness point.

Equation (2.16) also shows that the payoff advantage graph is the line segment that connects the point $(0, w_1)$ to the point $(1, -w_2)$ in $(s_A, \Delta w)$-space. If this line segment crosses the horizontal axis, it does so at the point $(s_A^*, 0)$. Examples are shown in Figure 2.2.

Recall from equation (2.12) that the sign of $\dot{s}_A$ is the same as the sign of Δw under continuous replicator dynamics. Using the current general notation, the continuous replicator equation is

$$\dot{s}_A = s_A(1 - s_A)\Delta w. \qquad \textbf{(2.17)}$$

Taking time derivatives of the expression $s_A + s_B = 1$, we see that $\dot{s}_A + \dot{s}_B = 0$, so $\dot{s}_B = -\dot{s}_A = -s_A(1 - s_A)\Delta w = -(1 - s_B)s_B \Delta w$.

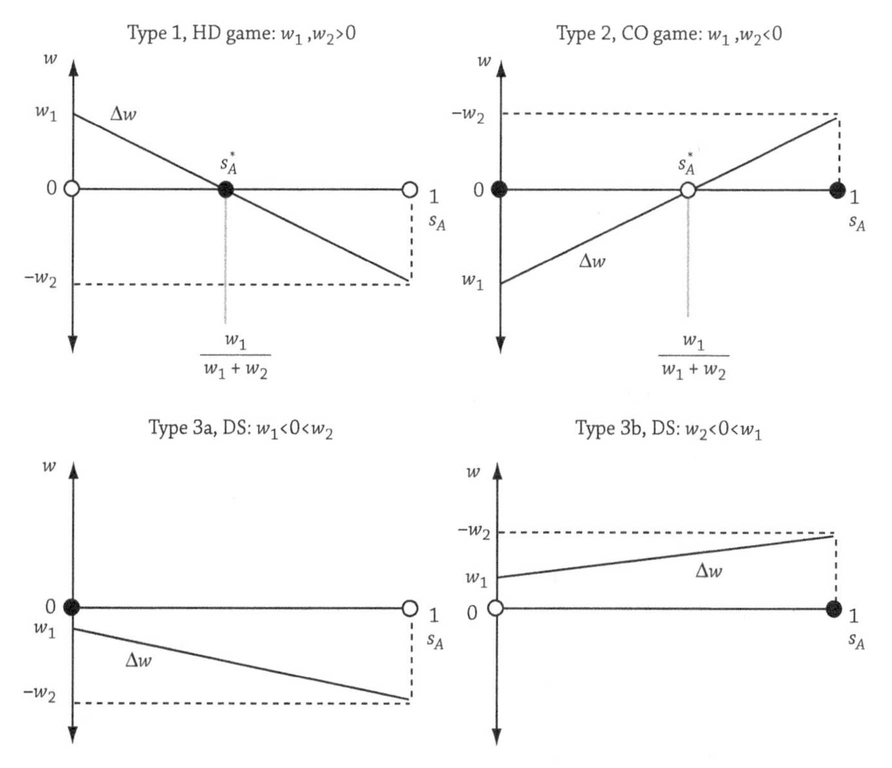

Figure 2.2 The three types of linear 1-dimensional games.

The upshot is that the signs of the w_i's tell the whole story. The signs determine whether the fitness difference line starts out positive or negative, and whether it crosses the horizontal axis. Then replicator dynamics determine behavior.

Consequently, as shown in Figure 2.2, there are three main kinds of 2×2 matrices: HD (or Type 1), CO (Type 2), and DS (Type 3).

HD type

When w_1 and w_2 are both positive, then mutants can invade either common type. As noted earlier, economists call this situation decreasing returns to scale or congestion, and biologists call it negative frequency dependence. Panel A of Figure 2.2 shows that if both w_i's are positive then:

- the only steady state is at $s_A^* = w_1/(w_1 + w_2)$, which is strictly between 0 and 1.
- The payoff advantage line Δw has negative slope $-(w_1 + w_2)$, and so s_A^* is a **downcrossing**. That is, $\Delta w > 0$ when $s_A < s_A^*$ and $\Delta w < 0$ when $s_A > s_A^*$.
- The steady state s_A^* is stable in the sense that all other states move towards s_A^* and converge to it under continuous replicator dynamics.

Downcrossings s_A^* are stable because the replicator equation (2.17) and the signs of Δw ensure that the state moves towards s_A^* from both sides.

We have encountered this sort of situation before, in our analysis of Hawk-Dove, so we refer to this case as HD. Now it is clear that the single, stable interior steady state is not necessarily a matter of having the cost of fighting greater than the value of the resource. More generally, it will occur whenever both pure strategies are invadeable for any reason.

CO type

What are the other possibilities? Of course, one or both of the w_i's might be zero. The chapter-end exercises invite you to analyze these knife-edge cases yourself.

Another possibility, labeled CO (or type 2) in Figure 2.2, is that the w_i's are both negative. That is, neither pure strategy can be invaded. It is an important case that we have not discussed previously. Biologists call it positive frequency dependence, and economists refer to it as increasing returns to scale or synergy.

Here is an example, based on some remarks by French Enlightenment philosopher Jean-Jacques Rousseau (1712–1778). The population consists of hunters in a village who seek to capture a great stag, *Cervus cervus*, a male with spectacular antlers. If the hunt is successful, everyone shares in the glory and their payoff is 10 each. But each hunter can be distracted from hunting the stag (strategy A) by hunting rabbits (strategy B). Hunting rabbits is quite reliable; no matter what other hunters do, it yields payoff 8. The stag is very good at escaping when some hunters are distracted (hunting rabbits makes a lot of noise), and so the chance of capturing it drops sharply when more hunters switch to strategy B.

This story is encapsulated in the following payoff table.

	Encounter rate:	
	s_A	$1-s_A$
Choice:	A	B
A	10	0
B	8	8

For these payoffs we have $w_1 = w_{AB} - w_{BB} = 0 - 8 = -8 < 0$ and $w_2 = w_{BA} - w_{AA} = 8 - 10 = -2 < 0$; both are negative, as advertised. The interior rest point is $s_A^* = w_1/(w_1 + w_2) = -8/(-8-2) = 8/10 = 0.80$. The payoff advantage $\Delta w = -8 + 10s_A$ is negative for $s_A < s_A^*$ and positive for $s_A > s_A^*$, so s_A^* is an **upcrossing**.

This example illustrates the following general features when both w_i's are negative:

- There are three steady states, the corners $s_A = 0$ and $s_A = 1$ as well as the interior steady state $s_A^* = w_1/(w_1 + w_2)$.
- The payoff advantage line Δw has positive slope $-(w_1 + w_2)$, and so s_A^* is a upcrossing.
- Consequently, the steady state s_A^* is unstable. It separates the basins of attraction of the two pure steady states $s_A = 0$ and $s_A = 1$.

The interpretation in terms of the Stag Hunt game is that if at first a larger fraction than $s_A^* = 0.80$ of the villagers focus on the stag, then under replicator dynamics, eventually all of them will choose strategy A, and the hunt will be successful. On the other hand, if a smaller fraction than 0.8 chooses A initially, then its popularity will decline and eventually everyone will end up hunting rabbits.

We refer to the general case of negative w_i's as coordination games (CO), because they have two pure strategy stable steady states in which each player has the incentive to coordinate behavior with the other players. The incentives come from positive frequency dependence.

Two final examples illustrate the general point. Choosing computer operating systems is a coordination game—the more people who use Linux, the better off are Linux users, and similarly for Windows users. Conventions like driving on the left side of the road or the right side have the same logic: you are better off doing what most other players do.

DS type

When w_1 and w_2 have opposite signs, then one of the strategies can invade the other, but can't itself be invaded. That strategy always has a higher payoff, and therefore game theorists call it a **dominant strategy**. We shall refer to this case as DS for short.

We encountered DS earlier, in the general parametrized Hawk-Dove example. Hawk is a dominant strategy in the extreme case $c < v$, that is, when the cost of fighting is less than the value of the resource.

To work out the general case, suppose that $w_1 > 0 > w_2$. Then A is the dominant strategy—it can invade B, but B can't invade it. As shown in the bottom-right panel of Figure 2.2, this case has the following features:

- The only steady state is at $s_A = 1$, where everyone plays the dominant strategy A.
- The payoff advantage line Δw lies entirely above the horizontal axis, and so there is no crossing point s_A^* between 0 and 1. Consequently,
- the steady state $s_A = 1$ is stable, and its basin of attraction is the entire state space $(0, 1)$.

The other DS possibility is $w_2 > 0 > w_1$. Here strategy B is dominant, Δw lies entirely below the horizontal axis, and the steady state $s_A = 0$ is the global attractor, as shown in the bottom-left panel of Figure 2.2. It really is the same as the first DS possibility with the strategy labels A and B switched.

Laissez-faire versus prisoner's dilemma

It might seem, then, that all DS cases are essentially the same. However, there is a crucial distinction regarding the effect of the dominant strategy on mean fitness $\bar{w}$.

As shown on the right side of Figure 2.3, we have a *social dilemma* when $\bar{w}$ is a decreasing function of the share s_A of the dominant strategy A. The more individuals who adopt that strategy, the lower is everyone's payoff. Often this subtype is referred to as the prisoner's dilemma (PD), and it is discussed further in the next section. The numerical values for the figure are given in exercise 3 at the end of the chapter.

In the other subtype of DS game, the mean payoff increases in the share of the dominant strategy, as on the left side of the figure. We refer to this sort of game as *laissez-faire* (LF) because the population spontaneously evolves towards the highest mean fitness. The more individuals who adopt the dominant strategy, the better off everyone is.

Two caveats are in order. First, the distinction between LF and PD is not captured in the composite payoff parameters w_1, w_2. Instead, one has to look at the underlying payoffs $w_{AA}, w_{AB}, w_{BA}, w_{BB}$. Second, $\bar{w}$ generally is a quadratic

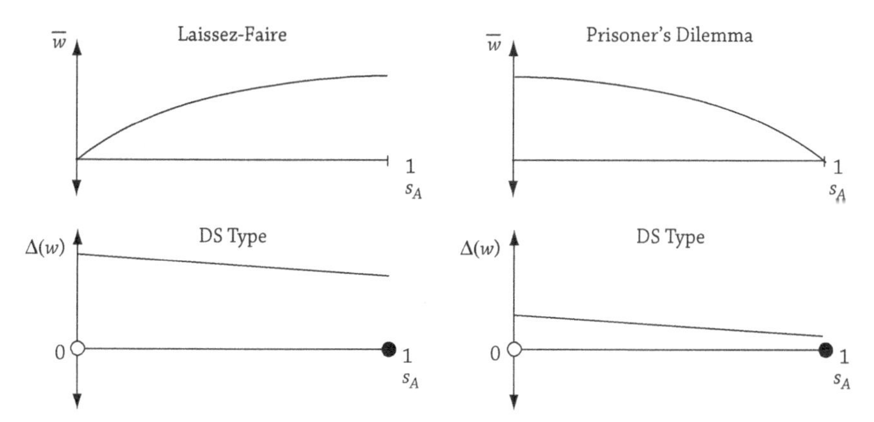

Figure 2.3 The two subtypes of DS games with dominant strategy A. In Laissez-Faire subtype (on the left) mean fitness $\bar{w}$ increases as s_A goes to fixation. In prisoner's dilemma subtype (on the right) $\bar{w}$ decreases as s_A goes to fixation.

function of the payoffs w_{ij}, not a linear function. Consequently, there are also mixed cases where a DS game is LF for s_A above (or below) a critical value $\hat{s}_A$ and is PD for s_A on the other side of the critical value. Chapter-end exercises invite the reader to explore both caveats.

2.4 DILEMMAS PLAYED BY VIRUSES AND EBAY SELLERS

The PD is perhaps the most famous 2×2 game. Instead of the usual contrived story about two prisoners who might or might not testify against each other, we will tell a true story involving two alternative behaviors of viruses. Then we'll tell a similar story about sellers on eBay. These examples have the same strategic tension as the standard PD, and they also illustrate how to estimate payoffs from available data.

Turner and Chao (1999) investigated viruses that are parasites of bacteria. Such viruses consist of gene-carrying strands of RNA wrapped in a protein coat that protects the RNA and allows the virus to stick to and enter its next host. Reproduction is asexual haploid, as in the ordinary replicator equation.

The interests of cloned viruses are aligned and they replicate without conflict. However, conflict arises when two or more genetically different viruses invade the same cell. Inside the bacterial host, the two viral types compete to replicate their own type. There is a cheater strategy that relies on a co-infecting viral genotype to produce the proteins which the cheater uses to build its coat. The cheater type foregoes the RNA that codes for coat proteins, thus gaining an advantage in replicating a shorter, more streamlined genome. All the cheater has to do is replicate its key RNA (i.e., the replication genes in its genome), and to grab the coat proteins created by the non-cheater viruses. When enough coat proteins are around, the cheater type can outcompete the non-cheater type that has to produce the coat proteins first and then replicate its own RNA.

Thus we have two viral morphs (or strategies), Defect (the cheater type), and Cooperate (the other type). Normalizing payoffs so that $W_{CC} = 1.0$, we see that the game has three unknown payoff parameters, as in the table.

	Encounter rate:	
	s_C	$1 - s_C$
Morph:	Cooperate	Defect
Cooperate	1	$1 - c_C$
Defect	$1 + v_D$	$1 - c_D$

Parameter estimates

Turner and Chao (1999) report experiments designed to provide estimates of the payoff parameters. They varied the initial shares of the two types of viruses, and measured how the progeny shares changed from the chosen initial frequency. The results are shown in Figure 2.4. Figure 2.4 shows that the Defect strategy always has higher fitness than Cooperate, the defining feature of DS-type games.

The payoff parameters are estimated as follows. The first parameter, v_D, represents the proportional advantage gained by the defector morph when the cooperator morph is common (i.e., $W_{DC} = 1 + v_D$). Figure 2.4 shows that the

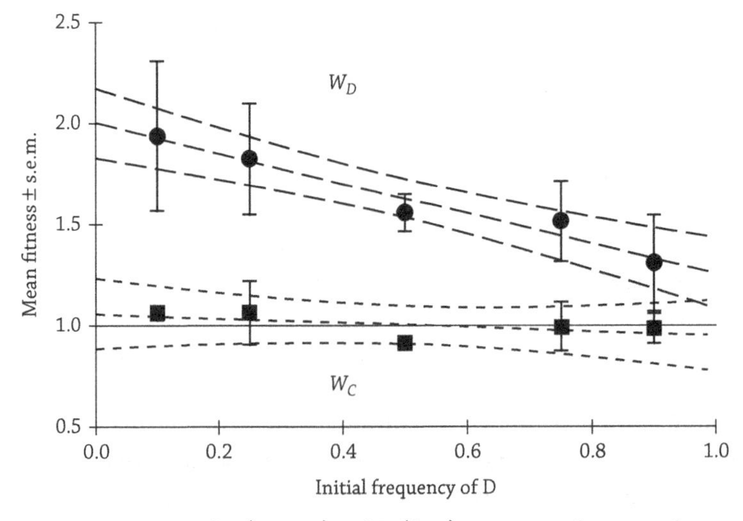

Figure 2.4 Fitness of C (squares) and D (dots) viruses as a function of $s_D = 1 - s_C$. Adapted from Turner and Chao (1999).

upper fitted line (for D) intercepts the vertical axis at $W_{DC} = 1.99$, yielding the estimate $v_D = 0.99$. The second parameter c_C represents the cost to the cooperator of being cheated of its proteins when the defector morph is common, as in the expression $W_{CD} = 1 - c_C$. The third parameter captures the cost to the defector of not having completely functional machinery for protein production, $W_{DD} = 1 - c_D$. The intercepts with the vertical line $s_D = 1$ in Figure 2.4 show that $W_{DD} > W_{CD}$, so $1 - c_D > 1 - c_C$ or $c_D < c_C$. This confirms that strategy D dominates C, so we indeed have a DS-type game.

In a separate experiment reported in the same paper, Turner and Chao (1999) measured $W_{D,D} = 0.83$ in a population of defector viruses compared against a pure population of cooperators, and also estimated that $W_{C,D} = 0.65$. Therefore, the empirical payoffs are:

	Cooperate	Defect
Cooperate	1.00	0.65
Defect	1.99	0.83

In the empirical payoff table, we see that $\overline{W} = W_{CC} = 1$ when $s_C = 1$. However, cheaters can invade because $W_{DC} = 1.99 > 1 = W_{CC}$. The invaders have a dominant strategy and so eventually the state evolves to $s_D = 1$ with lower average fitness $\overline{W} = W_{DD} = 0.83 < 1$—exactly as shown in the right side of Figure 2.3!

Essential conditions

A 2×2 game as in Table 2.3 is considered to be PD if the payoffs satisfy two conditions:

(a) $w_{DC} > w_{CC} > w_{DD} > w_{CD}$, and
(b) $w_{CC} > \frac{1}{2}(w_{DC} + w_{CD}) > w_{DD}$.

Condition (a) says that the highest payoff (often called the temptation) is for defecting on a cooperator, next is mutual cooperation, followed by mutual defection, and the lowest (often called the sucker's payoff) is for cooperating with a defector. This condition ensures that D is a strictly dominant strategy. The first part of condition (b) ensures that mutual cooperation has a higher mean payoff than the temptation-sucker combination, and the second part (which is overlooked in the literature we've seen) ensures that mutual defection has the lowest mean payoff.

The two conditions imply a strong tension between group interest (enhanced by cooperation according to condition b) and self interest (served by defection, the dominant strategy, by condition a). For that reason, PD-type games are true social dilemmas, and have attracted enormous attention.

An economic application

PD-type games turn up in politics (arms races are prime examples), literature (there is even a novel by that name!), and many other human endeavors. Here we mention a simple economic application suggested by a student.

eBay began to serve as an online auction platform in 1995, and by early 1997 it hosted millions of auctions every month. Unfortunately, in those early days, bidders on eBay couldn't be sure they were dealing with honest sellers. For example, sellers might create "shill" bidders to artificially push up the price, and this would give them an advantage over honest sellers. It is no longer easy to collect and analyze the historical data, but (for the sake of illustration), suppose that an analysis of early eBay sellers' profitability produced the following payoff table:

	Encounter rate:	
	s_H	$1-s_H$
Strategy:	Honest	Shill
Honest	5	0
Shill	6	1

That is, when your auction competes directly with another, your average profit is 5 if both auctions are honest but is 0 when you remain honest but the other seller attracts attention by using shills. The shill strategy is guesstimated to run slightly higher profits in both cases, 6 and 1 respectively.

It is easy to check that this game satisfies the PD conditions. The implication is that, even if 99% of sellers were honest when eBay first started, the share s_H would shrink, month by month and year by year, until it reached 0. This would greatly reduce average seller profitability from nearly 5 towards 1. That in turn, would be bad for eBay's business. To thrive, eBay had to change the game.

In 1998, with improvements in 1999, eBay introduced a reputation system that allowed bidders and sellers to rate each other. It was a game-changer, because sellers using shills (or engaging in other behavior to the detriment of buyers) risk bad ratings, which quickly lessen profit (Resnick and Zeckhauser 2002).

A guesstimate of the new game payoffs is

	Encounter rate:	
	s_H	$1-s_H$
Strategy:	Honest	Shill
Honest	5	3
Shill	6	0

That is, once the ratings system became established, honest sellers did much better against sellers using shills (payoff 3 instead of 0), and shill sellers did worse (0 instead of 1). Ratings have minimal impact (assumed zero) when almost all sellers are honest, but the impact increases with the share of dishonest sellers.

The new game clearly is no longer of type PD. What type is it? The composite parameters are $w_1 = 3-0 = 3$ and $w_2 = 6-5 = 1$. Since both are positive, the game is of type HD. Its unique steady state is at $s_H^* = \frac{w_1}{w_1+w_2} = \frac{3}{3+1} = 0.75$ and is stable. That is, if the numbers in the new table are correct, after things settled down, eBay participants should have expected to encounter 75% honest sellers and only 25% using shills. It is not hard to see that this implies that average payoff for sellers (of either kind) is 4.5. That is reasonably close to the original ideal value of 5, suggesting a very successful adaptation of eBay's business plan.

2.5 NONLINEAR FREQUENCY DEPENDENCE

In all examples presented so far, the fitness of each strategy i is given by an expression that is linear in the state $\mathbf{s} = (s_A, s_B)$. But sometimes fitness w_i is a nonlinear function of $\mathbf{s}$. For example, Chapter 12 will introduce a model of competition among two kinds of firms in a country. In that model, the fitness (i.e., profit) of firms of each kind is derived from the classic Cournot model and it turns out to be the quotient of fourth degree polynomials in the shares (s_A, s_B). Blume and Easley (1992) provides another economic example. In their model, fitness is the expected wealth of individual investors, and that turns out to be a nonlinear function of the alternative investment strategies that they consider.

To illustrate the main points of this section, consider the following simple example. Firms in the widget industry must adopt one of two alternative technologies. Technology A has decreasing returns to scale when rare and increasing returns when common; its profitability can be expressed as $w_A = 2s_A^2 - 2s_A + 1$, where s_A is the fraction of industry output produced using technology A. By contrast, technology B has moderately decreasing returns at all scales; its profitability is $w_B = 0.5(1.5 - s_B) = 0.5(s_A + 0.5)$ when $s_B = 1 - s_A$ of the output is produced using it.

Fitness functions that are nonlinear in $\mathbf{s}$ create fewer technical difficulties than one might suppose. Assume (as we have so far in this chapter) that there are two alternative strategies, so $\mathbf{s} = (s_A, s_B)$ where $s_B = 1 - s_A$. Write the payoff difference between the two available pure strategies as $\Delta w(s_A) = w_A(\mathbf{s}) - w_B(\mathbf{s})$. Although the fitnesses now are nonlinear, it is still true that the continuous replicator dynamic obeys equation (2.17). Thus $\dot{s}_A > 0$ (i.e., s_A increases) at interior states where $\Delta w(s_A) > 0$ and $\dot{s}_A < 0$ where $\Delta w(s_A) < 0$. Hence we have a steady state at any interior root s_A^* of the equation $\Delta w(s_A) = 0$.

In previous sections we established that the steady state s_A^* is stable if $\frac{d\Delta w(s_A^*)}{ds_A} < 0$, that is, if it is a downcrossing so Δw is downward sloping, because under this condition s_A decreases when it is above s_A^* and increases when it is below. By the same token, s_A^* is unstable if $\frac{d\Delta w(s_A^*)}{ds_A} > 0$, in other words, at an upcrossing, because s_A increases when it is above s_A^* and decreases when it is below. Recall also that the endpoint $s_A^* = 1$ is a stable steady state if $\Delta w(1) > 0$, and $s_A^* = 0$ is a stable steady state if $\Delta w(0) < 0$.

The logic is unchanged by the possibility that $\Delta w(s_1)$ is nonlinear. It is still true that states near a downcrossing move towards it, and states near an upcrossing move away. The only real novelty that nonlinearity brings is that the graph of Δw can cross the s_1-axis several times. Hence there can be several interior steady states. Of course, as in the linear case, either or both endpoints can also be stable steady states.

In the technology A versus B example, the fitness difference is $\Delta w(s_A) = w_A - w_B = 2s_A^2 - 2.5s_A + 0.75 = 2(s_A - 0.5)(s_A - 0.75)$, with roots $s_A^* = 0.5, 0.75$. As shown in Figure 2.5, the steady states $s_A^* = 0.5$ and $s_A = 1.0$ are stable because $\frac{d\Delta w}{ds_A}(0.5) = 4(0.5) - 2.5 = -0.5 < 0$ and $\Delta w(1) = 2(1)^2 - 2.5(1) + 0.75 = 0.25 > 0$. By contrast, the steady states $s_A^* = 0.75$ and $s_A = 0$ are unstable, since $\frac{d\Delta w(0.75)}{ds_A} = 4(0.75) - 2.5 = 0.5 > 0$ and $\Delta w(0) = 2(0)^2 - 2.5(0) + 0.75 = 0.75 > 0$.

Figure 2.5 is generic for 1-dimensional nonlinear dynamics in that downcrossings and upcrossings alternate. That is, the steady states are alternately stable and unstable, and the unstable steady states separate the basins of attraction of the stable steady states.

What do we mean by "generic"? The nice alternation property holds for any continuously differentiable fitness difference function Δw that has transverse crossings (i.e., $\frac{d\Delta w(s_1^*)}{ds_1} \neq 0$ at every interior root s_1^*) and nonvanishing endpoints (i.e., $\Delta w(0), \Delta w(1) \neq 0$). To be nongeneric, the fitness difference function must either be tangent to the horizontal axis at an interior root, or have a root at an endpoint. Such situations are very delicate; even the slightest change to the fitness

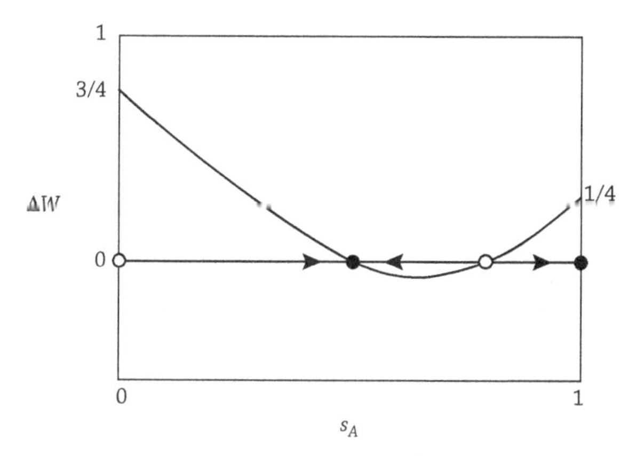

Figure 2.5 Graph of nonlinear fitness advantage $\Delta w = 2(s_A - 0.5)(s_A - 0.75)$. The unstable steady state $s_A^* = 0.75$ separates the basins of attraction for stable steady states $s_A^* = 0.5$ and 1.0.

function will break the tangency or move the root away from the endpoint. On the other hand, in the generic case, a sufficiently small change in the fitness function shifts the steady states only slightly and doesn't change their number or their stability. So generic here means robust as well as typical.

2.6 RPS AND THE SIMPLEX

The rest of the chapter will take a look at what can happen when players have three alternative strategies. We focus on the famous rock-paper-scissors game, using lowercase notation for continuous time games (e.g., rps with strategies r, p, s) and UPPERCASE for discrete time games (e.g., RPS with strategies R, P, S).

Building up from edge games

Let us begin with a very simple continuous rps game in which ties (e.g., when r meets r) get zero payoff, wins (e.g., for p when matched with r) get payoff 1, and losses (e.g., for r when matched with p) get −1. The payoff table for this game is:

encounter rate:	s_r	s_p	s_s
strategy :	r	p	s
r	0	−1	1
p	1	0	−1
s	−1	1	0

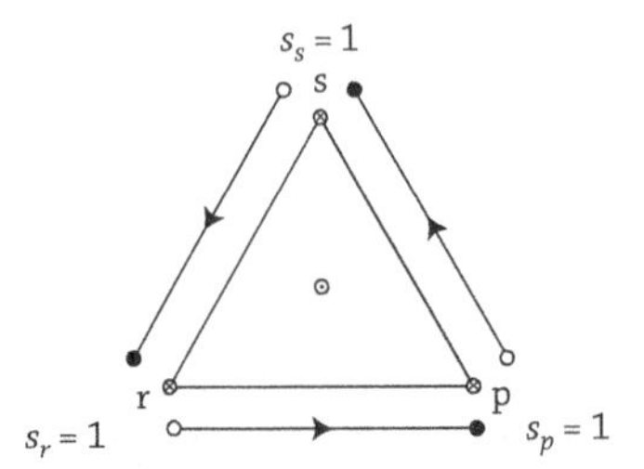

Figure 2.6 The three-strategy simplex exploded to show the three edge games for true-rps. Notice that each vertex is a source in one edge game and sink in another, and so each vertex in the full game is a saddle.

Recall from Chapter 1 that the state space is the simplex $\mathcal{S} = \{(s_r, s_p, s_s) \in \mathbb{R}^3_+ \mid s_r + s_p + s_s = 1\}$. The simplex is 2-dimensional because each state is pinned down by two of the three shares. For example, if $s_r \geq 0$ and $s_p \geq 0$ are known, then we also know that $s_s = 1 - s_r - s_p \geq 0$. The central triangle in Figure 2.6 represents $\mathcal{S}$; it is sometimes called a tenary plot, or a De Finetti diagram, or barycentric coordinates.

Note that each edge of $\mathcal{S}$—for example, the states for which $s_s = 0$ and so $s_r + s_p = 1$—is itself the state space of a game with only two alternative strategies. We will leverage the insights gained earlier in the chapter in order to better understand behavior in the three-strategy game.

In particular, the edge game of r versus p is given by the upper-left 2×2 block of the last payoff table:

encounter rate:	s_r	s_p
Player :	r	p
r	0	−1
p	1	0

To apply the analysis of Section 2.3, set $w_1 = w_{r,p} - w_{r,r} = -1 - 0 = -1$ and similarly $w_2 = w_{p,r} - w_{p,p} = 1 - 0 = 1$. Since $w_2 > 0 > w_1$, the game type is DS, with strategy p (paper) dominant over r (rock). Under replicator dynamics, paper will go to fixation at $s_p = 1$, as illustrated in the bottom line in Figure 2.6.

The edge game of p versus s is given by the lower–right 2×2 block of the rps payoff table:

encounter rate:	s_p	s_s
Player :	p	s
p	0	−1
s	1	0

The analysis of Section 2.3 quickly tells us that this edge game is also of type DS, with s dominant over p. Scissors will go to fixation, $s_s = 1$, as in the slanted line on the right side in Figure 2.6.

The final edge game, s versus r, has a payoff table obtained from the corners of the 3×3 rps table:

encounter rate:	s_r	s_s
Player :	r	s
r	0	1
s	−1	0

Once again, the game type is DS with r dominant over s. Rock will go to fixation, $s_r = 1$, as in the slanted line on the left side in Figure 2.6.

The payoffs in the game just analyzed are, of course, very special; there are lots of 3×3 payoff matrices $\mathbf{w} = ((w_{ij}))_{i,j=r,p,s}$ with the same basic cyclical DS structure. We will say that a 3×3 game $\mathbf{w}$ is *true-rps* if the following strict inequalities hold:

$$w_{s,r} < w_{r,r} < w_{p,r} \tag{2.18}$$

$$w_{r,p} < w_{p,p} < w_{s,p} \tag{2.19}$$

$$w_{p,s} < w_{s,s} < w_{r,s}. \tag{2.20}$$

That is, each strategy (when common) beats another strategy but is beaten by the remaining strategy, and the pairwise dominance relations are intransitive (i.e., cyclical).

2.7 REPLICATOR DYNAMICS FOR RPS

Intuitively, we can expect some sort of cycling behavior in true-rps games. When r is common, it loses share to p, which (when it becomes common) then loses share to s, which in turn loses share to r again. But do the cycles eventually die out? Or do they settle into a moderate amplitude, or perhaps become more extreme over time?

To figure it out, we turn once more to replicator dynamics. We begin with discrete time because, as you will see in the chapter-end exercises, it is easy to do such simulations with an electronic spreadsheet. Also, many organisms that express RPS dynamics actually have non-overlapping generations and for them discrete time is in fact a natural formulation.

Given a 3×3 payoff table with entries W_{ij}, $i,j = R,P,S$, the fitness of the three strategies at a given state $\mathbf{s} = (s_R, s_P, s_S) \in \mathcal{S}$ are:

$$W_R = s_R W_{R,R} + s_P W_{R,P} + s_S W_{R,S}\,, \tag{2.21}$$

$$W_P = s_R W_{P,R} + s_P W_{P,P} + s_s W_{P,S}\,, \tag{2.22}$$

and

$$W_S = s_R W_{S,R} + s_P W_{S,P} + s_S W_{S,S}\,. \tag{2.23}$$

Using these linear fitness functions and given initial shares $\mathbf{s}(0) = (s_R(0), s_P(0), s_S(0)) \in \mathcal{S}$, discrete time replicator dynamics are defined recursively for times $t = 1, 2, \ldots$ by:

$$s_i(t+1) = s_i(t)\frac{W_i(\mathbf{s}(t))}{\overline{W}(\mathbf{s}(t))},\ i = R,P,S, \tag{2.24}$$

where $\overline{W}(\mathbf{s}) = s_R W_R + s_P W_P + s_S W_S$.

If you run your spreadsheet simulator on the following RPS table, you will see that, beginning from almost any interior state, over time the state will spiral outward until it almost reaches the edges of the simplex:

encounter rate:	s_r	s_p	s_s
Player :	R	P	S
R	1	$\frac{1}{2}$	$\frac{3}{2}$
P	$\frac{3}{2}$	1	$\frac{1}{2}$
S	$\frac{1}{2}$	$\frac{3}{2}$	1

On the other hand, if you try the next payoff table, the cycle will die out and the state will spiral into the center of the simplex:

encounter rate:	s_r	s_p	s_s
Player :	R	P	S
R	1	$\frac{1}{2}$	4
P	4	1	$\frac{1}{2}$
S	$\frac{1}{2}$	4	1

Finally, apart from rounding or other errors, you should get cycling forever with the same amplitude when you use this payoff table:

encounter rate:	s_r	s_p	s_s
Player :	R	P	S
R	1	$\frac{1}{2}$	2
P	2	1	$\frac{1}{2}$
S	$\frac{1}{2}$	2	1

We'll say more about the stability properties of steady states in Chapter 4.

Lizard games

Male side-blotched lizards (*Uta stansburiana*) display three different mating strategies that are color coded as orange, blue, and yellow, by the colors on their throats (Sinervo and Lively 1996). The orange male strategy (which we'll designate as "rock") is fueled by high levels of testosterone, and involves aggressive attempts to exclude other males from very large territories and thus acquire a large female harem. The yellow male strategy ("paper") mimics female behavior. It is very effective when orange is common, because then yellow males can sneak in to copulate with many females while the orange males are busy fighting. The blue male strategy ("scissors") aggregates with other blue male neighbors to mutually defend adjacent territories against yellow, thereby thwarting the cheat strategy of yellow. But blues are no match for oranges in direct conflicts.

The side-blotched lizard is an annual with discrete one-year generation time. If we ignore genetics for the moment (and the interesting clumping behavior of blue males) and assume that the fitness of each type of lizard is due to a simple allele substitution, then we have a simple way to estimate fitness. Just measure the frequency of color alleles in each generation and use a technique called DNA paternity to figure out how many progeny were sired by each color type. Then measure fitness of each strategy each year as the number of progeny sired by that type divided by the number of adult males of that type.

Bleay et al. (2007) actually did this kind of genetic paternity testing, and they also tested paternity in an experiment reminiscent of Turner and Chao's work with viruses. Bleay et al. (2007) seeded a number of isolated neighborhoods (sandstone outcroppings) with lizard hatchlings with randomly assigned and quite varied initial shares of orange, blue, and yellow alleles. The investigators returned the next year, obtained eggs and hatchlings from all the reproductive females, and tested their paternity. They regressed the fitness of each male type (proportion of progeny sired in a female's clutch) against the frequency of each male allele, using data from all the isolated neighborhoods.

Figure 2.7 summarizes the results. The 3×3 array of panels shows how each male color type fares in competition against the other male types. The row i fitted line's intersection with the right edge of each panel (i.e., with the line $s_j = 1$ for

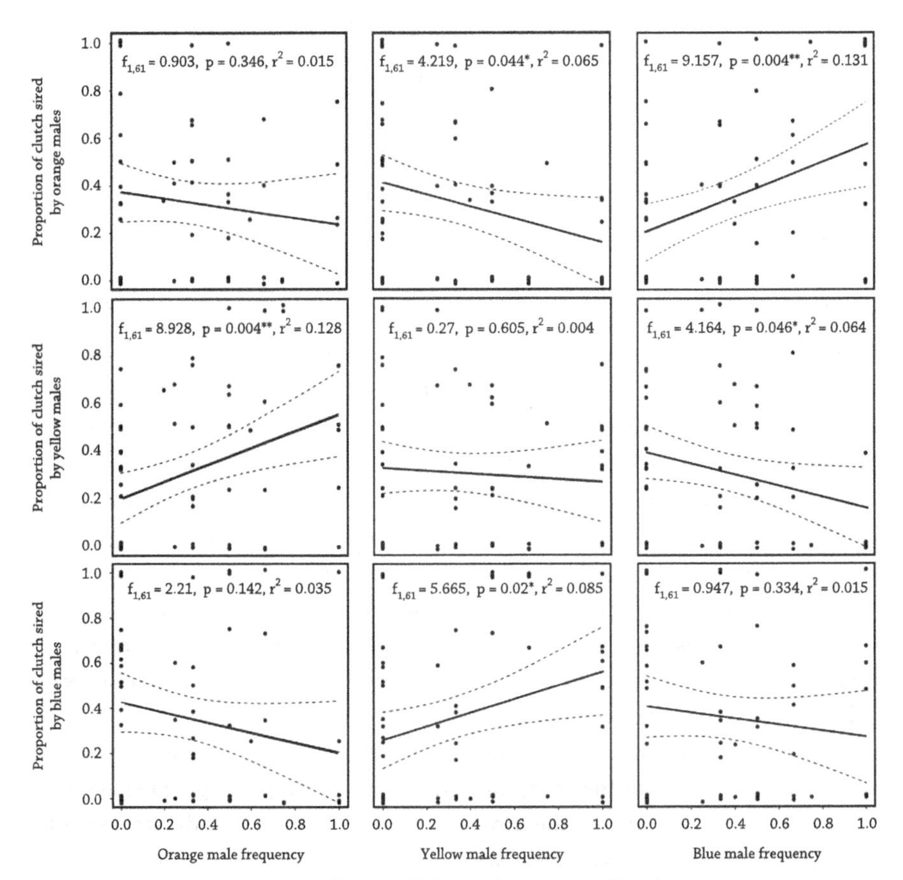

Figure 2.7 Fitness of orange, yellow, and blue males measured by the average proportion of the clutch that they sire, in neighborhoods of each color allele frequency.

panels in column j) estimates the fitness (W_{ij}, the proportion of progeny sired) of the color type in row i when color type j is common. While in the viral example of a 2×2 game we can simply use linear regression to estimate fitness payoffs, in 3×3 games we must use multiple regression to estimate the effects of each strategy, as in Sinervo and Lively (1996). (An appendix to Chapter 3 discusses the issue more broadly and offers more sophisticated statistical techniques for estimating 3×3 payoffs.)

The estimated payoffs are as follows.

$$\mathbf{W} = \begin{pmatrix} W_{oo} & W_{oy} & W_{ob} \\ W_{yo} & W_{yy} & W_{yb} \\ W_{bo} & W_{by} & W_{bb} \end{pmatrix} = \begin{pmatrix} 0.22 & 0.16 & 0.58 \\ 0.56 & 0.26 & 0.17 \\ 0.20 & 0.57 & 0.26 \end{pmatrix}.$$

As you will verify in a chapter-end exercise, replicator dynamics and steady states are unchanged when you multiply all entries in a given column by a positive constant. Here we normalize by dividing the numbers in each column by the diagonal element of that column, so that fitnesses are relative to the common type:

$$\mathbf{W} = \begin{pmatrix} 1.00 & 0.62 & 2.23 \\ 2.55 & 1.00 & 0.65 \\ 0.91 & 2.19 & 1.00 \end{pmatrix}.$$

Taking the natural logarithm, we obtain a very rough (see notes) continuous time approximation of the game:

$$\mathbf{w} = \begin{pmatrix} w_{oo} & w_{oy} & w_{ob} \\ w_{yo} & w_{yy} & w_{yb} \\ w_{bo} & w_{by} & w_{bb} \end{pmatrix} = \begin{pmatrix} 0 & -0.49 & 0.80 \\ 0.93 & 0 & -0.42 \\ -0.09 & 0.78 & 0 \end{pmatrix}$$

Do lizards play rps? To check, let's look at the 2×2 edge games. Indeed y dominates o:

$$\begin{pmatrix} w_{oo} & w_{oy} \\ w_{yo} & w_{yy} \end{pmatrix} = \begin{pmatrix} 0 & -0.49 \\ 0.93 & 0 \end{pmatrix},$$

and b dominates y:

$$\begin{pmatrix} w_{yy} & w_{yb} \\ w_{by} & w_{bb} \end{pmatrix} = \begin{pmatrix} 0 & -0.42 \\ 0.78 & 0 \end{pmatrix},$$

and the final edge game shows that o dominates b:

$$\begin{pmatrix} w_{oo} & w_{ob} \\ w_{bo} & w_{bb} \end{pmatrix} = \begin{pmatrix} 0 & 0.80 \\ -0.09 & 0 \end{pmatrix}.$$

Alternatively, one can check true-rps inequalities (2.18)–(2.20) in the estimated **W** or **w** matrices, relabeling strategies rps as oyb. All these inequalities are indeed satisfied. For another example (involving yeast), see the chapter-end exercises.

2.8 DISCUSSION

In this chapter we took our first look at strategic interdependence or frequency dependence. It is expressed via payoff or fitness functions that depend on the current shares of all strategies.

The simplest cases involve linear dependence among two alternative strategies, $i=1,2$. Here the state space is the 1-dimensional line segment $[0,1]$ representing the share s_1 of the first strategy; it is understood that $s_2 = 1 - s_1$. Linear dependence is fully specified by four numbers that we have shown in a table, but are most compactly shown in a 2×2 payoff matrix $\begin{pmatrix} w_{11} & w_{12} \\ w_{21} & w_{22} \end{pmatrix}$.

We will see in the next chapter how to use matrix algebra to express state-dependent payoffs of each strategy, and mean payoff.

The key to evolutionary analysis of these 2×2 games is the payoff difference function $\Delta w = (w_{11}s_1 - w_{12}s_2) - (w_{21}s_1 - w_{22}s_2)$. We observed that Δw graphs as a straight line whose intersections with the vertical axis $s_1 = 0$ and with the vertical line $s_1 = 1$ are composite payoff parameters that completely describe the game dynamics. We also saw that the continuous replicator equation can be written in terms of Δw (times a binomial variance factor), and that solutions of the equation $\Delta w = 0$ define interior steady states s_1^*.

Generically, there are only three types of 2×2 evolutionary games: HD type, where s_1^* is a downcrossing and hence stable; CO type, where it is an upcrossing and hence separates the basins of attractions of the two corner steady states; and DS type, where it is a no crossing, in other words, no solution in $[0,1]$ exists for the equation $\Delta w = 0$, so one of the strategies is dominant.

With these results in hand, we considered empirical work that estimates payoff matrices in various biological and economic applications with only two alternative strategies. We also showed how to leverage the linear results to deal with smooth nonlinear payoff functions. Such experimental approaches to estimating payoffs have recently been applied to cancer cell dynamics (Wu et al. 2014) and we suspect the number of such experiments to estimate payoff matrices will only accelerate in the coming years. In the last two sections we leveraged the linear 2×2 results to begin to analyze games with three alternative strategies. An appendix beginning on the next page takes it a step further.

The next chapter will build on the foundation laid here, and will develop tools to analyze more complicated evolutionary games.

APPENDIX: PAYOFF DIFFERENCES IN 3×3 GAMES

It will turn out to be useful to generalize the payoff difference expression Δw that served us so well in 2×2 games, and apply it to rps and other 3×3 games. Readers who enjoy algebra may appreciate the symmetries and nuances. Other readers may just want to skip to the bottom line and regard that as background information.

To begin, consider a general 3×3 payoff matrix in continuous time:

$$\mathbf{w}=\begin{pmatrix} w_{aa} & w_{ab} & w_{ac} \\ w_{ba} & w_{bb} & w_{bc} \\ w_{ca} & w_{cb} & w_{cc} \end{pmatrix}$$

As usual, the strategies have fitness given by:

$$w_a = w_{aa}s_a + w_{ab}s_b + w_{ac}s_c, \tag{2.25}$$

$$w_b = w_{ba}s_a + w_{bb}s_b + w_{bc}s_c, \tag{2.26}$$

$$w_c = w_{ca}s_a + w_{cb}s_b + w_{cc}s_c, \tag{2.27}$$

and average fitness is given by:

$$\overline{W} = w_a s_a + w_b s_b + w_c s_c. \tag{2.28}$$

Define the following pairwise payoff difference functions:

$$\Delta w_{a-b} = w_a - w_b = (w_{aa} - w_{ba})s_a + (w_{ab} - w_{bb)})s_b + (w_{ac} - w_{bc})s_c \tag{2.29}$$

$$\Delta w_{b-c} = w_b - w_c = (w_{ba} - w_{ca})s_a + (w_{bb} - w_{cb})s_b + (w_{bc} - w_{cc})s_c \tag{2.30}$$

$$\Delta w_{c-a} = w_c - w_a = (w_{ca} - w_{aa})s_a + (w_{cb} - w_{ab})s_b + (w_{cc} - w_{ac})s_c. \tag{2.31}$$

It is easy to confirm the following useful identities:

$$\Delta w_{a-b} + \Delta w_{b-c} + \Delta w_{c-a} = 0, \tag{2.32}$$

and

$$\Delta w_{i-j} = -\Delta w_{j-i}. \tag{2.33}$$

The continuous time replicator for this general 3×3 game reads

$$\dot{s_i}(t) = (w_i - \bar{w})s_i(t). \tag{2.34}$$

In particular, we have

$$\begin{aligned}
\dot{s}_c &= (w_c - \bar{w})s_c \\
&= (w_c - w_a s_a - w_b s_b - w_c s_c)s_c \\
&= s_c(w_c - w_c s_c - w_b(1 - s_a - s_c) - w_a s_a) \\
&= s_c(w_c - w_c s_c - w_b(1 - s_c) + w_b s_a - w_a s_a) \\
&= s_c(1 - s_c)(w_c - w_b) + s_c(w_b s_a - w_a s_a) \\
&= s_c(s_a + s_b)(w_c - w_b) + s_c s_a(w_b - w_a) \\
&= s_c s_a \Delta w_{c-b} + s_c s_b \Delta w_{c-b} + s_c s_a \Delta w_{b-a}.
\end{aligned} \tag{2.35}$$

Using the identities (2.32)–(2.33) we have

$$\begin{aligned}
\dot{s}_c &= s_c s_a \Delta w_{c-b} + s_c s_b \Delta w_{c-b} + s_c s_a \Delta w_{b-a} \\
&= s_c s_a(\Delta w_{a-b} + \Delta w_{c-a}) + s_c s_b \Delta w_{c-b} + s_c s_a \Delta w_{b-a} \\
&= -s_c s_a \Delta w_{b-a} + s_c s_a \Delta w_{c-a} + s_c s_b \Delta w_{c-b} + s_c s_a \Delta w_{b-a} \\
&= s_c s_a \Delta w_{c-a} + s_c s_b \Delta w_{c-b}.
\end{aligned}$$

The same sort of algebra yields:

$$\dot{s_a} = s_a s_b \Delta w_{a-b} + s_a s_c \Delta w_{a-c}. \tag{2.36}$$

$$\dot{s_b} = s_b s_a \Delta w_{b-a} + s_b s_c \Delta w_{b-c}. \tag{2.37}$$

That is, with continuous replicator dynamics, the growth rate $\frac{\dot{s}_i}{s_i}$ of each strategy share in any 3×3 game is a weighted sum of its payoff advantages Δw_{i-j} and Δw_{i-k} over the alternative strategies.

EXERCISES

1. Build a spreadsheet to simulate discrete replicator dynamics for general HD games. Have labeled cells near the top for entering the parameters v and c, and write formulas that display the resulting 2×2 matrix. Include another labeled cell for the initial H share. Then build up a row and column display similar to that used in the Chapter 1 spreadsheet to track H and D shares for 30 or more periods. Verify that you get convergence to the interior steady state v/c for appropriate values of the parameters.
2. Build a similar spreadsheet to simulate discrete replicator dynamics for RPS and other 3×3 matrix games. Check the claims made in Section 2.7 about stability and spiraling.

3. Consider the payoff matrix $\mathbf{P} = \begin{pmatrix} 0 & 5 \\ -1 & 3 \end{pmatrix}$.

   a. Verify that the first strategy (call it A) is dominant.
   b. Confirm that the mean payoff as a function of the share $x = s_A$ is $\overline{W}(x) = 3 - 2x - x^2$.
   c. Show that mean payoff is decreasing in x and conclude that game $\mathbf{P}$ has subtype PD.
   d. Now switch the diagonal entries 0 and 3 in $\mathbf{P}$ to obtain a new game matrix $\mathbf{Q}$. Verify that the first strategy A is still dominant, but that mean payoff is increasing in $x = s_A$ and so $\mathbf{Q}$ has subtype LF.

   Hint: Figure 2.3 is based on this exercise.
4. For an arbitrary 2×2 payoff matrix

$$\begin{pmatrix} w_{AA} & w_{AB} \\ w_{BA} & w_{BB} \end{pmatrix},$$

   obtain a quadratic expression for the mean payoff as a function of the share $x = s_A$ of the first strategy. Hint: Write s_B as $1 - x$ and use the definitions of $w_A(x), w_B(x)$ and $\bar{w}$.
5. (a) For which sets of payoff parameters in the previous exercise is mean payoff increasing in x? Decreasing in x? Hint: differentiate the quadratic expression.
   (b) Verify that essential condition (a) in Section 2.4 implies that $w_D(x) > w_C(x)$ for all values of $x = s_C$ (dominance), and that condition (b) implies that mean payoff is a decreasing function of $s_D = 1 - x$ (inefficiency). Conclude that these conditions are sufficient for a true social dilemma.
6. Weissing (1991) presents a putative yeast RPS uncovered by Paquin and Adams (1983) with normalized payoffs of three mutants as follows:

$$\mathbf{W} = \begin{pmatrix} 1.00 & 1.18 & 0.88 \\ 0.85 & 1.00 & 1.16 \\ 1.13 & 0.86 & 1.00 \end{pmatrix}.$$

   Paquin and Adams cited in Weissing (1991) estimated relative fitness of the three yeast strains (H_{30}, H_{203}, and H_{133} for $i,j = 1,2,3$) using a method similar to that used by Turner and Chao (1998). Weissing cautions that the payoffs were derived in pairwise competition, not in the key test where all three strategies are present in three-way competition. Despite this caveat,

and assuming the payoffs hold up in 3-way competition, is this true-RPS? Does the state spiral in, or out or neither?
Hint: To check RPS status, analyze the edge games. For dynamics, use the DEsolver in R.

7. For the general 2×2 matrix games presented in Section 2.3, what can you say about behavior when $w_1 = w_2 = 0$? How do you interpret this situation? What can you say about behavior when $w_1 > w_2 = 0$? How do you interpret this situation? What can you say about behavior when $w_1 < w_2 = 0$? How do you interpret this situation?
Remark. In a later chapter, we will discuss extensive form games. They often give rise to situations in which two or more w_{ij}'s are equal. Therefore, when the payoff table comes from an extensive form game, it is much more likely to have some $w_i = 0$.

NOTES

The stag hunt is a story that became a game. The story is briefly told by Jean-Jacques Rousseau, in *A Discourse on Inequality* (1754): "If it was a matter of hunting a deer, everyone well realized that he must remain faithful to his post; but if a hare happened to pass within reach of one of them, we cannot doubt that he would have gone off in pursuit of it without scruple ..." The game is a prototype of the social contract.

The rock-paper-scissors game has held the imagination of game theorists for decades. Maynard Smith (1982) was one of the first to point out that the rock-paper-scissors game might promote an endless cycle in the context of replicator dynamics. Actually, Sewall Wright probably made the same conjecture decades before (1968) when he described an unnamed 3×3 frequency dependent game that could cycle endlessly. Weissing (1991) provides a thorough analytical treatment of such games. On the empirical side, Buss and Jackson (1979) worked out intransitive relations in coral growth dynamics (eight different species!) and suggested ecosystem dynamics might be governed by intransitive dynamics. However, the first payoff matrix for the rock-paper-scissors game in a single population was derived for lizards in the context of mating system dynamics.

Chapter 7 will point out that since those early formulations, the number of natural systems in which rock-paper-scissors dynamics has been identified is growing exponentially. It is now recognized to be one of the apex social systems in many organisms including isopods, damselflies, lizards, rodents, fish, insects, bacteria, and even humans.

Passing from a discrete time payoff matrix W to its continuous time analogue w, or the reverse, is tricky. It is inaccurate simply to take the log or to exponentiate the matrix entries, for reasons alluded to in Appendix B of Chapter 1. We will see in the appendix to Chapter 4 how to use unit eigenvectors to transform a matrix into an (essentially) diagonal matrix where the log or exp operations can be performed, and then to transform it back.

We work through a continuous time rps example in the R workspace "RPS rK workpace Ch01.RData," where we introduce the DE solver of the R statistical package. This work space also gives the r-K game of Figure 1.1.

BIBLIOGRAPHY

Bleay, C., T. Comendant, and B. Sinervo. "An experimental test of frequency-dependent selection on male mating strategy in the field." *Proceedings of the Royal Society B: Biological Sciences* 274, no. 1621 (2007): 2019–2025.

Blume, Lawrence, and David Easley. "Evolution and market behavior." *Journal of Economic Theory* 58, no. 1 (1992): 9–40.

Buss, L. W., and J. B. C. Jackson. "Competitive networks: Nontransitive competitive relationships in cryptic coral reef environments." *American Naturalist* 113, no. 2 (1979): 223–234.

Maynard Smith, John. *Evolution and the theory of games*. Cambridge University Press, 1982.

Paquin, Charlotte, and Julian Adams. "Frequency of fixation of adaptive mutations is higher in evolving diploid than haploid yeast populations." *Nature* 302 (1983): 495–500.

Resnick, Paul, and Richard Zeckhauser. "Trust among strangers in internet transactions: Empirical analysis of eBays reputation system." *The Economics of the Internet and E-commerce* 11, no. 2 (2002): 23–25.

Sinervo, Barry, and C. M. Lively. "The rock-paper-scissors game and the evolution of alternative male strategies." *Nature* 380 (1996): 240–243.

Turner, Paul E., and Lin Chao. "Prisoner's dilemma in an RNA virus." *Nature* 398, no. 6726 (1999): 441–443.

Weissing, Franz J. "Evolutionary stability and dynamic stability in a class of evolutionary normal form games." In *Game equilibrium models I*, pp. 29–97. Springer Berlin Heidelberg, 1991.

Wright, Sewall. *Evolution and the genetics of populations, 4 vols*. University of Chicago Press, 1968.

Wu, Amy, David Liao , Thea D. Tlsty , James C. Sturm , Robert H. Austin. "Game theory in the death galaxy: Interaction of cancer and stromal cells in tumour microenvironment." *Interface Focus* 4 (2014): 20140028.

3

Dynamics in *n*-Dimensional Games

Players in some games have three or more alternative strategies. Fitness (or payoff) may be frequency dependent, but the dependence might be nonlinear. Many sorts of interactions involve players drawn from two or more distinct populations. Replicator dynamics are not always the best model of how the state adjusts over time.

This chapter will describe and analyze evolutionary models that incorporate such complications. It will note numerous applications and suggest empirical approaches.

3.1 SECTORING THE 2-D SIMPLEX

We begin by deriving simple analytic techniques for analyzing any evolutionary game with a 3×3 payoff matrix. These techniques center on the delta (payoff difference) functions introduced in the previous chapter. Here we use them to chop up the simplex into sectors, and show that dynamics have tight bounds within each sector. Then, by piecing the sectors back together, we get a nice picture of the overall dynamics.

Consider an arbitrary linear game with three strategies played by a single population. Such games were described in the previous chapter via a payoff table with rows labeled by the pure strategies (call them a, b, c), columns labeled by their shares (s_a, s_b, s_c), and entries like w_{ab} specifying the fitness. In this chapter we use more compact notation, and put those entries into a 3×3 payoff matrix:

$$\mathbf{w} = \begin{pmatrix} w_{aa} & w_{ab} & w_{ac} \\ w_{ba} & w_{bb} & w_{bc} \\ w_{ca} & w_{cb} & w_{cc} \end{pmatrix}. \tag{3.1}$$

As before, the fitness w_i of strategy i is its weighted average fitness using the current shares s_j, $j \in \{a,b,c\}$ as weights, so $w_i(s_a,s_b,s_c) = s_a w_{ia} + s_b w_{ib} + s_c w_{ic}$ for $i = a,b,c$. The mean fitness is the weighted average fitness across strategies, $\bar{w}(s_a,s_b,s_c) = s_a w_a + s_b w_b + s_c w_c$. In matrix notation (see the Math Glossary if you are rusty), this takes the compact form

$$\bar{w}(\mathbf{s}) = \mathbf{sw} \cdot \mathbf{s} = \sum_{i,j=a,b,c} s_i s_j w_{ij}. \tag{3.2}$$

Recall from both previous chapters that the continuous replicator equation equates the growth rate of each strategy share to its relative fitness. Writing this out in present notation gives

$$\dot{s}_i = s_i(w_i(\mathbf{s}) - \bar{w}(\mathbf{s})),\ i = a,b,c. \tag{3.3}$$

Finally, recall that an appendix to the previous chapter showed that (by expanding $\bar{w}(\mathbf{s})$ and using a little algebra) the first continuous replicator equation can be rewritten in terms of payoff differences $\Delta w_{i-j} = w_i(\mathbf{s}) - w_j(\mathbf{s})$ as follows:

$$\dot{s}_a = s_a s_b \Delta w_{a-b} + s_a s_c \Delta w_{a-c} \tag{3.4}$$

and that the other two equations have similar re-expressions.

With this as background, we use a little more algebra to isolate the deltas. Re-written in terms of growth rates, equation (3.4) is

$$\frac{\dot{s}_a}{s_a} = s_b \Delta w_{a-b} + s_c \Delta w_{a-c}, \tag{3.5}$$

and analogously we have

$$\frac{\dot{s}_b}{s_b} = s_a \Delta w_{b-a} + s_c \Delta w_{b-c} \tag{3.6}$$

and

$$\frac{\dot{s}_c}{s_c} = s_a \Delta w_{c-a} + s_b \Delta w_{c-b}. \tag{3.7}$$

Now consider the set of points where the right-hand side of (3.5) is equal to the right hand side of (3.6). The expressions are linear so that set of points is a straight line, along which the growth rates of strategies a and b are equal, $\frac{\dot{s}_a}{s_a} = \frac{\dot{s}_b}{s_b}$. Similarly, by equating the right-hand sides of (3.6) and (3.7), we get a straight line where b and c have equal growth rates, and by equating (3.5) and (3.7) we get the line where a and c have equal growth rates or payoffs.

At the intersection of two of these straight lines, of course, all three growth rates are the same, by transitivity. What is that common growth rate? It can't be positive, since that would soon lead to a point s with $s_a + s_b + s_c > 1$, above the simplex. Nor can it be negative, since that would take us below the simplex. Hence the common growth rate must be zero, and the intersection therefore is a steady state.

So much for the intuition. The following algebra confirms the intuition, and shows that the equal growth lines have a surprisingly simple and useful characterization. The first equal growth line is the set of points $s = (s_a, s_b, s_c)$ in the simplex such that

$$0 = \frac{\dot{s}_a}{s_a} - \frac{\dot{s}_b}{s_b} = s_b \Delta w_{a-b} - s_a \Delta w_{b-a} + s_c \Delta w_{a-c} - s_c \Delta w_{b-c}. \tag{3.8}$$

Recall the simple identities mentioned in Chapter 2:

$$0 = \Delta w_{a-b} + \Delta w_{b-c} + \Delta w_{c-a} \tag{3.9}$$

and

$$\Delta w_{a-b} = -\Delta w_{b-a}. \tag{3.10}$$

Inserting (3.10) and then (3.9) into equation (3.8) we have:

$$\begin{aligned}
\frac{\dot{s}_a}{s_a} - \frac{\dot{s}_b}{s_b} &= s_b \Delta w_{a-b} + s_a \Delta w_{a-b} + s_c(\Delta w_{a-c} - \Delta w_{b-c}) \\
&= (s_b + s_a)\Delta w_{a-b} + s_c \Delta w_{a-b} \\
&= (s_a + s_b + s_c)\Delta w_{a-b} \\
&= \Delta w_{a-b}.
\end{aligned} \tag{3.11}$$

Similar derivations show that:

$$\frac{\dot{s}_b}{s_b} - \frac{\dot{s}_c}{s_c} = \Delta w_{b-c} \tag{3.12}$$

and

$$\frac{\dot{s}_c}{s_c} - \frac{\dot{s}_a}{s_a} = \Delta w_{c-a} \tag{3.13}$$

Thus the equal growth rate lines $\frac{\dot{s}_i}{s_i} = \frac{\dot{s}_j}{s_j}$ in the simplex coincide with the equal fitness lines $\Delta w_{i-j} = 0$! These lines break the simplex up into sectors (or wedges) in each of which the fitnesses of the three strategies has constant ordering, for example, $w_a > w_b > w_c$ in one sector and $w_b > w_c > w_a$ in another. Of course, if the lines intersect within the simplex, then as noted earlier, we have a steady state at which all three strategies have equal fitness.

To illustrate, consider the following payoff matrix, which is inspired by the European common lizard (expressed in continuous time with i,j = r,s,p):

$$\mathbf{w} = \begin{pmatrix} -c & 1 & 0 \\ -1 & 0 & 1 \\ 1 & 1 & -k \end{pmatrix} \tag{3.14}$$

Readers who want to streamline the algebra can substitute $c = k = \frac{1}{2}$ in the expressions to follow, with very little loss of insight.

Use the substitution $s_s = 1 - s_r - s_p$ to see that the delta functions are :

$$\Delta w_{r-s} = (1-c)s_r + s_s - s_p = 1 - cs_r - 2s_p \tag{3.15}$$

$$\Delta w_{s-p} = -2s_r - s_s + (k+1)s_p = -1 - s_r + (k+2)s_p \tag{3.16}$$

$$\Delta w_{p-r} = (1+c)s_r - ks_p = (1+c)s_r - ks_p \tag{3.17}$$

Notice that equations (3.15)–(3.17) sum to 0, as they must given the definitions of the deltas. Equating the last three expressions to zero yields the equal growth rate (and equal fitness) lines:

$$cs_r + 2s_p = 1 \tag{3.18}$$

$$-s_r + (k+2)s_p = 1 \tag{3.19}$$

$$(1+c)s_r - ks_p = 0. \tag{3.20}$$

To find the steady state $s^* = (s_r^*, s_s^*, s_p^*)$, we first solve the two equations by eliminating s_r via $[3.18] + c[3.19]$, to obtain

$$(2+2c+ck)s_p = 1 + c \Longrightarrow s_p^* = \frac{1+c}{2+2c+ck}. \tag{3.21}$$

Next we insert result (3.21) into (3.20):

$$(1+c)s_r = ks_p \Longrightarrow s_r^* = \frac{k}{1+c}s_p^* = \frac{k}{2+2c+ck}. \tag{3.22}$$

Finally we insert (3.21)–(3.22) into $s_s = 1 - s_r - s_p$, to obtain:

$$s_s^* = \frac{1+c-k+ck}{2+2c+ck}. \tag{3.23}$$

The $\Delta = 0$ lines are drawn in Panel A of Figure 3.1 for the values $c = k = 0.5$. For example, equation (3.18) describes a very steep line $\Delta w_{r-s} = 0$ where r and s have equal growth and fitness. To the left of this line, s_p is higher, so equation (3.15) tells us that $\Delta w_{r-s} < 0$ and $w_s > w_r$. To the right of this line, $\Delta w_{r-s} > 0$

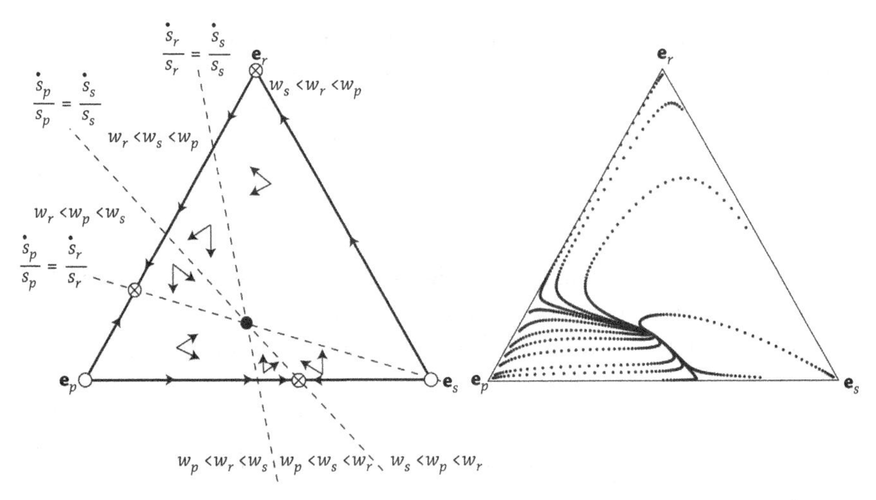

Figure 3.1 Asymmetric rsp game of the European common lizard with $c = k = \frac{1}{2}$. Panel A: Phase portrait for monotone dynamics. The interior steady state $(s_r^*, s_s^*, s_p^*) = (2,5,6)/13$ is globally stable. There are two edge steady states, both of them saddles: one at $(0,3,2)/5$ and the other at $(1,0,3)/4$. Panel B: Numerical solution of replicator dynamics. Notice the trajectories connecting the edge saddles to the stable interior equilibrium.

and $w_r > w_s$. The line $\Delta w_{p-s} = 0$—or $\frac{\dot{s}_p}{s_p} = \frac{\dot{s}_s}{s_s}$ as in the figure label—similarly separates the part of the simplex where $w_p > w_s$ from the other part where $w_s > w_p$. Of course, the remaining line separates the regions with different fitness orderings for r and p. Putting it all together, we get the six sectors shown in the figure with all the possible fitness orderings.

A last feature of Panel A is the directional arrows that pertain to a wide class of adjustment dynamics, not just to replicator dynamics. For example, in the sector containing the p corner, we have the ordering $w_p < w_r < w_s$. We should have $\dot{s}_p < 0$ since p has lowest fitness in this region, and $\dot{s}_s > 0$ since s has highest fitness. This means that the direction of adjustment should be somewhere between moving directly away from the p vertex (i.e., directly towards the s-r edge) and moving directly towards the s vertex (i.e., directly away from the p-r edge). The arrows show the range of possible adjustment directions in each sector, for any dynamics that respect the fitness ordering. We will discuss this wider class of dynamics, called monotone dynamics, later in this chapter.

The final details are the solution of edge equilibria which we can obtain by solving the 2×2 games on the edges. This is even simpler than the internal steady state, and we obtain the general formula for the $s_s = 0$ edge steady state of $(\frac{k}{k+1+c}, 0, \frac{1+c}{k+1+c})$ and the $s_r = 0$ edge steady state of $(0, \frac{k+1}{k+2}, \frac{1}{k+2})$. Inspection of the Δs_{r-p} and Δw_{s-p} equations reveals that both are downcrossings and stable equilbria along the edge but that each is actually a saddle given that the third strategy can invade.

Any 3×3 game can be analyzed in a similar fashion. The first exercise at the end of this chapter invites you to give it a try.

Panel B of Figure 3.1 shows that the interior steady state is stable under replicator dynamics. There is a general tendency for the state to move counterclockwise around the interior steady state, but not to cross the saddle path leading from the edge steady state (a saddle point) to the interior steady state (a sink). We used the *deSolve* package in [R] to produce the ternary plot (package *vcd*) with code shown at the end of the chapter. If you subscribe to Mathematica, you could also use a very nice package called Dynamo by Sandholm et al. (2012).

3.2 ESTIMATING 3×3 PAYOFF MATRICES

The 3×3 matrix specified in (3.14) is a mix of a priori beliefs (that the lizards play rsp) and auxiliary parameters (c,k). Rather than simply picking convenient parameter values, such as $c=k=0.5$ in Figure 3.1, it is preferable to let the data identify them. Sinervo et al. (2007) reports data that would allow c and k to be estimated. It would be better still, data permitting, to estimate the entire payoff matrix, and thus refine our a priori beliefs.

How do you estimate an entire payoff matrix? Chapter 2 suggested a simple approach for male lizards: define W_{ij} empirically as the fitness observed (progeny per capita) for phenotype i when phenotype j has nearly full share, that is, estimate it as the right edge intercept of the mean fitness line. However, as explained in the appendix below, this technique uses only a small part of the relevant data and has other shortcomings. In this section we sketch a more sophisticated approach to estimating 3×3 payoff matrices. The appendix collects more precise specifications and tests, and additional details can be found in Friedman et al. (2016) (article and supplement).

Figure 3.2 shows the data we will use to estimate the payoff matrix—shares of the three lizard morphs described in the previous chapter, based on annual hand counts of the male lizards in four separate locations. We code morph O as $i=1$, B as $i=2$, and Y as as $i=3$. There are 21 years of data from the "Main world," a large set of rock outcroppings in a location west of Los Baños, California, plus 12 years of data from each of three other locations, denoted O, B, and Y worlds for the (experimentally manipulated) initial frequencies of orange, blue, and yellow male morphs. The data also include population density for each year and each location, and (because it affects survival of young lizards) a summer temperature series.

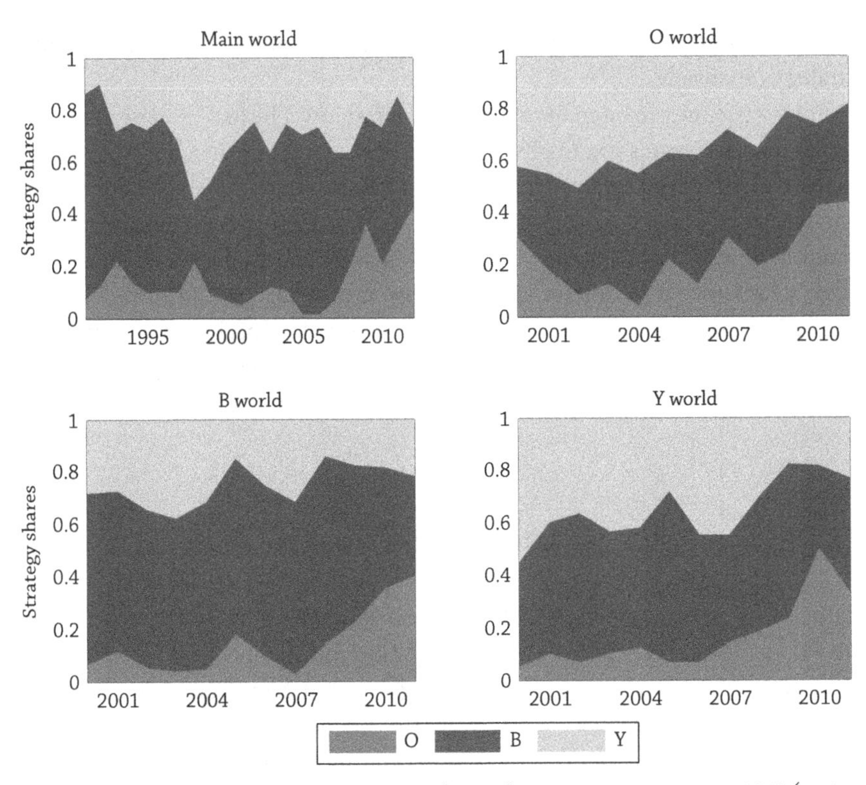

Figure 3.2 Annual allele shares in *Uta stansburia*, four separate sites, 1991–2011 (main site) and 1999–2011 (three smaller sites with selected alleles boosted in the first year, Sinervo et al. 2001). (Sinervo unpub. data).

The basic idea is to find nine fitness values W_{ij}, $i,j = 1,2,3$ that best account for the observed share dynamics, assuming that they obey a noisy version of the discrete time replicator equation. Confining our attention for now to the Main world, let $S_j(t)$ be the observed share of morph $j = 1,2,3$ at time t. Given any proposed fitness values W_{ij}, we compute the fitness $W_i(t) = W_{i\cdot} \cdot S(t) \equiv \sum_j W_{ij}S_j(t)$ of morph $i = 1,2,3$ at time t as the usual weighted average fitness against current population frequency. Thus mean fitness at time t is $\bar{W}(t) = S(t)W \cdot S(t) \equiv S(t) \cdot (W_1(t), W_2(t), W_3(t)) \equiv \sum_i \sum_j W_{ij}S_i(t)S_j(t)$.

Discrete time replicator dynamics give the next period population state $\mathbf{Z}(t+1) \in \mathcal{S}$ that would arise absent stochastic disturbances ("noise"), as follows:

$$Z_i(t+1) = \frac{W_i(t)}{\bar{W}(t)} S_i(t),\ i = 1,2,3. \tag{3.24}$$

Of course, the actual state $\mathbf{S}(t+1)$ will differ from the prediction $\mathbf{Z}(t+1)$ due to sampling fluctuations, good or bad luck for individuals, or unforeseen weather or predators or disease. To capture such disturbances, we could simply tack on

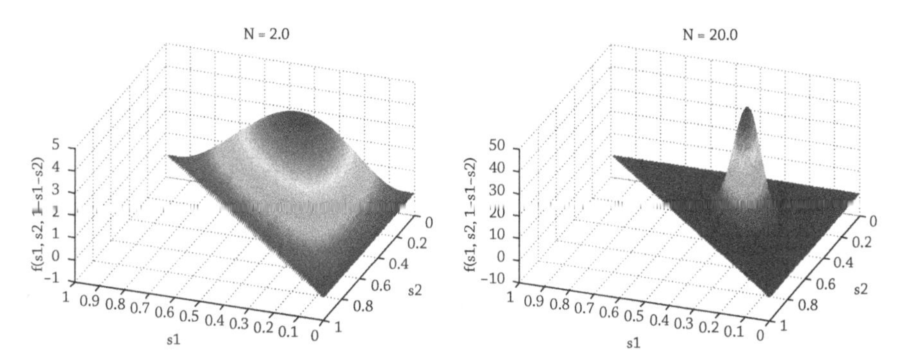

Figure 3.3 The Dirichlet density with mean and mode Z at the centroid of the simplex. Panel A has $N = 2 \cdot 3 = 6$, creating a more uniform density than in Panel B, where $N = 20 \cdot 3 = 60$ leads to a very sharp peak (note the compressed vertical scale).

an independent normally distributed error term, as is done routinely in many applications. But that would misspecify dynamics on the simplex. Except in the zero-probability event that the three such terms summed to 0, those terms would take us out of the simplex because the three shares would no longer add to 1.0. We could impose a correlated error structure to ensure a zero sum, but then we'd still face the problem that sometimes a share might end up negative or larger than 1.0. There are ways to repair this, but they are messy and tend to create new problems.

Fortunately, there is an off-the-shelf distribution function tailor-made for the simplex, the Dirichlet distribution. As explained in the appendix, we complete the structural model via $S(t+1) \sim Dir(N\mathbf{Z}(t+1))$, where $N > -1$ is the effective sample size of breeding adult males. The Dirichlet distribution Dir has density zero outside the simplex, eliminating the problems just mentioned. It also ensures that each $S_i(t+1)$ is a random variable with mean $Z_i(t+1)$ as specified in the deterministic model, and variance $\frac{Z_i(t+1)[1-Z_i(t+1)]}{N+1}$. Thus N can be interpreted as a precision parameter, inversely related to the variance, as illustrated in Figure 3.3.

The idea now comes down to finding values of the parameters N, W_{ij} that are most likely to produce the observed data. These parameters are subject to several constraints. First, as explained in Chapter 1, fitnesses in the discrete replicator can't be negative, so each matrix entry must satisfy $W_{ij} \geq 0$. We also must normalize the matrix entries because they appear in both the numerator and denominator of equation (3.24), so there is no effect when we multiply all matrix entries by an arbitrary constant $c > 0$. To resolve that indeterminacy, we impose the constraint $\sum_i \sum_j W_{ij} = 1$—that is, we normalize the original matrix by multiplying by $c = \frac{1}{\sum_i \sum_j W_{ij}}$.

The appendix derives exact expressions for the likelihood that the data would be as observed, conditional on Dirichlet/replicator dynamics with the proposed

Table 3.1. ESTIMATES OF THE NORMALIZED FITNESS MATRIX **W**, WITH BOOTSTRAPPED STANDARD ERRORS IN PARENTHESES.

0.0000 (0.0352)	0.2462 (0.0420)	0.0207 (0.0472)
0.1383 (0.0469)	0.0600 (0.0232)	0.1941 (0.0456)
0.1701 (0.0527)	0.1705 (0.0395)	0.0000 (0.0319)

payoff matrix $\mathbf{W} = ((W_{ij}))$ and precision parameter N. A standard numerical algorithm searches the 9-dimensional constrained parameter space of proposals, and finds a parameter vector that maximizes the likelihood of the observed share data. This vector is called the maximum likelihood estimate (MLE) for the model.

Applying these procedures to all the data shown in Figure 3.2 (see the appendix for details on how to combine data from the four different "worlds"), we obtain the fitness matrix estimates shown in Table 3.1, together with a precision parameter estimate of $\hat{N} = 21.7 \pm 3.4$, which is analogous to the population genetics concept of effective population size, N_e.

Note that in a predominantly Orange state (first column) the fittest strategy at 0.17 is the sneaker strategy Yellow (third row), while Blue (center row) at 0.19 is fittest against Yellow (third column) and Orange at 0.25 is fittest against Blue (center column). Hence the data confirm that the lizards play a generalized RPS game. (It is not an true-RPS game since each morph does worse against itself than against the third strategy.) The estimated value of N is much smaller than actual male population, typically around 100 in each world. This suggests considerable background noise, likely due to idiosyncratic events such as unusually hot or cold weather, or a large or small influx of lizard-eating snakes. A complementary explanation in terms of the RPS game itself is that in fact, only a fraction of males have progeny that survive to be counted the next year, selection culls the actual number of alleles that reproduce, similar to the effects of selection on effective population size.

The appendix extends the approach sketched here to account for two important covariates—population density (which also captures some of the impact of female mating strategies) and temperature (essentially, the number of summer afternoons too hot for young lizards to venture out from under a rock). It turns out that these covariates shift the payoff matrix considerably, sometimes to the detriment of Orange (e.g., in 1994–97 and again in 2002 and 2004–05 in Main world) and sometimes to the detriment of Yellow (e.g., 2007–08 and 2010–11 in Main world). The covariate results suggest that distributions become entrained in all four worlds—even though O, B, and Y worlds were started with very different morph shares in 1999, within five or six years they tended to move more in synch. As explained in our (2016) paper, these results also suggest different sorts of morph loss at the extreme northern and southern ends of this species' range. Section 14.3 discusses the implications for forming new species.

3.3 MORE STRATEGIES

It is conceptually straightforward to work with any finite number of alternative strategies. The logic is exactly the same as with two or three strategies, but we need more sophisticated notation since now we are working with n-vectors and $n \times n$ matrices.

Suppose now that our single-population model has strategies $\{1,2,\ldots,n\}$. As always, we denote by $s_i = \frac{N_i}{N}$ the share of strategy i, that is, the number N_i of individuals using that strategy as a fraction of the population size N. The state vector is $\mathbf{s} = (s_1,\ldots,s_n)$, in other words, the state space $\mathcal{S}$ consists of row vectors whose components s_i are non-negative and sum to 1.

Fixation on strategy i is indicated by the vector $\mathbf{s} = \mathbf{e}_i = (0,\ldots,0,1,0,\ldots,0)$, that is, $s_i = 1$ and the other shares are 0. For example, in the rps game, $n = 3$ and the state space $\mathcal{S}$ is the equilateral triangle. Its three vertices $\mathbf{e}_1 = (1,0,0)$, $\mathbf{e}_2 = (0,1,0)$, and $\mathbf{e}_3 = (0,0,1)$ denote respectively the three pure strategies r, p, and s.

Cason et al. (2010) analyze an $n = 4$ strategy game called rock-paper-scissor-dumb (rpsd) with fitness matrix

$$\mathbf{w} = \begin{pmatrix} 9 & 0 & 12 & 2 \\ 12 & 9 & 0 & 2 \\ 0 & 12 & 9 & 2 \\ 9 & 9 & 9 & 0 \end{pmatrix}.$$

Despite its name, the last strategy is not especially stupid—it is the second-best response to the other strategies, albeit its own worst response. More to the point, the state space is the 3-dimensional simplex spanned by the vertices $\mathbf{e}_1 = (1,0,0,0)$, $\mathbf{e}_2 = (0,1,0,0)$, $\mathbf{e}_3 = (0,0,1,0)$, and $\mathbf{e}_4 = (0,0,0,1)$. A state $\mathbf{s} = (s_1,\ldots,s_4) = \sum_{i=1}^{4} s_i \mathbf{e}_i$ is a mix of these four pure strategies. This state space can be visualized as a tetrahedron, as shown in Figure 3.4.

It is hard to visualize simplicies when $n \geq 5$, but it is straightforward to write out the algebraic expressions. The most general finite strategy games with one population and linear frequency dependence can be summarized by a payoff

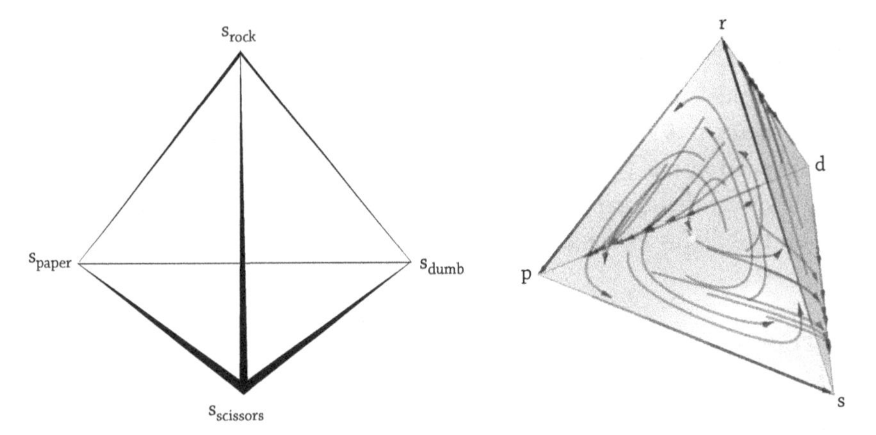

Figure 3.4 (a) The state space for the rpsd game **w**, with strategies (1,2,3,4) labeled respectively (rock, paper, scissors, dumb). (b) A few trajectories under replicator dynamics for **w**, drawn by Dynamo (Sandholm et al. 2015).

matrix

$$\mathbf{w} = ((w_{ij})) = \begin{pmatrix} w_{11} & w_{12} & \dots & w_{1n} \\ w_{21} & w_{22} & \dots & w_{2n} \\ \vdots & \vdots & \dots & \vdots \\ w_{n1} & w_{n2} & \dots & w_{nn} \end{pmatrix}. \tag{3.25}$$

The fitness w_i of each pure strategy i at the current state $\mathbf{s} \in \mathcal{S}$ can be written compactly in matrix notation:

$$w_i(\mathbf{s}) = \mathbf{e}_i \mathbf{w} \cdot \mathbf{s} \equiv \sum_{j=1}^{n} w_{ij} s_j. \tag{3.26}$$

As usual, the equation just says that the fitness of strategy i is its weighted average fitness against all stategies j and the weights are their current shares s_j.

For example, at the centroid state $\mathbf{s} = (1,1,1,1)/4$ in the rpsd game, we have $w_1 = w_2 = w_3 = (9+0+12+2)/4 = 23/4$ and $w_4 = (9+9+9+0)/4 = 27/4$. The first three strategies have the standard rps feature that w_i is highest (=12) against the preceding strategy $\mathbf{s} = \mathbf{e}_{i-1}$ (using the mod 3 convention that 1 follows 3) and is lowest (=0) against the next strategy $\mathbf{e}_{i+1}$ (mod 3). Strategy 4, called d or dumb, has fitness $w_4 = 9$ against the other pure strategies but fares badly against itself: $w_4(\mathbf{e}_4) = 0$.

Continuous time replicator dynamics have a familiar look even with with n alternative strategies. Using the abbreviation $\mathbf{s} = (s_1, \dots, s_n)$ for $\mathbf{s}(t) = (s_1(t), \dots, s_n(t))$, we have

$$\dot{s_i} = (w_i - \bar{w}) s_i, \; i = 1, \dots, n, \tag{3.27}$$

where $\bar{w} = \sum_{j=1}^{n} w_j s_j = \sum_{i=1}^{n} \sum_{j=1}^{n} w_{ij} s_i s_j \equiv \mathbf{sw} \cdot \mathbf{s}$ is average fitness. As usual, the equation simply says that each strategy i's growth rate is its fitness relative to the population average.

Discrete time replicator dynamics can also be expressed nicely using matrix notation. For a time step of $\Delta t > 0$, use the abbreviation $\mathbf{s}' = (s'_1, \ldots, s'_n)$ for $\mathbf{s}(t + \Delta t) = (s_1(t + \Delta t), \ldots, s_n(t + \Delta t))$, and express each fitness matrix element in discrete terms via $W_{ij} = \exp(w_{ij} \Delta t)$. (Unless otherwise stated, we shall assume $\Delta t = 1$.) Then current average fitness is

$$\bar{W} = \mathbf{sW} \cdot \mathbf{s} \equiv \sum_{i=1}^{n} \sum_{j=1}^{n} W_{ij} s_i s_j \tag{3.28}$$

and the discrete replicator equation is

$$s'_i = \frac{W_i}{\bar{W}} s_i, \; i = 1, \ldots, n. \tag{3.29}$$

Is there a higher-dimensional generalization of the way we sectored the simplex in 3×3 games? To get started on this question, we return to continuous time replicator dynamics, and rewrite equation (3.27) as

$$\begin{aligned} \dot{s}_i / s_i = w_i - \bar{w} = w_i \sum_{j=1}^{n} s_j - \sum_{j=1}^{n} w_j s_j = \sum_{j=1}^{n} (w_i - w_j) s_j \\ = \sum_{j=1}^{n} s_j \Delta w_{i-j}. \end{aligned} \tag{3.30}$$

Of course, we can rewrite the formula as

$$\dot{s}_i = \sum_{j=1}^{n} s_i s_j \Delta w_{i-j}, \; i = 1, \ldots, n, \tag{3.31}$$

a direct generalization of equation (3.3). But equation (3.30) itself provides a useful insight: the growth rate of any strategy's share, $\dot{s}_i / s_i$, comes at the expense of every other strategy's share. Strategy i grows at the expense of strategy j to the extent that j has a large share s_j and a disadvantage Δw_{i-j} that is large and positive.

Where will two growth rates be equal? In the 2-dimensional simplex for 3×3 games it was along a straight line. In the 3-dimensional simplex (the tetrahedron) for the 4×4 game rpsd, the linear equation equating two growth rates will define a 2-dimensional plane. There are $4 \cdot 3/2 = 6$ ways to chose two different strategies to compare growth rates, so we might hope that these six planes chop the tetrahedron up into nice regions, and that if enough of them intersect, they do so at a steady state. We might even hope that they can be described by simple fitness differences, Δw_{i-j}, or at least some not-too-complex expression involving the deltas.

Putting hopes and fears aside for the moment, let's attack the question directly. Use (3.30) to write the growth rate for strategy k and subtract it from the given growth rate for strategy i. The result is

$$\dot{s}_i/s_i - \dot{s}_k/s_k = \sum_{j=1}^{n} s_j \Delta w_{i-j} - \sum_{j=1}^{n} s_j \Delta w_{k-j} \tag{3.32}$$

$$= \sum_{j=1}^{n} s_j (\Delta w_{i-j} - \Delta w_{k-j}) \tag{3.33}$$

$$= \sum_{j=1}^{n} s_j \Delta w_{i-k} = \Delta w_{i-k} \sum_{j=1}^{n} s_j \tag{3.34}$$

$$= \Delta w_{i-k}, \tag{3.35}$$

where (3.33) uses the simple identity $\Delta w_{i-j} - \Delta w_{k-j} = w_i - w_j - w_k + w_j = w_i - w_k = \Delta w_{i-k}$, and (3.34) uses the basic simplex identity $\sum_{j=1}^{n} s_j = 1$. The conclusion is as pretty as can be: as in the 3×3 case, the growth rate difference between any two strategies i and k is just their payoff difference Δw_{i-k}.

Thus we can chop up the n-simplex, an $n-1$ dimensional object, with a bunch of $n-2$ dimensional hyperplanes defined by the linear equations $\Delta w_{i-k} = 0$. For each pair $i \neq k$ we get a hyperplane, and there are $n(n-1)/2$ possible pairs. But do they all intersect the simplex? A moment's thought shows that the $\Delta w_{i-k} = 0$ hyperplane does cut through the simplex as long as there are points s such that $\Delta w_{i-k}(s) > 0$ and other points in the simplex such that $\Delta w_{i-k}(s) < 0$ (i.e., as long as neither strategy dominates the other). This will be the case as long as we have already eliminated strictly dominated strategies from among the n strategies under consideration.

We get a steady state at the intersection of any collection of $\Delta w_{i-k} = 0$ hyperplanes as long as every $j = 1, \ldots, n$ appears somewhere in the list of i's and k's. The reasoning is familiar: at such an intersection, all fitnesses must be the same, and their common value must be zero for the state to remain in the simplex.

In principle we can assess the stability of the steady state by looking at the directional cones in each region adjoining the steady state. This works very nicely, as we have seen, for $n = 3$ and sometimes it is helpful for $n = 4$. For higher dimensions it seems impractical. An appendix to Chapter 4 presents classic techniques (involving matrix eigenvalues) that continue to work well in higher dimensions.

3.4 NONLINEAR FREQUENCY DEPENDENCE

In the previous chapter, we saw that generic 1-dimensional (i.e., two alternative strategy) games with nonlinear strategic interdependence looked locally like linear games, and that you can apply the linear analysis to each piece of the state space to get the full picture. Can we do something similar in higher dimensions?

When fitness is a nonlinear function of $n = 3$ alternative strategies, the state space is still the 2-dimensional simplex and we can still draw the three $\Delta w_{i-j} = 0$ lines. But now the lines are usually not straight—they are curves that may intersect in more than one place in the interior of the simplex. (Again, if two of them cross at a point, so must the third, by transitivity.) The three lines again chop up the simplex into regions with constant fitness ordering. At each transversal intersection of these lines (and of the simplex edges), we can assess the local stability in exactly the same way as in the linear case. We can usually ignore the exceptional or nongeneric cases in which the lines are tangent to each other or intersect a vertex of the simplex.

In higher dimensions, the $\Delta w_{i-j} = 0$ loci for $i \neq j$ are nonlinear. They are generically $n - 2$ dimensional manifolds, that is, curved 2-dimensional surfaces in the 3-dimensional simplex when $n = 4$ as in Figure 3.5. For general n they are curved analogues of hyperplanes. Exactly as in the linear case, these loci chop up the simplex into sectors with constant ordering of the fitnesses w_i, for $i = 1, \ldots, n$. Each such ordering defines a cone of possible directions in which the system can evolve.

Any interior intersection of all the $\Delta w_{i-j} = 0$ loci is a steady state whose (local) stability can be assessed by looking at the cones in neigboring sectors. As we will see, this sort of assessment is not always conclusive. The eigenvalue techniques presented in Chapter 4 can be used in such cases.

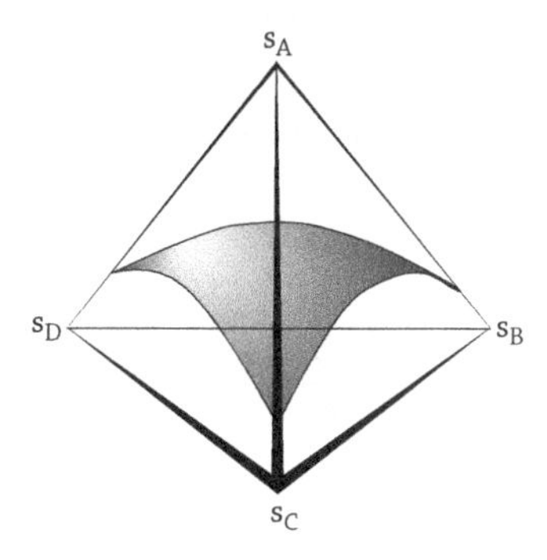

Figure 3.5 A nonlinear 2-dimensional $\Delta w_{i-j} = 0$ locus in the 4-simplex.

3.5 TWO-POPULATION GAMES: THE SQUARE

So far we have only considered strategic interaction within a single (large) population: every individual faces the same state $\mathbf{s}$ and obtains the same fitness $w_i(\mathbf{s})$ by choosing any particular strategy i from the same set of possible strategies. But in many interesting situations, individuals belonging to one population interact with individuals belonging to a different population with different alternative strategies and different fitness functions. For example,

- Buyers interact with sellers, and the buyers' payoffs depend mainly on what the sellers are doing. The sellers have different alternative strategies than buyers and a different payoff function.
- Ants sometimes maintain colonies of aphids, and the ants' fitness depends largely on what the aphids do, and of course the aphids' fitness depends on ant strategies.
- The fitness of a male strategy in many biological games depends on what the females are doing, and vice versa.
- The fitness of a guild of pollinators depends on what the flower species are doing, and vice versa.

To model such situations, we need to consider two distinct populations. Let population 1 have $n \geq 2$ alternative strategies with shares $\mathbf{p} = (p_1, p_2, \ldots, p_n)$ and population 2 have $m \geq 2$ alternative strategies with shares $\mathbf{q} = (q_1, q_2, \ldots, q_m)$. For example, buyers may $(i = 1)$ or may not $(i = 2)$ test milk before purchase, and sellers may $(j = 1)$ or may not $(j = 2)$ dilute the milk and add melamine. Here $n = m = 2$, the simplest case.

Suppose that the fitness of population 1 depends only on behavior in the second population (i.e., on $\mathbf{q}$), and fitness in population 2 depends only on behavior in the first population (i.e., on $\mathbf{p}$). Suppose also that the fitness functions are linear. This is known in the biological literature as "the asymmetric case."

In this case we can write the fitness functions in terms of a $n \times m$ matrix $\mathbf{u}$ and a $m \times n$ matrix $\mathbf{v}$:

$$u_i(\mathbf{q}) = \mathbf{e}_i \mathbf{u} \cdot \mathbf{q} = \sum_{k=1}^{m} u_{ik} q_k, \quad i = 1, \ldots, n, \tag{3.36}$$

and

$$v_j(\mathbf{p}) = \mathbf{e}_j \mathbf{v} \cdot \mathbf{p} = \sum_{k=1}^{n} v_{jk} p_k, \quad j = 1, \ldots, m. \tag{3.37}$$

Game theorists often write the two matrices together, as the $n \times m$ bimatrix $\mathbf{w} = [\mathbf{u}, \mathbf{v}^T] \equiv ((u_{ij}, v_{ji}))$. In the bimatrix, the first population's strategies define the rows, and the second population's strategies define the columns. This convention is the same as for the matrix $\mathbf{u}$ but is the opposite of the convention used for $\mathbf{v}$. Therefore the bimatrix transposes the second population's fitness matrix $\mathbf{v}$.

It's time for a simple, concrete example. Let $\mathbf{u} = \begin{pmatrix} 1 & -1 \\ -4 & 2 \end{pmatrix}$ and $\mathbf{v} = \begin{pmatrix} -8 & 2 \\ 0 & 1 \end{pmatrix}$ in the buyer-seller interaction just mentioned. The bimatrix $\mathbf{w}$ is $\begin{pmatrix} 1,-8 & -1,0 \\ -4,2 & 2,1 \end{pmatrix}$. Thus, if a buyer tests ($i = 1$), then (due to the expense and delay of testing) she gets $u_{12} = -1$ and the seller gets $v_{21} = 0$ if the seller is honest ($j = 2$), whereas she gets the "whistleblower" payoff $u_{11} = 1$ and the seller gets the penalty payoff $v_{11} = -8$ if the seller cheats ($j = 1$). Buyers' expected payoffs are given by (3.36), for example, buyers who test get $u_1(\mathbf{q}) = u_{11}q_1 + u_{12}q_2 = 1q_1 + (-1)(1 - q_1) = 2q_1 - 1$, and buyers who don't test get $u_2(\mathbf{q}) = u_{21}q_1 + u_{22}q_2 = -4q_1 + 2(1 - q_1) = 2 - 6q_1$. The expected payoff advantage to the buyer's first strategy is $\Delta u(\mathbf{q}) = u_1 - u_2 = 8q_1 - 3$. Thus it is advantageous for a buyer to test before purchasing when at least 3/8 of the sellers cheat, and not otherwise.

Similar calculations using (3.37) for the seller population yield fitness (or expected payoff) $v_1(\mathbf{p}) = v_{11}p_1 + v_{12}p_2 = -8p_1 + 2(1 - p_1) = 2 - 10p_1$ and $v_2(\mathbf{p}) = v_{21}p_1 + v_{22}p_2 = 0p_1 + 1(1 - p_1) = 1 - p_1$. The expected payoff advantage to the seller's first strategy is $\Delta v(\mathbf{s}) = v_1 - v_2 = 1 - 9p_1$, which is positive (i.e., it is advantageous for sellers to cheat) when fewer that 1/9 of the buyers test before purchase.

The state space $\mathcal{S}$ has a nice representation for two-population games with $n = m = 2$ alternative strategies for each population. It is the unit square $\{(p_1, q_1) \in [0,1] \times [0,1]\}$, as shown in Figure 3.6. Not shown explicitly, the alternative strategy shares of course are $p_2 = 1 - p_1$ for buyers and $q_2 = 1 - q_1$ for sellers.

With linear fitness functions, average payoffs are $\bar{u} = \mathbf{pu} \cdot \mathbf{q} = \sum_{i=1}^{n} \sum_{j=1}^{m} u_{ij}p_iq_j$ in population 1 and $\bar{v} = \mathbf{qv} \cdot \mathbf{p} = \sum_{j=1}^{m} \sum_{i=1}^{n} v_{ij}p_iq_j$ in population 2. Replicator dynamics are still given by equation (3.29) in discrete time and (3.27) in continuous time. The main novelty here is that the dynamics of the two populations are intertwined, since the first population's fitnesses u_i now depend on the state q of the second population and the second population's fitnesses v_j now depend on the state p of the first population. That is, the populations co-evolve.

To see how coevolution works, consider again the buyer-seller example. We know that the share p_1 of buyers who test increases when $\Delta u(\mathbf{q}) = 8q_1 - 3 > 0$, that is, when the fraction of cheating sellers is $q_1 > 3/8$, and the share of testers decreases when $q_1 < 3/8$. Hence the equal fitness line $\Delta u(\mathbf{q}) = 0$ or $q_1 = 3/8$ divides the unit square into two rectangles, and p_1 decreases in the lower rectangle and increases in the upper rectangle. By the same logic, the equal fitness line $\Delta v(\mathbf{p}) = 0$, here $p_1 = 1/9$, divides the square into two rectangles; in the left (or west) rectangle, the first strategy has the advantage and the share q_1 of cheaters increases, while it decreases in the right (or east) rectangle. The equal fitness lines intersect at the steady state $(p_1^*, q_1^*) = (\frac{1}{9}, \frac{3}{8})$.

Stability analysis

Is this steady state stable? To check, first note that it is the corner of four subrectangles defined jointly by the two equal fitness lines, as shown in Figure 3.6. In the large northeastern rectangle, as we have just seen, the trajectory is southeast: q_1 is decreasing and p_1 is increasing. Hence eventually the state must leave the northeastern rectangle (and since it can't cross the eastern boundary $p_1 = 1$) the state must enter the southeastern rectangle. There its trajectory is south*west* because, with $q_1 < 3/8$, the fraction of testers begins to decrease. Eventually the state must enter the small southwestern rectangle, where the trajectory is *north*west. Then the state must enter the northwestern rectangle, whence it moves north*east* into the northeastern rectangle where we began this story.

The upshot is that the state spirals clockwise around the steady state. But does it spiral inward, so that the steady state is stable? Or outward, implying instability? The analysis so far is inconclusive.

One way to check is to simply pick some initial point, say $\mathbf{p} = \mathbf{q} = (1,1)/2$, and to iteratively apply the discrete replicator equation. This can be done with a spreadsheet and the result will be an outward spiral, for any initial point different from the steady state (and not already on the boundary of the square).

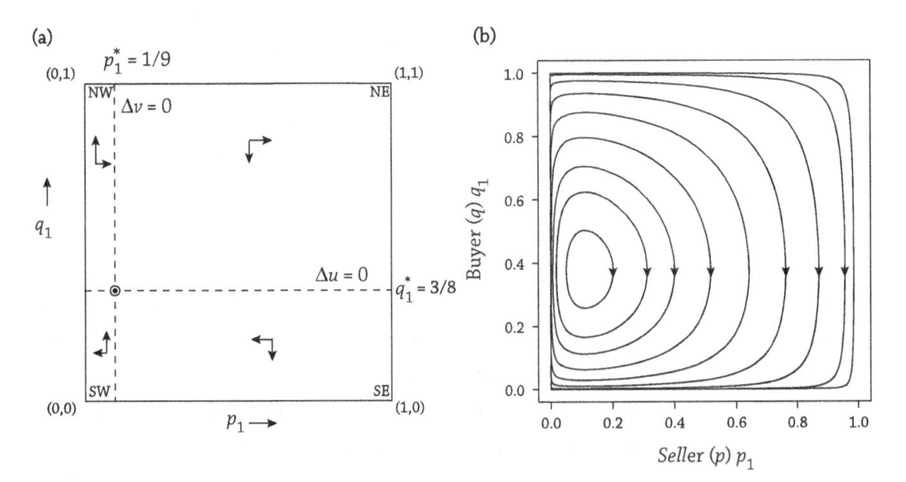

Figure 3.6 (a) Loci $\Delta v(p) = 0$ and $\Delta u(q) = 0$, a closed loop, and trajectory directions, for the Buyer-Seller game. (b) Numerical solution of the game confirms the closed orbits depending on where the game starts.

Alternatively, one can simulate the continuous time replicator dynamics,

$$\dot{p}_1 = p_1(\mathbf{e_1 u \cdot q} - \mathbf{pu \cdot q}) = p_1(\mathbf{e_1} - \mathbf{p})\mathbf{u \cdot q} = p_1(1-p_1)\Delta u(\mathbf{q}) \tag{3.38}$$

$$\dot{q}_1 = q_1(\mathbf{e_1 v \cdot p} - \mathbf{qv \cdot p}) = q_1(\mathbf{e_1} - \mathbf{q})\mathbf{v \cdot p} = q_1(1-q_1)\Delta v(\mathbf{p}). \tag{3.39}$$

Most simulators will show a very slow outward spiral. We invite the reader to test the R package using the routines buyer_seller() and plot_buyer_seller() in the section on R code for various continuous evolutionary games.

It seems then, from numerical simulation, that the steady state is unstable. But things are not quite what they seem. We will see in Chapter 4 how to use more advanced analytical techniques—eigenvalues of the Jacobian matrix and Lyapunov functions—to properly assess stability. We'll see that in the present case that the steady state is not unstable after all; the trajectories actually are closed loops as in Figure 3.6. It turns out that the outward spiral is due entirely to numerical round-off error! Intuitively, the problem is that the numerical solution follows short tangent lines, but tangents to a circle always lead out of the circle, to a wider circle, and similarly for loops as in the Figure 3.6.

3.6 HAWK-DOVE WITH TWO POPULATIONS

Recall the Hawk-Dove example from the previous chapter, with $value = 4, cost = 6$ and fitness matrix $\mathbf{u} = \begin{pmatrix} -1 & 4 \\ 0 & 2 \end{pmatrix}$. Suppose now that there are two distinct populations, that individuals in one population interact only with individuals in the other, and that both populations have the same payoff matrix. Then $\mathbf{u}=\mathbf{v}$ and the bimatrix is $\mathbf{u}, \mathbf{v}^T = \begin{pmatrix} -1,-1 & 4,0 \\ 0,4 & 2,2 \end{pmatrix}$.

It is straightforward to check that $\Delta u(\mathbf{q}) = 2 - 3q_1$ and $\Delta v(\mathbf{p}) = 2 - 3p_1$. Hence, the interior steady state is $(p_1^*, q_1^*) = (\frac{2}{3}, \frac{2}{3})$.

As shown in Figure 3.7, the square again breaks into four rectangles. Trajectories in the small northeastern rectangle head southwest. There is one trajectory, called the saddle path, that leads to the steady state. All other trajectories must eventually exit into either the northwestern rectangle or the southeastern rectangle. Trajectories in the large southwestern rectangle are northeast. Again, except for initial conditions that lie on the saddle path, the trajectories all must eventually exit into either the northwestern or southeastern rectangle. Trajectories in the northwestern rectangle head northwest, and can terminate only in the extreme corner $(p_1^*, q_1^*) = (0,1)$. Trajectories in the southeastern rectangle head southeast, and can terminate only in the extreme corner $(p_1^*, q_1^*) = (1,0)$.

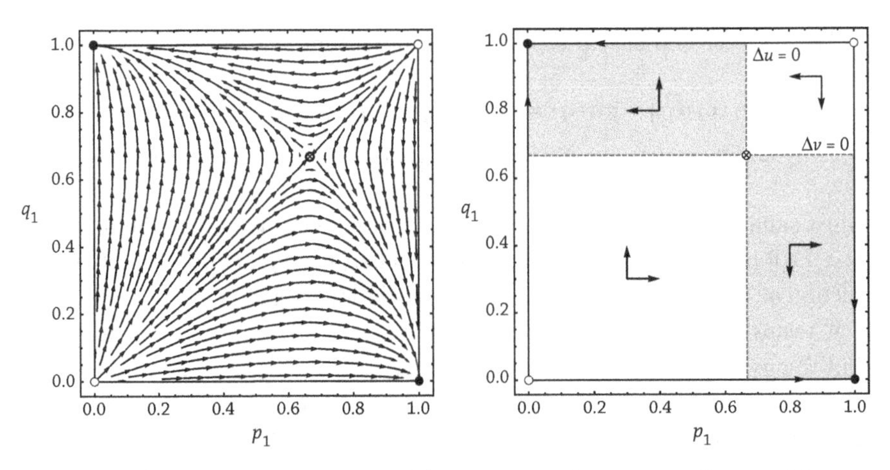

Figure 3.7 (A) Replicator dynamics and (B) Trajectory directions and loci $\Delta u(q) = 0$ and $\Delta v(p) = 0$, for the Hawk-Dove game played between two populations.

Thus there is a saddle path that runs from the corners $(0,0)$ and $(1,1)$ to the interior steady state, $(p_1^*, q_1^*) = (\frac{2}{3}, \frac{2}{3})$. The trajectory beginning at any point northwest of the saddle path ultimately terminates at the corner $(p_1^*, q_1^*) = (0,1)$, while the trajectory beginning at any point southeast of the saddle path ultimately terminates at the other corner $(p_1^*, q_1^*) = (1,0)$. We conclude that the saddle path separates the basins of attraction for these two corner steady states, and that the interior steady state $(p_1^*, q_1^*) = (\frac{2}{3}, \frac{2}{3})$ is a saddle point.

At first, this might seem puzzling. The interior steady state was stable in the one-population game. Here it is unstable; almost every trajectory, even those beginning very close to it, moves away. Eventually, these nonexceptional trajectories converge to a state in which everyone in one population plays hawk and everyone in the other plays dove. But that makes sense when there are two populations that can have different destinies.

The one-population game corresponds to the exceptional case in the two population game, because the fractions of hawks in "both populations" must be the same if there really is only one population. Thus the one-population game occurs precisely on the saddle path, and so the share vector converges to the interior steady state. In the two-population game, by contrast, any initial imbalance between the two populations grows over time. The population with more hawks eventually fixates on all hawk, and the other population fixates on all dove.

In a paper that began as a class project, Oprea et al. (2011) recruited 12 human subjects to play a Hawk-Dove game, with payoffs representing dollar payments. In this laboratory experiment, play converged closely to the symmetric mixed Nash equilibrium when player matchings were within a single population. But when the experimenters made an unannounced switch to a two-population matching procedure (the six players in one population interacted only with those in the other population) the state consistently showed clear movement towards one of the corner steady states. So the theory can work pretty well—even with not especially large populations of humans!

3.7 OWN-POPULATION EFFECTS

In the previous two sections there are two distinct roles, but the fitness u_i of each strategy in the first population depends directly only on the state $\mathbf{q}$ of the second population and not on the state $\mathbf{p}$ of its own population. Likewise, the second population's fitnesses v_j depend only on $\mathbf{p}$ and not on $\mathbf{q}$.

In many realistic situations, fitnesses of each strategy depend on the state of both populations. For example, the efficacy of tending aphid colonies depends on what other ants are doing as well as what the aphids are doing. Or the buyer strategy of waiting until closing time and then asking for a discount will work better when sellers are willing to bargain and also when there are not too many other buyers hanging around at closing time.

A simple example

To illustrate, we return again to the Hawk-Dove example with $v = 4, c = 6$, and basic fitness matrix $\mathbf{u} = \begin{pmatrix} -1 & 4 \\ 0 & 2 \end{pmatrix}$. Suppose again that there are two distinct populations and that the matrix $\mathbf{u}$ describes interactions with the other population. But now suppose that the matrix $a\mathbf{u} = \begin{pmatrix} -a & 4a \\ 0 & 2a \end{pmatrix}$ describes interactions within each population, and overall fitness is the sum of fitness obtained in all interactions. The previous section analyzed the case $a = 0$. When $0 \le a < 1$, the interactions with the other population are more important, and when $a > 1$ the interactions within each population are more important.

Now the fitness of strategy i in the first population is

$$u_i = \mathbf{e}_i a\mathbf{u} \cdot \mathbf{p} + \mathbf{e}_i \mathbf{u} \cdot \mathbf{q} = \mathbf{e}_i \mathbf{u} \cdot (a\mathbf{p} + \mathbf{q}), \quad i = 1,2, \tag{3.40}$$

and the fitness difference is $\Delta u(\mathbf{p},\mathbf{q}) = u_1 - u_2 = (\mathbf{e}_1 - \mathbf{e}_2)\mathbf{u} \cdot (a\mathbf{p} + \mathbf{q})$. Of course, $(\mathbf{e}_1 - \mathbf{e}_2) = (1,-1)$ so for the given matrix $\mathbf{u}$ we have $\Delta u = (1,-1)\mathbf{u} \cdot (a\mathbf{p} + \mathbf{q}) = (-1,2) \cdot (a\mathbf{p} + \mathbf{q}) = 2(1+a) - 3ap_1 - 3q_1$. As a increases from 0, the equal fitness line $\Delta u = 0$ rotates from the horizontal position clockwise around the point $(p_1^*, q_1^*) = (\frac{2}{3}, \frac{2}{3})$. A similar calculation shows that the equal fitness line $0 = \Delta v = 2(1+a) - 3aq_1 - 3p_1$ rotates from the vertical position counterclockwise around the same point.

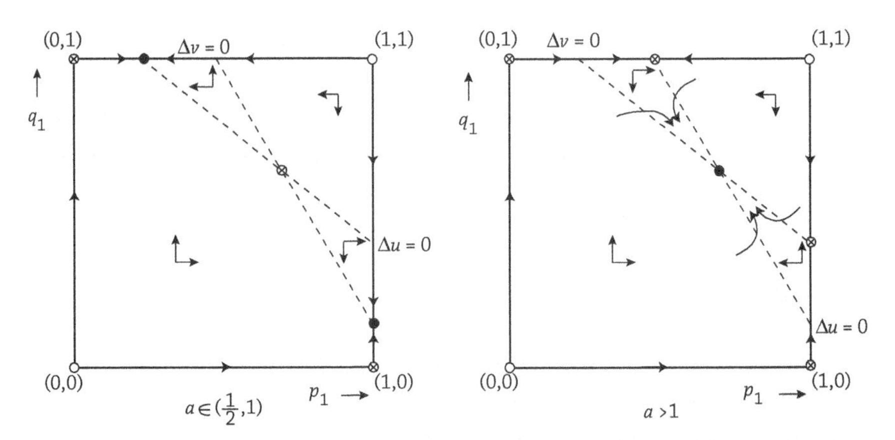

Figure 3.8 Loci $\Delta u = 0$ and $\Delta w = 0$, and trajectory directions, for the Hawk-Dove game played between two populations with weak (Panel A) and strong (Panel B) own-population effects.

For small positive values of the own-population interaction amplitude a, there is no qualitative effect—the populations still evolve (along slightly distorted paths) until one is all hawk and the other is all dove, and the basins of attraction are still separated by a saddle path along the diagonal. However, when $a = 0.5$ the equal fitness line $\Delta w = 0$ intersects the stable steady state $(0,1)$ and the equal fitness line $\Delta u = 0$ intersects the stable steady state $(1,0)$. As shown in Figure 3.8, for values $0.5 < a < 1.0$, there are still two symmetric stable steady states, but they take the form $(1, q(a))$ and $(q(a), 1)$ (i.e., all hawks in one population and mostly doves in the other except for some small fraction $q(a)$ of hawks).

That fraction increases with a until, at $a = 1$, the two equal fitness lines coincide. Then the entire (double) equal fitness line is neutrally stable. For $a > 1$, the two equal fitness lines again divide the square into four regions, and the direction of trajectories in the northeast region is still southwest, and the southwestern region still has trajectories headed northeast. However, now the trajectories in the southeastern region point northwest (instead of southeast) and those in the northwestern region point southeast (instead of northwest). The upshot is that the interior steady state $(\frac{2}{3}, \frac{2}{3})$ is now stable instead of a saddle point. That is, as shown in Figure 3.8B, from any initial state in the interior of the square, the state evolves towards a mix of 2/3 hawks and 1/3 doves in both populations!

Generic games on the square

Of course, this example is quite special. In general it takes two separate matrices to define payoffs in the first population—call them $\mathbf{u}_a$ for own-population interaction and $\mathbf{u}$ for interactions across populations—and another two matrices for the other population, say $\mathbf{v}_a$ and $\mathbf{v}$.

Steady states in the square (i.e., in the case that there are only two alternative actions for players in either population) can be found as follows. Compute the fitness difference functions

$$\Delta u = (1,-1)(\mathbf{u_a} \cdot \mathbf{p} + \mathbf{u} \cdot \mathbf{q}) \tag{3.41}$$

and

$$\Delta v = (1,-1)(\mathbf{v_a} \cdot \mathbf{q} + \mathbf{v} \cdot \mathbf{p}). \tag{3.42}$$

Steady states occur wherever both (a) $\dot{\mathbf{p}} = 0$, and (b) $\dot{\mathbf{q}} = 0$. Condition (a) is satisfied for replicator dynamics on (and only on) three lines: the vertical edges $p_1 = 0$ and $p_1 = 1$ of the square, and the equal fitness line $\Delta u = 0$. Condition (b) is satisfied precisely on on the horizontal edges $q_1 = 0$ and $q_1 = 1$, and on the equal fitness line $\Delta v = 0$. Hence steady states consist of the intersections of one of the first set of three lines with one of the second set. These intersections include all four corners of the square, plus any interior intersection of the two equal fitness lines. An intersection of an equal fitness line with an edge represents a steady state if the two are complementary, for example, $\Delta v = 0$ [from set (b)] with a vertical edge [from set (a)].

The stability of each steady state can be determined as follows. The equal fitness lines divide the square into four regions. Each region has a distinctive direction for its trajectories: northeast (i.e., $\dot{p}_1 > 0, \dot{q}_1 > 0$), northwest ($\dot{p}_1 > 0, \dot{q}_1 < 0$), southwest, or southeast. Such trajectories must either exit their region to a specific neighboring region, or remain in that region forever after. The geometry of the regions and their directions then determine the stability of each steady state.

We have already seen several different configurations. In the two-population Hawk-Dove game with no own-population effects, the interior steady state is a saddle, the diagonally opposite corners $(p_1, q_1) = (0,0)$ and $(1,1)$ are sources (unstable: every nearby point moves away), and the other corners $(1, 0)$ and $(0, 1)$ are sinks (stable: every nearby point moves in). The intersections of the equal fitness lines with the edges are not steady states.

We saw a second configuration in the buyer-seller game: the corners were all saddle points and the interior steady state was neutrally stable, as trajectories looped around it.

There is only one other generic configuration for linear games with no own-population effects (i.e., with $\mathbf{u_a} = \mathbf{v_a} = \mathbf{0}$). It occurs when one of the populations has a dominant strategy, for example, $w_1 > w_2$ at every state $(\mathbf{p}, \mathbf{q})$. In that case, every trajectory eventually hits the edge corresponding to fixation on the dominant strategy, and along that edge the trajectory moves towards fixation in the other population on the strategy that provides greater fitness against the dominant strategy. In this case there is a source at one corner, a sink at the diagonally opposite corner (corresponding to the fixations just mentioned), saddles at the other two corners, and no interior steady state.

Own-population effects allow three variants of the configurations already mentioned. The Hawk-Dove game with $0.5 < a < 1.0$ (moderate own population effects) is a slight variant of the first configuration: the sinks move off the corners to intersections of edges with the equal fitness lines, and the corners become saddles. In the other variants, the closed loops in the second configuration can turn into either inward spirals (so the interior equilibrium is stable) or outward spirals (so the interior equilibrium is unstable).

Own-population effects also allow two generic configurations that are qualitatively different from those just discussed. The first was illustrated in the Hawk-Dove game with strong own-population effects ($a > 1$). It features an interior sink, while all four corners are sources, and other intersections of the equal fitness lines with edges are saddles. The other new configuration is the opposite: all four corners are sinks, the interior steady state is a source, and the intersections of the equal fitness lines with edges are saddles.

How about nonlinear games? Here the fitness functions still have the same state space, the square, as long as each of two populations has two alternative strategies. The linear analysis still covers all the generic configurations in terms of local behavior. But nonlinearity allows more interior steady states, and therefore more complex combinations.

The analysis still proceeds by drawing the equal fitness loci $\Delta u = 0$ and $\Delta v = 0$, which chop the square up into regions in which the signs of $\dot{p}_1$ and $\dot{q}_1$ are constant. Generically four such regions with differing sign combinations surround each interior intersection of the lines, and direction of adjustment (e.g., northwest) in these regions determines the local stability of the steady state exactly as in the linear analysis. Similarly for intersections of the equal fitness lines with the edges of the square, and for the corners of the square.

Behavior away from the steady states can be a bit more complicated than in the nonlinear case. For example, there can be a stable limit cycle—a closed loop towards which nearby trajectories spiral. The notes tell you where to find explanations of such behavior.

3.8 HIGHER-DIMENSIONAL GAMES

So far in this chapter we have seen two different 2-dimensional state spaces: the square (for two populations, each with two alternative strategies) and the triangle (for one population with three alternative strategies).

Several important applications have 3-dimensional state spaces, of which there are three different kinds. The first is the cube, representing three populations, each with two alternative strategies. The state space is $\{(p_1,q_1,r_1) \in [0,1]^3\}$, the unit cube. For example, take the Buyer-Seller game and add a third population of regulators who can either be aggressive or accommodating with the sellers. Rowell et al. (2006) feature a different example in which a signaler population chooses the degree of honesty in [0, 1] (e.g., male frogs can croak at a pitch reflecting their actual size or at a lower pitch that greatly exaggerates their size), and two different populations of listeners (e.g., female frogs and predators) each choose a degree of belief in [0, 1].

The second state space of three dimension is the prism $\{(r_1,s_1,s_2,s_3) : r_1,s_i \in [0,1], s_1+s_2+s_3=1\}$. This applies to two populations, one with two alternative strategies and the other with 3. We will see an example in Chapter 7, in which a predator (such as an insect-eating bird) has two alternative behaviors while the prey (e.g., butterflies) have three alternative defenses. A student project developed a different example—an evolutionary model of the video game *StarCraft II* in which the game developer (Blizzard Entertainment) can choose the degree of change in rules $r_1 \in [0,1]$, while the player population can develop relative fighting expertise $\mathbf{s}$ among the three species (Terran, Zerg, and Protoss).

The third 3-d model has as its state space the tetrahedron $\{(s_1,s_2,s_3,s_4) : (s_i \in [0,1], s_1+s_2+s_3+s_4=1\}$. It pertains to a single population with four alternative strategies. An example is the rpsd game considered earlier. Another example, from another student project, is the competition for gamblers' time and monetary stakes among four kinds of poker: Texas Hold-Em, five-card stud, and five- and seven-card draw.

Of course, there is no shortage of models of dimension 4 and higher. The higher the dimension, however, the more difficult it is to visualize the state space and to analyze the model. Usually it is better to simplify the application enough to keep the state space dimension below 4.

3.9 ALTERNATIVE DYNAMICS

So far we have focused on replicator dynamics, especially the haploid and continuous time versions, but sometimes we mentioned sign preserving dynamics. In later chapters that work out applications in the social and virtual worlds, we will encounter several alternative varieties of dynamics. To set the stage, in this section we provide an overview and preview, together with a few remarks on when the specific variety of evolutionary dynamics can make a difference.

Perhaps the most basic choices in modeling how the state evolves are whether to regard the process as stochastic or deterministic, and whether to use discrete or continuous time.

How should you make these choices? We advise you to seek the simplest model that can address your questions with the available data. For convenience, our discussion below will focus on deterministic, continuous time models, but the lizard application in Section 3.2 illustrated a discrete stochastic modification, and applications in later chapters will illustrate other possibilities.

Continuous deterministic dynamics are convenient for at least three reasons. First, we can get such dynamics from the other kinds by taking an appropriate limit as the time increment and the scale of random fluctuations go to zero and the number of players goes to infinity in each population (e.g., Binmore and Samuelson 1995). Second, continuity in the state over time satisfies a basic principle of evolution, captured in Darwin's dictum "natura non facit saltum." Third, such dynamics take the relatively tractable form of a system of ordinary differential equations (ODEs).

The essence of biological evolution is that fitter genotypes increase relative to less-fit genotypes. The economic analogue is that (due to imitation and learning or resource redistribution or entry and exit) higher payoff strategies increase relative to lower payoff strategies in every population. This principle, called *monotonicity*, implies some relationship between the relative fitness functions $w_i(s) - \bar{w}(s)$ and the corresponding dynamics $\dot{s}_i = F_i(s)$.

Monotonicity comes in six major varieties. To see them clearly, consider a single population that has available alternative strategies $i = 1, \ldots, n$. One can impose monotonicity on the rates of change $\dot{s}_i$ or else on the growth rates $\dot{s}_i / s_i$. In either case, one can specify that the $\dot{s}_i$'s (or the $\dot{s}_i / s_i$'s)

(a) are *proportional* to relative fitness; or
(b) have the same *rank-order* as relative fitness; or
(c) are *positively correlated* with relative fitness.

Replicator dynamics are specification (a) applied to growth rates. Choosing the time scale so that the constant of proportionality is 1.0 and equating each growth rate $\dot{s}_i / s_i$ to relative fitness $w_i(s) - \bar{w}(s)$, one gets the familiar system of ordinary differential equations.

Replicator dynamics have a compelling biological justification—genetic fitness is *measured* as the relative growth rate. One can also get replicator dynamics from certain sorts of imitation processes. For example, suppose that individuals are rematched pairwise at random times (with a Poisson distribution), and the player with lower payoff switches to the other player's strategy with probability proportional to its payoff advantage. It has been shown that, in the large population limit, the state obeys the replicator ODE.

An alternative version of specification (a) just equates the rate of change $\dot{s}_i$ (not the growth rate) to relative fitness in the interior of the state space. This makes more sense for memes when the entire population has access to all possible alternatives, because in this specification a very rare strategy with high fitness can gain share quickly. A caveat: for reasons explained in Friedman (1991), when working with rates of change, it is natural to define the relative fitness in terms of the simple average of nonextinct strategies rather than the population weighted average. Sandholm (2010, Chapter 6) refers to this version of specification (a) as a projection dynamic, and obtains two slightly different versions dubbed BNN dynamics (for Brown, Nash, and von Neumann) and Smith dynamics.

Specification (b), often referred to as monotone dynamics (Nachbar 1990; Samuelson 1991), has become increasingly popular in social world applications because of its greater generality. The simplex sectoring and state space chopping featured in this chapter are the essence of monotone dynamics, which are defined by the directional cones, like northwest, that we assigned to each region.

In some applications even (b) may be too strong. It might turn out that the population fraction using third-best action sometimes grows more rapidly (or shrinks more slowly) than that for the second-best action. In such cases version (c) might be useful. The geometric interpretation is as follows. At any given point s, draw the relative fitness vector $(w_1 - \bar{w}, \ldots, w_n - \bar{w})$ and the velocity vector $(\dot{s}_1, \ldots, \dot{s}_n)$. Then the dot product of the two vectors must be positive (i.e., the angle between the two vectors must be acute). See Friedman (1991) for further discussion of (c); and see Weibull (1995) for yet another variant in which each rate of change must have the same sign as its relative fitness. All versions have plausible arguments in their favor and also have attractive asymptotic properties.

Best response dynamics are a specific example of monotone dynamics satisfying (c), but not quite (b), when $n > 2$. The idea, which originates in Cournot (1838), is that players make a best response to the current state. Such dynamics can be written in continuous time as $\dot{s} = \alpha(B(s) - s)$, where $\alpha > 0$ is an adjustment rate parameter and, at each state s, the function B picks the strategy with maximal payoff (or any one of them when there are several). That is, $B(s) = \mathbf{e}_i$ if $w_i(s) \geq w_j(s)$ for all alternative pure strategies j. Best response dynamics, then, increase the share of the currently fittest strategy at the expense of all other strategies. For example, in a model with more than two pure strategies, the second best strategy (if not already extinct) loses share even if the second best strategy is almost as good as the best strategy and far better than the other $n-2$ strategies.

Of course, (b-c) coincide when there are only $n = 2$ alternative strategies in each population. In this case, all six dynamic specification lead to the same qualitative phase portraits (i.e., they have exactly the same stable equilibria and basins of attraction). This case is often called "sign-preserving dynamics."

In higher dimensions, the interior steady states are the same for all six varieties of dynamics, as is their classification as a source or sink or saddle. However, centers may be (locally asymptotically) stable for some specifications and neutral or unstable for others. For the three varieties involving growth rates, the vertices of the simplex are always steady states, and trajectories that begin in an edge or face always remain there. However, faces (including vertices) are not stable for these growth rate dynamics unless they are also stable for the corresponding rate of change dynamics.

In later chapters we will introduce specific dynamics, and usually will mention whether they satisfy (a), (b), or (c) in rates of change or in growth rates. Following common practice, when we just say "monotone dynamics" without further qualification, we will mean variety (b) in rates of change.

3.10 DISCUSSION

Readers who have absorbed the main ideas of this chapter are ready to begin applying evolutionary game theory to new questions. Although the rest of the book contains many additional techniques and insights, the basic tools and perspective are already covered.

To summarize briefly, the main steps in applying basic evolutionary game theory are the same for applications in the social and virtual worlds as they are for applications in the natural world:

- Identify a small number of strategically distinct populations (typically one or two, occasionally three or more) and their relevant alternative strategies (usually two or three). This defines the model's state space. The idea is to capture the more important aspects of the application in a tractable (low-dimensional) model, not to strive for perfect correspondence between model and reality.
- Specify how the fitness (or payoff) of each strategy depends on the current state. If necessary, include stochastic elements and density (population size) dependence.
- Identify steady states and, if possible, their stability properties (source, sink, saddle, or center). Often the delta/sectoring techniques featured in this chapter are helpful.

- Use behavior in the neighborhood of steady states to help piece together the full picture of how the state evolves over time in your model.
- The dynamic specification can help resolve stability questions (especially for centers) and may be of interest in its own right. The application should dictate the proper specification, for example, replicator dynamics for many biological applications or a particular form of imitation or noisy best response for some social or virtual world applications. Chapter 8 and other later chapters will have much more to say about this step.

A parting thought. Does it really matter which dynamic specification you choose? Sometimes no. In the next chapter we will study shortcuts that help you find steady states quickly, and that identify good candidates for stable steady states. However, we will see that a shortcut (called ESS) that many researchers rely on to identify stable steady states can fail for even monotone dynamics. An exercise in Chapter 8, inspired by Golman and Page (2010), shows that most of the simplex can shift from one basin to another depending on the dynamics. As explained in the last note, there are even situations where most trajectories never converge to any steady state. So sometimes the choice of dynamics makes all the difference, and that is why our recipe for applications gives it such prominence.

APPENDIX: ESTIMATING 3×3 PAYOFF MATRICES

Here we collect details on how we specify the 3×3 Dirichlet/replicator model and fit it to the lizard data. The first step is to take raw census counts of adult males by morph allele N_o, N_b, N_y and $N_{tot} = N_o + N_b + N_y$ and to compute the relative frequencies $S_1 = \frac{N_o}{N_{tot}}$, $S_2 = \frac{N_b}{N_{tot}}$, and $S_3 = \frac{N_y}{N_{tot}}$. Thus the main data input is a $T \times 3$ matrix that lists the state vector $S(t) = (S_1(t), S_2(t), S_3(t))$ in the simplex $\mathcal{S}$ each period $t = 1, \ldots, T$.

As mentioned in the text, we assume that

$$Z_i(t+1) = \frac{W_i(t)}{\bar{W}(t)} S_i(t) \tag{3.43}$$

defines the deterministic part of the dynamics, and that the observed frequency in the next generation $S(t+1)$ is a realization from $Dir(NZ(t+1))$, where $N > -1$ is the effective sample size or precision parameter. That distribution is defined as follows. Given an arbitrary probability vector over three outcomes, $p = (p_1, p_2, p_3) \in \mathcal{S}$, the multinomial distribution with N independent draws gives the probabilities of all possible counts of the three outcomes. The Dirichlet distribution $Dir(Nz)$ is the conjugate prior, that is, given any count vector

$Nz = (Nz_1, Nz_2, Nz_3)$, it gives the probability that the underlying p-vector was $x \in \mathcal{S}$. The Dirichlet density function is

$$f(x_1, x_2, x_3 | N, z_1, z_2, z_3) = K(NZ) x_1^{Nz_1 - 1} x_2^{Nz_2 - 1} x_3^{Nz_3 - 1} I_{x \in \mathcal{S}}, \tag{3.44}$$

where I is the indicator function, which is 0 outside the simplex and 1 on the simplex, and the normalizing constant $K(NZ)$ ensures that density integrates to 1. Textbooks confirm that $x \in \mathcal{S}$ is a random variable with mean z, and that the variance of x_i is $\frac{z_i(1-z_i)}{N+1}$.

Two covariates capture time- or world-specific shifts in the fitnesses. The first is $\tau_k(t) = h_r(t) - \bar{h}_r^k$, a measure of temperature restriction the previous summer (when progeny are recruiting) normalized for each world k by subtracting the historical mean value $\bar{h}_r^k$ over the life of that world. For world-specific shifts we take $\nu_k(t)$, the male population size in world k in year t, normalized by subtracting its world-specific mean value.

We assume that these covariates shift fitness multiplicatively, with the proportional change in fitness coefficient a linear function of each variable. Thus we will estimate the fitness matrices $\Omega_k = ((\Omega_{ijk}))_{i,j=1,\ldots,3}$ where

$$\Omega_{ijk}(t) = W_{ij} \exp\left(\beta_{ij}^{\nu} \cdot \nu_k(t) + \beta_{ij}^{\tau} \cdot \tau_k(t)\right). \tag{3.45}$$

The normalization of ν and τ ensures that we have $\Omega_k(t) = W$ when both variables are at their average levels. In the most complete specification, therefore, in addition to the 3×3 fitness matrix W we are also estimating two other 3×3 coefficient matrices: β^{ν} for population density effects and β^{τ} for temperature effects.

As noted in the text, to resolve the scale indeterminacy of the discrete replicator equation, we impose the constraint $\sum_i \sum_j W_{ij} = 1$ in forming the Z vector. For the same reason, the β matrices need normalization, and a natural choice is that the coefficients sum to zero. Thus estimation of equation (3.45) is subject to the restrictions $W_{ij} \geq 0$, $\sum_{i,j} W_{ij} = 1$, $\sum_{i,j} \beta_{ij}^{\nu} = 0$, $\sum_{i,j} \beta_{ij}^{\tau} = 0$.

The basic model (ignoring the β parameters for the moment) has parameter vector $\theta = (N, W_{11}, W_{12}, \ldots, W_{33})$ to be estimated. Given the previous state $S(t-1)$, the conditional density of the current state is

$$f(S(t)|S(t-1)) = \frac{\Gamma(N)}{\prod_{i=1}^{3} \Gamma(NZ_i(t))} \prod_{i=1}^{3} S_i(t)^{NZ_i(t)-1}. \tag{3.46}$$

This expression spells out the normalizing constant $K(NZ)$ in (3.44), using the gamma function $\Gamma(z) \equiv \int_0^{\infty} y^{z-1} e^{-y} dy$.

Summing the log of the right-hand side of (3.46) over $t = 1,\ldots,T$, the conditional log-likelihood function of the sample is:

$$\ln\ell = T\ln\Gamma(N) + \sum_{t=1}^{T}\sum_{i=1}^{3}[-\ln\Gamma(NZ_i(t)) + \ln S_i(t)(NZ_i(t)-1)] \quad (3.47)$$

The $t = 1$ terms do not contribute to the sum since, with no previous period to which to apply replicator dynamics, their likelihood is 1. We maximized the right-hand side of (3.47) numerically over the set of parameter values $\theta = (N, W_{11}, W_{12}, \ldots, W_{33})$ satisfying $N > -1$, $W_{ij} \geq 0$, and $\sum_{i,j=1}^{3} W_{ij} = 1$. The numerical maximization is via Matlab's active set algorithm `fmincon`, using as initial values 1/9 for each entry W_{ij} of the payoff matrix and 20 for N. The algorithm converged reasonably quickly to the values reported in Table 3.1 for a range of other initial values, for example, for N between 15 and 30. The Matlab code can be found in Magnani, Friedman, and Sinervo (2015).

Extending the basic model to include covariates, panel data is conceptually straightforward. To deal with four separate locations ("worlds") we write the conditional log-likelihood function for location k as:

$$\ln\ell_k = T_k\ln\Gamma(N) + \sum_{t=1}^{T_k}\sum_{i=1}^{3}[-\ln\Gamma(NZ_{ik}(t)) + \ln S_{ik}(t)(NZ_{ik}(t)-1)]. \quad (3.48)$$

Note that N and the W_{ij}'s are not location-specific because, to keep the number of free parameters small relative to overall sample size, we assume that the underlying stochastic structural model is the same in all locations. Of course stochastic realizations can and do vary across locations. Adding the covariates ν and τ affects the likelihood function only through the (deterministic) replicator dynamic equation for $Z_{ik}(t)$, where the payoff matrix W is replaced by the augmented fitness matrix $\Omega_k(t)$ defined in equation (3.45) of the text.

We now assume that stochastic realizations are conditionally independent, that is, conditional on $S_{ik}(t-1)$ and other covariates, the $S_{ik}(t)$'s are independent across k. Then the augmented model has log-likelihood function

$$\sum_{k\in\{MW,OW,BW,YW\}} \ln\ell_k, \quad (3.49)$$

where $\ln\ell_k$ is defined in (3.48) using the augmented fitness matrices Ω_k in the replicator expression defining $Z_{ik}(t)$. In addition to the coefficient constraints on W mentioned earlier, we also imposed $\sum_{i,j}\beta_{ij}^{\nu} = 0, \sum_{i,j}\beta_{ij}^{\tau} = 0$, and used the agnostic initial values $\beta_{ij}^{\nu} = \beta_{ij}^{\tau} = 0$. To avoid regions of the

Table 3.2. ESTIMATES OF THE $(W, \beta_\nu, \beta_\tau)$ MODEL (WITH BOOTSTRAPPED STANDARD ERRORS IN PARENTHESES). IN PANEL (A) THE COVARIATES ARE SUPPRESSED, AND THE ESTIMATED EFFECTIVE SAMPLE SIZE IS $\hat{N} = 21.7 \pm 3.4$. THE REMAINING PANELS SHOW ESTIMATES OF THE FULL MODEL, FOR WHICH $\hat{N} = 34.2 \pm 1.2$. COVARIATES ARE JOINTLY HIGHLY SIGNIFICANT ACCORDING TO THE STANDARD LIKELIHOOD RATIO TEST (NOT SHOWN).

0.0000 (0.0352)	0.2462 (0.0420)	0.0207 (0.0472)
0.1383 (0.0469)	0.0600 (0.0232)	0.1941 (0.0456)
0.1701 (0.0527)	0.1705 (0.0395)	0.0000 (0.0319)

(a) Estimated W in simple model

0.0052 (0.0216)	0.2583 (0.0548)	0.0273 (0.0477)
0.0578 (0.1058)	0.0401 (0.0363)	0.3304 (0.1064)
0.0634 (0.0765)	0.2174 (0.0704)	0.0000 (0.0380)

(b) Estimated W in full model

−2.5329 (4.6263)	0.0456 (2.5650)	4.7470 (2.7592)
−3.0065 (4.8190)	3.4059 (5.5519)	1.1364 (2.9938)
−3.3406 (3.0572)	2.1167 (3.1126)	−2.5715 (4.5553)

(c) Estimated β_ν

3.4187 (4.1768)	−1.4064 (2.6704)	−1.9101 (3.8363)
−0.3363 (3.6359)	−1.5314 (4.0852)	−1.8428 (3.3206)
−1.1370 (3.0220)	−1.5874 (2.7406)	6.3326 (4.5957)

(d) Estimated β_τ

parameter space where the algorithm failed to converge we also imposed $\beta_{ij}^\nu, \beta_{ij}^\tau \in [-10, 10]$, but these constraints never bind once the algorithm approaches convergence.

The estimated matrices have implications for dynamic behavior. If the temperature and density covariates were always fixed at their average levels, then the estimates in Table 3.2b imply that from any interior initial condition (including the large experimental perturbations to OW, BW, and YW) the state would converge in finite time to the steady state $S^* = \lambda W^{-1}\mathbf{1} \approx (0.38, 0.40, 0.22)$. Of course, covariates are not fixed, and indeed vary stochastically from year to year and across locations. Climate directly influences morph payoffs via progeny recruitment during hot versus cool summers, and population density has an indirect influence. To assess the impact, we compute the effective payoff matrix $\hat{\Omega}_t^k$ for each World ($k = M, O, B, Y$) and each year t by amplifying or diminishing each payoff entry according to the density and temperature variables experienced in that world that year. The dots in Figure 3.9 show drastic year-to-year changes in the Main world steady state for the payoff matrix $\hat{\Omega}_t^M$. In many years the equilibrium lies on an edge, indicating under the conditions experienced that year

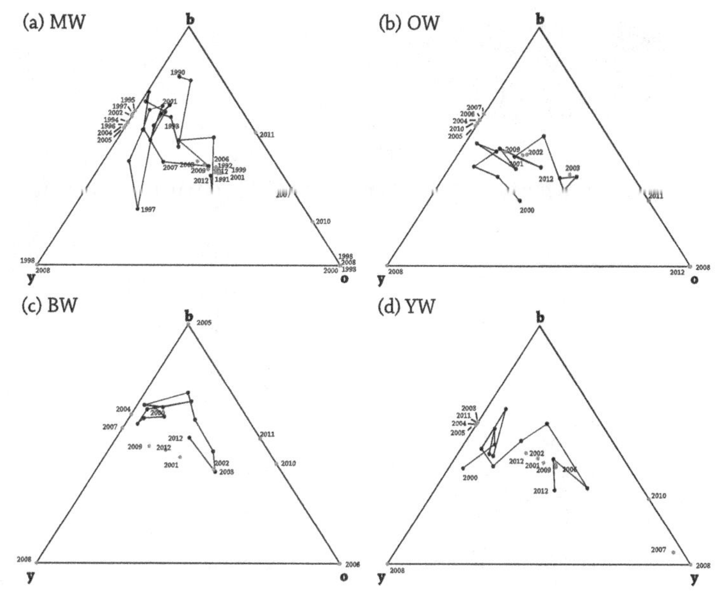

Figure 3.9 Simplex and states in all worlds. Black points and lines are time series $S(t)$, while the points labeled by years are the Nash equilibrium for $\hat{\Omega}_t^{xW}$.

1 morph (the one fixated at the corner of the simplex opposite that edge) is in danger of local extinction.

Simulation validity checks

Finally, to test estimation techniques, we create simulated data and see the extent to which we can recover the true parameters used in the simulations. We specify a payoff matrix $\mathbf{W}$ and precision N, and run the Dirichlet/replicator model for 60 periods. A first vanilla example is

$$\mathbf{W} = \begin{bmatrix} 0.24 & 0.09 & 0.67 \\ 0.67 & 0.24 & 0.09 \\ 0.09 & 0.67 & 0.24 \end{bmatrix} \tag{3.50}$$

with extremely high precision $N = 2000$. Figure 3.10 plots one simulation of 1000 time periods.

It turns out that the bivariate analysis of Chapter 2 has difficulties with these data. Figure 3.11 shows the scatter plots of $S_i(t+1)$ on $S_j(t)$, including the fitted line $S_i(t+1) = \alpha_{ij} + \beta_{ij} S_j(t)$ in each panel. Table 3.3 reports the fitted values α_{ij}, β_{ij} as well as the corresponding estimate $\gamma_{ij} \equiv \alpha_{ij} + \beta_{ij}$ of the the

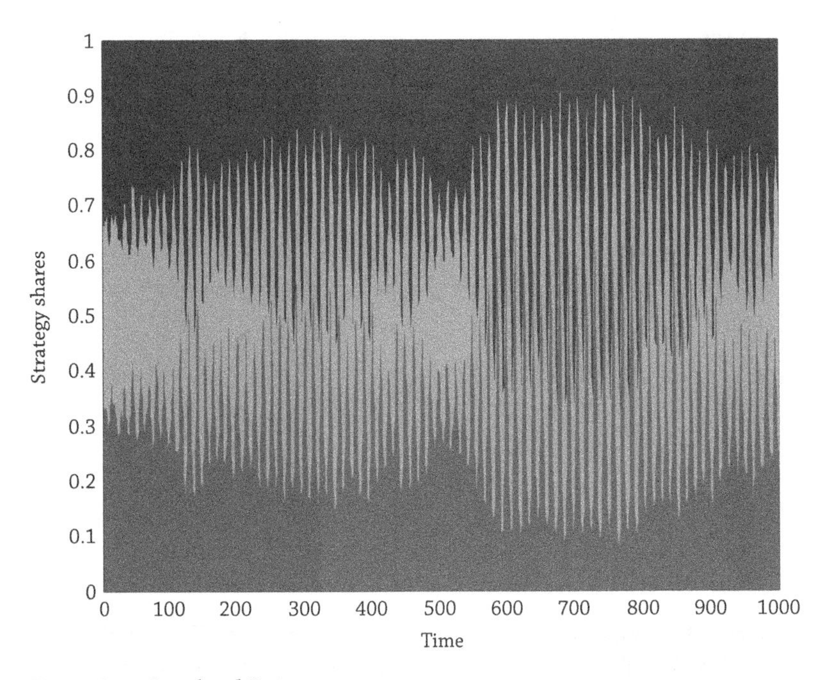

Figure 3.10 Simulated Series

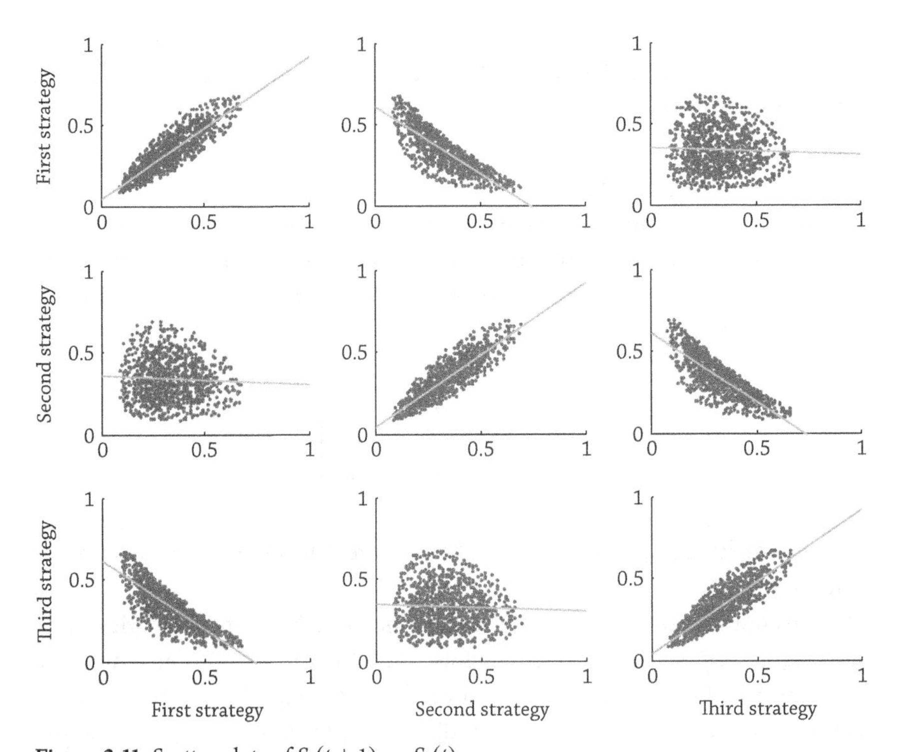

Figure 3.11 Scatter plots of $S_i(t+1)$ on $S_j(t)$

Table 3.3. BIVARIATE ANALYSIS SUMMARY

$\alpha_{11} = 0.87$	$\alpha_{12} = -0.82$	$\alpha_{13} = -0.04$
$\beta_{11} = 0.04$	$\beta_{12} = 0.61$	$\beta_{13} = 0.35$
$\gamma_{11} = 0.91$	$\gamma_{12} = -0.21$	$\gamma_{13} = 0.31$
$\alpha_{21} = -0.04$	$\alpha_{22} = 0.87$	$\alpha_{23} = -0.82$
$\beta_{21} = 0.35$	$\beta_{22} = 0.04$	$\beta_{23} = 0.61$
$\gamma_{21} = 0.31$	$\gamma_{22} = 0.91$	$\gamma_{23} = -0.21$
$\alpha_{31} = -0.82$	$\alpha_{32} = -0.04$	$\alpha_{33} = 0.87$
$\beta_{31} = 0.61$	$\beta_{32} = 0.35$	$\beta_{33} = 0.04$
$\gamma_{31} = -0.21$	$\gamma_{32} = 0.31$	$\gamma_{33} = 0.91$

payoff matrix element W_{ij}. Clearly the procedure does not recover the true payoff matrix (3.50).

By contrast, as reported in the supplement to Friedman et al. (2016), the MLE procedure does a decent job of recovering matrix and precision parameters from a simulated data series of length only $T = 60$ when the matrix has a steady state is not too close to an edge or corner.

Overlapping generations

The discrete time approach just described works well with annual creatures (which these lizards generally are), but there would seem to be a problem for creatures whose lifestyle allows adults to be active for more than one time period.

To generalize the model, assume that some fraction $\delta \in [0,1)$ of adults survives to breed again the next season, and that all adults, no matter their age, have the same average rates of survival and fecundity. Let $\Delta t > 0$ represent the time elapsed from breeding until the new generation to reaches adulthood, and let $\mathbf{s}' = (s_1', \ldots, s_n') = \mathbf{s}(t + \Delta t) = (s_1(t + \Delta t), \ldots, s_n(t + \Delta t))$ denote next period's state (i.e., the subsequent shares of the n alternative strategies). Given a continuous fitness matrix $\mathbf{w} = ((w_{ij}))$, write the discrete fitness matrix as $\mathbf{W} = ((W_{ij}))$, where $W_{ij} = \exp(w_{ij} \Delta t)$. Recall that at current state $\mathbf{s}$, average fitness is

$$\bar{W} = \mathbf{sW} \cdot \mathbf{s} \equiv \sum_{i=1}^{n} \sum_{j=1}^{n} W_{ij} s_i s_j. \tag{3.51}$$

The expected number of new recruits per capita for strategy i is $W_i = \sum_{j=1}^{n} W_{ij} s_j$, while the corresponding expected number of surviving adults is δ. To get the new shares, we must normalize the sum $W_i+\delta$ using the total

$$\sum_{i=1}^{n}(W_i+\delta)s_i=\sum_{i=1}^{n}([\sum_{j=1}^{n}W_{ij}s_j]s_i+\delta s_i)=\bar{W}+\delta.$$

Thus the new share for strategy i is

$$s_i'=\frac{W_i+\delta}{\bar{W}+\delta}s_i,\ i=1,\ldots,n. \tag{3.52}$$

Note that in the special case $\delta=0$, the expression (3.52) reduces to $s_i'=\frac{W_i}{\bar{W}}s_i$, the same as considered before. Likewise, when fertility is very low per generation time Δt, then the fraction in (3.52) is close to $\frac{\delta}{\delta}=1$, so s_i hardly changes in one generation.

Overlapping generations have no impact on empirical replicator dynamics in continuous time. The growth rate of strategy i is its own net rate of increase $\mathbf{e_i w}\cdot\mathbf{s}-\delta$ minus the average rate $\mathbf{sw}\cdot\mathbf{s}-\delta$, so the continuous replicator equation $\dot{s}_i=[(\mathbf{e_i w}\cdot\mathbf{s}-\delta)-(\mathbf{sw}\cdot\mathbf{s}-\delta)]s_i=[\mathbf{e_i w}\cdot\mathbf{s}-\mathbf{sw}\cdot\mathbf{s}]s_i$ is unchanged.

EXERCISES

1. Sector the triangle, and find all steady states, when fitness is given by Table 3.1. Alternatively, pick any interesting matrix from Bomze (1983) and perform the same exercise.
2. Find own-population effects $\mathbf{u}_a$ and/or $\mathbf{v}_a$ in the Buyer-Seller game that change the closed loops into inward spirals (so the interior equilibrium is stable). Then (probably just changing a few signs), find other own-population effects that produce outward spirals (so the interior equilibrium is unstable).
3. State spaces of dimension 4 include the tesseract (or hypercube) and the 5-simplex. What are the corresponding numbers of populations and strategies? Try to find three other state spaces of dimension 4.
4. Develop a one-line proof that $\Delta_{k-j}=0$ loci coincide with the equal growth rate loci for replicator dynamics. Hint: Write the usual replicator equation $\dot{s}_i=(w_i-\bar{w})s_i$, divide both sides by s_i, and subtract the equation for $i=j$ from the equation for $i=k$.

NOTES

Our techniques for using delta loci to sector the simplex combine known ideas, but we have not seen them exposited systematically in previous literature.

The original RSP game for common lizards (Sinervo et al. 2007) was a discrete time game between overlapping generations in which juveniles recruited as a function of frequency of adult males. The game specified in matrix 3.14 is a continuous time analogue, altered slightly to illustrate the flow from edge saddles.

The subsection on overlapping generations assumes constant survival and fecundity rates. This is a good approximation for many animals and plants, but not for humans. To model populations in which fertility and survival rates differ greatly among adults of different ages, we need a much higher dimensional state space to keep track of the age structure. Chapter 9 takes a step in this direction when it examines life-cycle games.

Stories, mainly from China, of milk tainted with melamine have appeared in western media since 2007; see for example Xin and Stone (2008). Of course, the issue of product quality is always present for experience and credence goods (e.g., auto repairs). Only when private and/or public watchdogs (e.g., the Food and Drug Administration in the United States) fail in their mission does the issue become prominent in modern markets for standardized goods.

Own population effects can be important in applications with two or more populations. For example, Chapter 12 develops a model of international trade, in which firms in the home country compete in goods markets with foreign firms, but also with other home country firms.

The analysis of how the intensity parameter a affects the phase portrait is an application of bifurcation theory. Some other applications, and pointers to the technical literature, will be presented in Chapter 14. Benndorf et al. (2015) independently developed theoretically an equivalent Hawk-Dove example, and found that it did a good job of predicting human behavior in the laboratory.

The section on dynamics develops the perspective introduced in Friedman (1998). We understand that Schlag (1998) was the first to derive replicator dynamics from imitation behavior, although the point is best known from Björnerstedt and Weibull (1999); also see Binmore et al. (1995), Blume and Easley (1992), and Börgers and Sarin (1997). See Sandholm (2010, Table 5.1, p. 150) and surrounding text in his Chapters 4, 5, and 6 for a recent and very helpful survey of evolutionary dynamics.

As explained in numerous books on nonlinear dynamics, such as Sprott (2003), ODE solutions in state spaces of dimension 2 and higher can converge to limit cycles. In dimensions 3 and higher they can converge to "strange attractors" and exhibit extreme sensitivity to initial conditions. See also the Chapter 7 notes.

The material in the appendix was developed largely by Jacopo Magnani, and it is now part of our joint article Friedman et al. (2016). We appreciate the willingness of Dr. Magnani, and the fourth coauthor of that article, Dhanashree Paranjpe, to share that material.

BIBLIOGRAPHY

Benndorf, Volker, Ismael Martinez-Martinez, and Hans-Theo Normann. "Intra- and Intergroup Conflicts: Theory and Experiment in Continuous Time." DICE manuscript, Heinrich Heine University, Dusseldorf, Germany, October 2015.

Binmore, Kenneth G., Larry Samuelson, and Richard Vaughan. "Musical chairs: Modeling noisy evolution." *Games and Economic Behavior* 11, no. 1 (1995): 1–35.

Björnerstedt, Jonas, and Weibull, Jörgen. "Weibull (1999), "Nash equilibrium and evolution by imitation." In The Rational Foundations of Economic Behavior, K. Arrow, E. Colombatto, M. Perlaman, and C. Schmidt, eds., 155–171. Palgrave Macmillan, 1999.

Blume, Lawrence, and David Easley. "Evolution and market behavior." *Journal of Economic Theory* 58, no. 1 (1992): 9–40.

Bomze, Immanuel M. "Lotka-Volterra equation and replicator dynamics: A two-dimensional classification." *Biological Cybernetics* 48, no. 3 (1983): 201–211.

Börgers, Tilman, and Rajiv Sarin. "Learning through reinforcement and replicator dynamics." *Journal of Economic Theory* 77, no. 1 (1997): 1–14.

Cason, Timothy N., Daniel Friedman, and Ed Hopkins. "Testing the TASP: An experimental investigation of learning in games with unstable equilibria." *Journal of Economic Theory* 145, no. 6 (2010): 2309–2331.

Cournot, Antoine-Augustin. *Recherches sur les principes mathématiques de la théorie des richesses par Augustin Cournot.* chez L. Hachette, 1838.

Friedman, Daniel. "Evolutionary games in economics." *Econometrica: Journal of the Econometric Society* 59 (1991): 637–666.

Friedman, Daniel. "On economic applications of evolutionary game theory." *Journal of Evolutionary Economics* 8, no. 1 (1998): 15–43.

Friedman, Daniel, Dhanashree Paranjpe, Jacopo Magnani, and Barry Sinervo. "Estimating Payoff Matrices from Time Series Data: The Case of the Lizard Rock-Paper-Scissors Oscillator." (2016).

Golman, Russell, and Scott E. Page. "Individual and cultural learning in stag hunt games with multiple actions." *Journal of Economic Behavior and Organization* 73, no. 3 (2010): 359–376.

Magnani, Jacopo, Daniel Friedman, and Barry Sinervo. Maximum Likelihood Estimate of Payoffs from Time Series. UC Santa Cruz, 2015. Software. doi:10.7291/d1h59d.

Nachbar, John H. "Evolutionary" selection dynamics in games: Convergence and limit properties." *International Journal of Game Theory* 19, no. 1 (1990): 59–89.

Oprea, Ryan, Keith Henwood, and Daniel Friedman. "Separating the Hawks from the Doves: Evidence from continuous time laboratory games." *Journal of Economic Theory* 146, no. 6 (2011): 2206–2225.

Rowell, Jonathan T., Stephen P. Ellner, and H. Kern Reeve. "Why animals lie: How dishonesty and belief can coexist in a signaling system." *The American Naturalist* 168, no. 6 (2006): E180–E204.

Samuelson, Larry. "Limit evolutionarily stable strategies in two-player, normal form games." *Games and Economic Behavior* 3, no. 1 (1991): 110–128.

Sandholm, William H. *Population games and evolutionary dynamics*. MIT Press, 2010.

Schlag, Karl H. "Why imitate, and if so, how?: A boundedly rational approach to multi-armed bandits." *Journal of Economic Theory* 78, no. 1 (1998): 130–156.

Sinervo, Barry, Benoit Heulin, Yann Surget-Groba, Jean Clobert, Donald B. Miles, Ammon Corl, Alexis Chaine, and Alison Davis. "Models of density-dependent genic selection and a new rock-paper-scissors social system." *The American Naturalist* 170, no. 5 (2007): 663–680.

Sinervo, Barry, Colin Bleay, and Chloe Adamopoulou. "Social causes of correlational selection and the resolution of a heritable throat color polymorphism in a lizard." *Evolution* 55, no. 10 (2001): 2040–2052.

Sprott, Julien Clinton. *Chaos and time-series analysis*. Vol. 69. Oxford: Oxford University Press, 2003.

Weibull, Jörgen W. *Evolutionary game theory*. MIT Press, 1995.

Xin, Hao, and Richard Stone. "Tainted milk scandal. Chinese probe unmasks high-tech adulteration with melamine." *Science* 322, no. 5906 (2008): 1310–1311.

4

Equilibrium

Steady states, especially stable steady states, are crucial in applied work because it is much easier to collect consistent observations and to learn something about the system when the state has settled down. Economists, biologists, and engineers therefore pay special attention to steady states, which they usually call equilibrium states, or just equilibria.

If you are mainly interested in equilibria, the approach so far is rather cumbersome. You have to specify replicator or other dynamics, work out their trajectories, find basins of attraction, and so on. It would be wonderful to have shortcuts that could help you find the equilibria (and tell you something about stability) just from looking at the fitness function.

Such shortcuts do exist. Indeed, there are many different static definitions of equilibrium, and this chapter will explore two of the most important. The first is *Nash equilibrium* (NE for short), named after 1994 Nobel laureate John Nash, whose 1950 PhD dissertation proved that NE exist for quite general fitness functions. The second is *evolutionarily stable state* (ESS for short, and sometimes also called "evolutionary stable strategy"), named by John Maynard Smith, and developed jointly with George Price in the 1970s.

This chapter's primary theme is the connection between these two static equilibrium concepts and dynamic behavior. To oversimplify, a NE is a steady state of any reasonable dynamic process, and an ESS is typically a dynamically stable steady state. The chapter will note when there are steady states that are not NE, when there are stable steady states that are not ESS, and when other caveats are in order. Along the way, it will mention techniques for assessing stability of equilibrium.

The last part of the chapter looks at "runaway" dynamics in the context of Kirkpatrick's (1982) model of sexual selection. Runaways offer a perspective on equilibrium that can benefit most researchers, while sexual selection is a complication often important in biological applications and occasionally in social science applications. Appendix succinctly presents three standard techniques for assessing stability of equilibrium, based on the eigenvalues of Jacobian matrices, on Lyapunov functions, and on Liouville's theorem.

4.1 EQUILIBRIUM IN 1 DIMENSION

Let us begin with the simplest case: a single population with two alternative strategies. The state space is 1-dimensional, since every state can be written $\mathbf{s} = (s_1, 1 - s_1)$ with $s_1 \in [0,1]$. As usual, assume that the fitness functions $w_1(\mathbf{s})$ and $w_2(\mathbf{s})$ are continuously differentiable, and possibly linear.

Recall from previous chapters how to check whether a point $\mathbf{s}^* = (s_1^*, 1 - s_1^*)$ is a steady state. If the point is interior (i.e., if $0 < s_1^* < 1$), then $\mathbf{s}^*$ is a steady state of any monotone dynamic if and only if it is a solution of the equation $\Delta w(\mathbf{s}) = 0$. By definition, $\Delta w = w_1 - w_2$, so the steady state equation can be written

$$w_1(\mathbf{s}^*) = w_2(\mathbf{s}^*). \tag{4.1}$$

When fitness is linear, there is a 2×2 matrix $\mathbf{w} = ((w_{ij}))$ such that $\Delta w(\mathbf{s}^*) = w_1(\mathbf{s}^*) - w_2(\mathbf{s}^*) = \mathbf{e}_1\mathbf{w}\cdot\mathbf{s} - \mathbf{e}_2\mathbf{w}\cdot\mathbf{s}$, so in this case (4.1) can be rewritten as

$$\Delta w(\mathbf{s}^*) = \mathbf{e}_\Delta \mathbf{w}\cdot\mathbf{s} = 0, \tag{4.2}$$

where $\mathbf{e}_\Delta = \mathbf{e}_1 - \mathbf{e}_2 = (1, -1)$.

But what if the point is not interior? A corner state $\mathbf{s} = \mathbf{e}_i$, that is, a state such that $s_i^* = 0$ or 1, is always a steady state for replicator dynamics. For monotone dynamics, $\mathbf{e}_i$ is a steady state if, and only if, it has the higher payoff when prevalent. That is, it must satisfy

$$w_i(\mathbf{e}_i) \geq w_j(\mathbf{e}_i), \tag{4.3}$$

for all j.

Nash found a way to capture both corner and interior steady states in a more unified manner. His idea was that, at equilibrium, no player can increase her fitness by changing her strategy. Formally, a state $\mathbf{s}^* = (s_1^*, 1 - s_1^*)$ is a NE of a 1-dimensional population game if

$$s_1^* > 0 \Longrightarrow w_1(s^*) \geq w_2(s^*), \text{and} \tag{4.4}$$

$$s_1^* < 1 \Longrightarrow w_2(s^*) \geq w_1(s^*). \tag{4.5}$$

That is, if anyone is using the first strategy then it must be at least as good as the alternative, and likewise, if anyone is using the second strategy then it must be at least as good as its alternative.

Comparing the last three equations, you can see that a corner is a NE (called a "pure strategy NE") if and only if it is a steady state for all monotone dynamics. Comparing the last two equations to (4.1), you can see that an interior point is a NE if and only if it is a steady state. Thus any NE is automatically a steady state for any monotone dynamic.

How about stability? A corner steady state is stable if (4.3) holds as a strict inequality for $j \neq i$, since (by definition) monotone dynamics then push towards the corner $\mathbf{e}_i$. As for interior steady states, recall from Chapter 2 that the key condition is whether it is a *downcrossing*, defined as a solution $\mathbf{s}^* = (s_1^*, 1 - s_1^*)$ of $\Delta w(\mathbf{s}) = 0$ such that $d\Delta w(\mathbf{s}^*)/ds_1 < 0$. When the fitness function is linear, equation (4.2) tells us that the condition is

$$\mathbf{e}_\Delta \mathbf{w} \cdot \left[\frac{d\mathbf{s}^*}{ds_1} \right] = \mathbf{e}_\Delta \mathbf{w} \cdot \mathbf{e}_\Delta < 0, \qquad (4.6)$$

since $\frac{d\mathbf{s}^*}{ds_1} = (\frac{ds_1}{ds_1}, 0 - \frac{ds_1}{ds_1}) = (1, -1) = \mathbf{e}_\Delta$. Recall that the interior steady state in Coordination-type games was an upcrossing (i.e., $d\Delta w(\mathbf{s}^*)/ds_1 > 0$) and unstable, while the interior steady state in Hawk-Dove-type games was a downcrossing and dynamically stable.

The stability conditions are encapsulated in the following tentative definition. A state $\mathbf{s}^* = (s_1^*, 1 - s_1^*)$ is an ESS if either

1. $\mathbf{s}^* = \mathbf{e}_i$ and $w_i(\mathbf{s}^*) > w_j(\mathbf{s}^*)$ for $j \neq i$, or
2. $w_i(\mathbf{s}^*) = w_j(\mathbf{s}^*)$ and $d\Delta w(\mathbf{s}^*)/ds_1 < 0$.

The first part is simply the strict version of (4.3), and the second part is just the downcrossing property for an interior equilibrium.

To illustrate the computations, recall the basic Hawk-Dove game matrix $\mathbf{w} = \begin{pmatrix} \frac{v-c}{2} & v \\ 0 & \frac{v}{2} \end{pmatrix}$, where $c > v > 0$. An interior equilibrium $\mathbf{s}^* = (s_1^*, 1 - s_1^*)$ solves $0 = \Delta w(\mathbf{s}) = \mathbf{e}_\Delta \mathbf{w} \cdot \mathbf{s} = (\frac{v-c}{2}, \frac{v}{2}) \cdot (s_1, 1 - s_1) = \frac{1}{2}(v - cs_1)$, so $s_1^* = \frac{v}{c}$ and $\mathbf{s}^* = (\frac{v}{c}, \frac{c-v}{c})$.

To verify that $\mathbf{s}^*$ is a NE, we compute $w_1(\mathbf{s}^*) = \mathbf{e}_1 \mathbf{w} \cdot \mathbf{s}^* = (\frac{v-c}{2}, v) \cdot (\frac{v}{c}, \frac{c-v}{c}) = \frac{v(c-v)}{2c} = (0, \frac{v}{2}) \cdot (\frac{v}{c}, \frac{c-v}{c}) = \mathbf{e}_2 \mathbf{w} \cdot \mathbf{s}^* = w_2(\mathbf{s}^*)$; so equations (4.4)–(4.5) are indeed both satisfied (with $=$, not $>$). To verify that $\mathbf{s}^*$ is actually an ESS, we now only need to check the downcrossing property. By (4.6) we have $d\Delta w(\mathbf{s}^*)/ds_1 = \mathbf{e}_\Delta \mathbf{w} \cdot \mathbf{e}_\Delta = (\frac{v-c}{2}, \frac{v}{2}) \cdot (1, -1) = -\frac{c}{2} < 0$. Hence $\mathbf{s}^* = (\frac{v}{c}, \frac{c-v}{c})$ is indeed a downcrossing and hence satisfies part 2 of the ESS definition.

This theoretical exercise gives us a prediction: for any monotone dynamic and any interior initial condition, the state will converge over time towards $\mathbf{s}^*$ because it is an ESS and is the only ESS. Thus we predict that the observed population will be near $\mathbf{s}^*$.

Does this prediction pan out in actual data? Oprea, Henwood, and Friedman (2012) ran eight laboratory sessions, each including 10 two-minute periods in which a dozen human subjects, matched in a single population, could continuously switch between hawk and dove play. As illustrated in Figure 4.1, from the initial $\mathbf{s} = (\frac{1}{2}, \frac{1}{2})$ mix, the players' average choice converged within a few seconds very closely to the predicted mix $\mathbf{s}^* = (\frac{v}{c}, \frac{c-v}{c}) = (\frac{2}{3}, \frac{1}{3})$, and rarely moved more than 0.03 away from this point.

What if the environment changes? Suppose, for example, that the value v of the contested resource increases. Instead of trying to discover the actual dynamic process and plotting the trajectory from the old equilibrium to the new equilibrium, we can use a shortcut that biologists and engineers call *sensitivity analysis* and economists call *comparative statics*. We just compute how the equilibrium changes (i.e., we compute $\frac{d\mathbf{s}^*}{dv}$).

In the Hawk-Dove game, the calculation is straightforward. For $\mathbf{s}^* = (\frac{v}{c}, \frac{c-v}{c})$, we have $\frac{d\mathbf{s}^*}{dv} = (\frac{1}{c}, \frac{-1}{c})$. Thus the fraction of hawks should increase, and the fraction of doves decrease, by an amount proportional to the change in v, with the proportion being the reciprocal of the combat cost c.

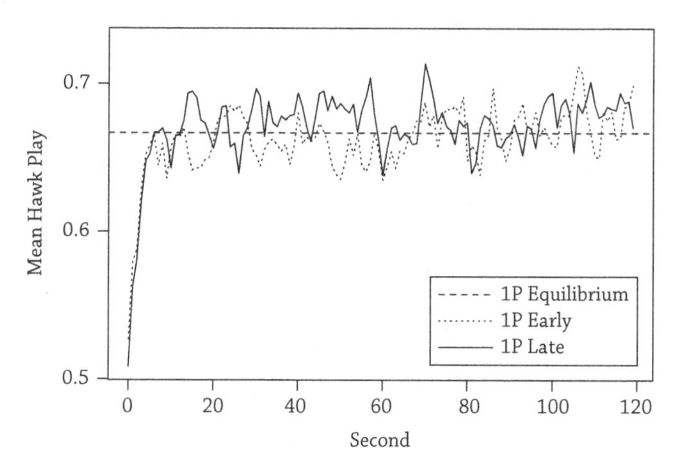

Figure 4.1 Hawk share over time. Averages taken each second over the first five ("Early") and last five ("Late") periods of eight sessions, each with 12 human subjects playing the standard single population ("1P") HD game for cash, with $v = 12$ and $c = 18$. From Oprea, Henwood, and Friedman (2011).

4.2 NASH EQUILIBRIUM WITH n STRATEGIES

How well do the equilibrium definitions and results generalize to higher dimensions and to nonlinear fitness functions? Pretty well, it turns out, but the explanation will be easier to follow with new notation.

In this section we consider strategic interactions of a single population with alternative strategies $i = 1, 2, \ldots, n$. Let $\mathbf{a} = (a_1, a_2, \ldots, a_n)$ be any real vector; for example, with $n = 3$ we could set $\mathbf{a} = (-1.5, 2, -0.1)$. The $\mathbf{a}$-weighted fitness at a given state $\mathbf{s}$ is

$$f(\mathbf{a}, \mathbf{s}) = \sum_{i=1}^{n} a_i w_i(\mathbf{s}). \tag{4.7}$$

That is, we construct f to be linear in its first argument $\mathbf{a}$ but not necessarily in its second argument $\mathbf{s}$. Of course, if f turns out to be linear in both arguments then there is some matrix $\mathbf{w}$ such that f is just matrix multiplication: $f(\mathbf{a}, \mathbf{s}) = \mathbf{aw} \cdot \mathbf{s}$. But the new notation is especially handy when the fitnesses $w_i(\mathbf{s})$ are possibly nonlinear in the state $\mathbf{s}$.

Using the new notation, continuous replicator dynamics can be written as

$$\dot{s}_i = (f(\mathbf{e}_i, \mathbf{s}) - f(\mathbf{s}, \mathbf{s}))s_i,\ i = 1, \ldots, n \tag{4.8}$$

since $f(\mathbf{e}_i, \mathbf{s})$ is simply the fitness $w_i(\mathbf{s})$ of strategy i in the new notation, and $f(\mathbf{s}, \mathbf{s})$ is mean fitness. Let $A(\mathbf{s}) = \{i : s_i > 0\}$ denote the active (nonextinct) strategies at state $\mathbf{s}$, and recall that $\mathbf{s}$ is a replicator steady state if and only if $\dot{s}_i = 0,\ i = 1, \ldots, n$. It is obvious from (4.8) that $\mathbf{s}$ is a replicator steady state if and only if

(i) $f(\mathbf{e}_i, \mathbf{s}) = f(\mathbf{s}, \mathbf{s})$ for all $i \in A(\mathbf{s})$.

Steady states of monotone (in rates of change) dynamics require the additional proviso that no extinct strategy has higher fitness:

(ii) $f(\mathbf{e}_i, \mathbf{s}) \leq f(\mathbf{s}, \mathbf{s})$ for all $i \notin A(\mathbf{s})$.

Let $\mathcal{S}$ denote the n-simplex. In the new notation, the definition of NE for a single-population game is very tidy:

Definition. A state $\mathbf{s} \in \mathcal{S}$ is an NE if, for all $\mathbf{x} \in \mathcal{S}$,

(i') $f(\mathbf{s}, \mathbf{s}) \geq f(\mathbf{x}, \mathbf{s})$.

To unpack the definition, note first that, by (4.7), $\bar{w}(\mathbf{s}) = f(\mathbf{s}, \mathbf{s}) = \sum_{i=1}^{n} s_i w_i(\mathbf{s}) = \sum_{i=1}^{n} s_i f(\mathbf{e}_i, \mathbf{s})$. That is, the fitness $\bar{w}$ of the mixed strategy $\mathbf{s}$ is always the weighted average of the fitnesses $f(\mathbf{e}_i, \mathbf{s})$ of the active strategies i. But no strategy i can have greater fitness than a NE mix $\mathbf{s}$—to see this, just insert $\mathbf{x} = \mathbf{e_i}$ into (i'). If there were a strategy $i \in A(\mathbf{s})$ with fitness lower than $\bar{w}$, then you could exclude it from the mix and raise fitness, contrary to (i'). Hence at NE, the fitnesses are equal: $\bar{w} = f(\mathbf{s}, \mathbf{s}) = f(\mathbf{e}_i, \mathbf{s})$ for all $i \in A(\mathbf{s})$.

This simple argument proves a point worth emphasizing:

Proposition 1. *At Nash equilibrium, each active strategy has equal and maximal fitness.*

Thus the current NE definition really does generalize the tentative definition for one dimension given in the previous section. The intuition is the same for both versions—at NE, no player can gain by unilaterally changing her strategy.

With Proposition 1 in mind, it is easy to see that condition (i') actually implies the steady state conditions (i) and (ii). Thus we have another general result:

Proposition 2. *A Nash equilibrium is a steady state for every monotone dynamic.*

Of course, the conditions simplify at interior points, where all $i \in A(\mathbf{s})$. Then condition (i) guarantees that condition (i$'$) holds with equality. For future reference, we record this simple result as

Proposition 3. *If $\mathbf{s} \in \mathcal{S}$ is an interior steady state under monotone dynamics, then it is also a Nash equilibrium.*

As noted earlier, John Nash's dissertation proved that NE exist for a very broad class of games. An existence result general enough for our purposes follows.

Proposition 4. *Let $f : \mathbb{R}^n \times \mathcal{S} \to \mathbb{R}$ satisfy (4.7) and be continuous on $\mathcal{S}$. Then there is at least one point $\mathbf{s} \in \mathcal{S}$ that satisfies condition (i$'$), Hence f has at least one Nash equilibrium.*

Nash's proof used Brouwer's fixed point theorem, a mainstay of topology. Most recent graduate textbooks contain a shorter proof using Kakutani's variant of Brouwer's theorem, and those proofs cover Proposition 4.

4.3 ESS WITH n STRATEGIES

So far, so good, but what can we say about stability? A pure NE $\mathbf{s}^* = \mathbf{e}_i$ is dynamically stable if we have a strict inequality in (i'), or, equivalently, in (4.3). But such strict NE are not possible in the interior, nor even on faces and edges of the n-simplex, by Proposition 1. Hence a more subtle analysis is necessary for mixed NE.

The 1-dimensional case offers a clue: there the downcrossing condition ensures dynamic stability. The geometric intuition is that, at any downcrossing, deviations from equilibrium shrink over time. How can we generalize that idea to higher dimensions?

Equation (4.6) provides a hint on how to proceed. It says that, for a 2×2 matrix $\mathbf{w}$ to produce a downcrossing, it must satisfy $\hat{w} = \mathbf{hw} \cdot \mathbf{h} < 0$, where $\mathbf{h} = (1, -1)$. But the same inequality works for the vector $\mathbf{h} = (b, -b)$ for any $b \neq 0$, since $(b, -b)\mathbf{w} \cdot (b, -b) = b(1, -1)\mathbf{w} \cdot (1, -1)b = b^2\hat{w} < 0$. Recall that a matrix $\mathbf{w}$ is called *negative definite* if the inequality $\mathbf{yw} \cdot \mathbf{y} < 0$ holds for all $\mathbf{y} \neq 0$. Thus the downcrossing condition can be restated as follows: $\mathbf{w}$ is negative definite on the space of deviations $(b, -b)$ from a given state $\mathbf{s}$.

Now we have a condition that generalizes! Even better, it still says (in fancier language) that deviations shrink over time. To spell it out, write an arbitrary deviation $\mathbf{h} = (\mathbf{s} - \mathbf{x}) = (s_1 - x_1, \ldots, s_n - x_n)$ from an equilibrium $\mathbf{s}$ to another point $\mathbf{x}$ in the simplex, and note that the components of $\mathbf{h}$ must sum to $1 - 1 = 0$. Let $\mathbb{R}_0^n$ denote the subspace of deviations, i.e., all n-vectors $\mathbf{h}$ whose components sum to 0.

The key stability condition for an $n \times n$ fitness matrix $\mathbf{w}$ is that it is negative definite on $\mathbb{R}_0^n$:

$$0 > \mathbf{hw} \cdot \mathbf{h} = (\mathbf{s} - \mathbf{x})\mathbf{w} \cdot (\mathbf{s} - \mathbf{x}) = \mathbf{sw} \cdot \mathbf{s} - \mathbf{sw} \cdot \mathbf{x} - \mathbf{xw} \cdot \mathbf{s} + \mathbf{xw} \cdot \mathbf{x}. \tag{4.9}$$

(The explanation of the eigenvalue method in Appendix mentions why a negative definite condition on $\mathbb{R}_0^n$ implies that deviations shrink over time.) Using the f notation, inequality (4.9) can be written more generally as

$$f(\mathbf{s}, \mathbf{s}) - f(\mathbf{s}, \mathbf{x}) - f(\mathbf{x}, \mathbf{s}) + f(\mathbf{x}, \mathbf{x}) < 0 \tag{4.10}$$

for all $\mathbf{h} \neq 0 \in \mathbb{R}_0^n$, (i.e., for all $\mathbf{s} \neq \mathbf{x}$ in the simplex).

This gets us pretty close to a definition of ESS, but a couple of subtleties remain. When f is nonlinear, the inequality need not hold everywhere; local stability requires only that it hold for $\mathbf{x}$ in some neighborhood of $\mathbf{s}$. Using the notation $\mathbf{s}_{\epsilon,x} = (1 - \epsilon)\mathbf{s} + \epsilon\mathbf{x}$ for nearby states, we are led to the following *definition*: A state $\mathbf{s} \in \mathcal{S}$ is an ESS if there is some $\bar{\epsilon} > 0$ such that, for all other states $\mathbf{x} \neq \mathbf{s} \in \mathcal{S}$,

$$f(\mathbf{s}, \mathbf{s}_{\epsilon,x}) > f(\mathbf{x}, \mathbf{s}_{\epsilon,x}) \quad \text{for all} \quad \epsilon \in (0, \bar{\epsilon}]. \tag{4.11}$$

Writing out $\mathbf{s}_{\epsilon,x}$, it is obvious that (4.11) is closely related to (4.10) when f is linear. The intuition again is that any small mutation (of size $\epsilon > 0$) fails to take hold because, at the post-invasion state $\mathbf{s}_{\epsilon,x}$, the strategies that are part of the ESS state $\mathbf{s}$ have higher fitness, and so any small deviation from $\mathbf{s}$ shrinks over time. For a careful discussion of these matters, see Sandholm (2010, Section 8.3), and see the notes for a cookbook method to check whether a NE is an ESS.

This definition of ESS is valid for linear as well as nonlinear fitness functions, but some authors prefer to use the original definition, which is valid only for linear fitness functions. It reads as follows:

Original Definition. A state $\mathbf{s} \in \mathcal{S}$ is an ESS if, for all other states $\mathbf{x} \neq \mathbf{s} \in \mathcal{S}$, either

(i) $f(\mathbf{s},\mathbf{s}) > f(\mathbf{x},\mathbf{s})$, or
(ii) $f(\mathbf{s},\mathbf{s}) = f(\mathbf{x},\mathbf{s})$ and $f(\mathbf{s},\mathbf{x}) > f(\mathbf{x},\mathbf{x})$.

This original version spells out the reasons an ESS state s resists any invasion—either (i) the mutants $\mathbf{x}$ are less fit, or (ii) they are equally fit at the current state (when they are rare) but less fit when they become prevalent.

Comparing these last two conditions to those that define NE, it follows immediately that ESS are automatically NE. Also, the first condition translates immediately as stating that $\mathbf{s}$ is a strict NE. Hence, for the record, we have

Proposition 5. *For any given linear fitness function, (a) any ESS* $\mathbf{s} \in \mathcal{S}$ *is also a Nash equilibrium, and (b) any strict Nash equilibrium is also an ESS.*

ESS is an acronym for evolutionarily stable state, but in what sense is ESS a dynamic stability condition? Our first result, known for a long time, is that it is a sufficient condition for replicator dynamics. More formally,

Proposition 6. *Let* $\mathbf{s} \in \mathcal{S}$ *be an ESS for fitness function* f. *Then* $\mathbf{s}$ *is a locally stable steady state under replicator dynamics.*

A recent proof can be found in Sandholm (2010, Section 8.4), together with similar results for several important alternative dynamics.

Unfortunately, there are evolutionary dynamics that can destabilize an ESS. If one uses the weakest notion of monotonicity—variety (c) in Section 9 of Chapter 3, that there is only a positive correlation of fitnesses and rates of change—then it is not especially difficult to construct examples (for $n = 3$) of such dynamic instability. Perhaps more surprisingly, one can also construct variety (b) monotone dynamics—fitnesses have the same rank order as rates of change—that destabilize an ESS. See Counterexample 2 of Friedman (1991).

The rest of this section will use some specific examples to illustrate NE, ESS, and stable steady states.

Zeeman's 3 x 3 game

For replicator dynamics (and many other sorts of dynamics but not all), we know from the preceding discussion that ESS is a sufficient condition for dynamic stability. But is it a necessary condition?

The answer is no. Zeeman (1980) showed that there are interior NE that are not ESS, and yet are dynamically stable under replicator (and many other) dynamics, with large basins of attraction. Zeeman illustrated this point using the 3×3 matrix game

$$\mathbf{w}=\begin{pmatrix} 0 & 6 & -4 \\ -3 & 0 & 5 \\ -1 & 3 & 0 \end{pmatrix}. \tag{4.12}$$

We begin a systematic analysis of (4.12) by sectoring the simplex as in the previous chapter. The payoff advantage of the first strategy s_1 over the second s_2 is

$$\Delta_{12}(\mathbf{s})=(1,-1,0)\mathbf{w}\cdot\mathbf{s}=(3,6,-9)\cdot\mathbf{s}=3(s_1+2s_2-3s_3). \tag{4.13}$$

As shown in Panel A of Figure 4.2, the line $\Delta_{12}(\mathbf{s})=0$ intersects the $s_1=0$ (or bottom) edge of the simplex at $(0,3,2)/5$ and the $s_2=0$ (or right) edge at $(3,0,1)/4$. Hence $w_1>w_2$ at points in the simplex to the left of the straight line connecting these points. Similarly,

$$\Delta_{13}(\mathbf{s})=(1,0,-1)\mathbf{w}\cdot\mathbf{s}=(1,3,-4)\cdot\mathbf{s}=s_1+3s_2-4s_3, \tag{4.14}$$

and $\Delta_{13}(\mathbf{s})=0$ intersects the bottom edge at $(0,4,3)/7$ and the right edge at $(4,0,1)/5$, so $w_1>w_3$ at points in the simplex to the left of the straight line connecting these points. Also,

$$\Delta_{23}(\mathbf{s})=(0,1,-1)\mathbf{w}\cdot\mathbf{s}=(-2,-3,5)\cdot\mathbf{s}=-2s_1-3s_2+5s_3, \tag{4.15}$$

and $\Delta_{23}(\mathbf{s})=0$ intersects the bottom edge at $(0,5,3)/8$ and the right edge at $(5,0,2)/7$, so $w_2>w_3$ at points in the simplex to the right (not left, since (4.15)

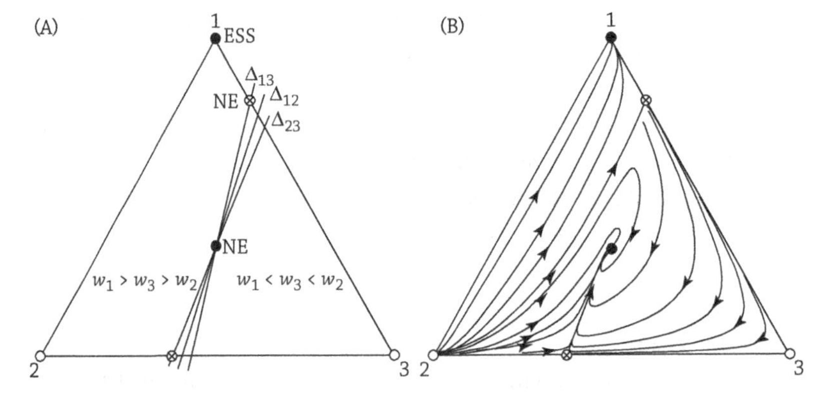

Figure 4.2 Zeeman game (4.12). Panel A: Sectoring. Panel B: Replicator dynamics.

has coefficients with signs on s_i opposite to those of the other two Δ expressions) of the straight line connecting these points.

In view of equation (4.8) and subsequent item (i), the only candidates for equilibrium (or any kind of steady state) are intersections of the edges and the $\Delta_{ij}=0$ lines. Corners are where two edges intersect, and they are automatically steady states for replicator dynamics. However, only the first corner $\mathbf{e}^1=(1,0,0)$ satisfies the pure Nash property (i′), by virtue of the fact that the first entry in the first column of $\mathbf{w}$ exceeds the other entries in that column. Indeed, this fact ensures that it is also an ESS, using either the preliminary or the official definition.

We have already found the six intersections of $\Delta_{ij}=0$ lines with edges. Of these, only $(4,0,1)/5$ is a NE because it satisfies $w_1=w_3$ (since it is on the $\Delta_{13}(s)=0$ line) and also satisfies $w_1>w_2$ (since it is to the left of the $\Delta_{12}(s)=0$ line which hits that edge at $(3,0,1)/4$). The other five intersections either fail to equate the two payoffs pertaining to their edge, or else the remaining payoff is higher. For example, $(0,5,3)/8$ does equate the pertinent payoffs w_2 and w_3, but w_1 is higher. Hence it is not a NE, but it is a saddle point for replicator dynamics, with unstable saddle path leading into the interior of the simplex, as can be seen in Panel B of Figure 4.2.

The final candidate for equilibrium is the intersection of any pair (hence all three) $\Delta_{ij}=0$ lines. This intersection occurs at $(1,1,1)/3$, the center of the simplex, as is easy to verify from (4.13)–(4.15). Since all three payoffs are equal at this point, it is by definition an interior NE. To check stability under replicator dynamics, we compute the Jacobian matrix $((\frac{\partial \dot{s}_i}{\partial s_j}))$, evaluate it at $(1,1,1)/3$, and multiply it by the projection matrix $\mathbf{P}=\frac{1}{3}\begin{pmatrix} 2 & -1 & -1 \\ -1 & 2 & -1 \\ -1 & -1 & 2 \end{pmatrix}$, as explained in the Appendix. The result is $\mathbf{J}=\frac{1}{9}\begin{pmatrix} 4 & 9 & -13 \\ -5 & -9 & 14 \\ 1 & 0 & -1 \end{pmatrix}$. The characteristic equation reduces (after more tedious algebra) to $3\lambda^3+2\lambda^2+\lambda=0$, with roots (i.e., eigenvalues) 0 (corresponding to the eigenvector (1,1,1) normal to the simplex, hence irrelevant) and $\frac{-1\pm\sqrt{-2}}{3}$. Since the real parts of the two relevant eigenvalues are negative (at $-1/3$), we conclude that this NE is indeed a stable steady state for replicator dynamics.

However, $(1,1,1)/3$ is not an ESS. To see this, consider an "invasion" of the first strategy (i.e., set $\mathbf{x}=\mathbf{e}^1=(1,0,0)$ in the last version of the ESS definition). Since we are at an interior state, we must have $f(\mathbf{s},\mathbf{s})=f(\mathbf{x},\mathbf{s})$ so part (i) does not apply. By part (ii), then, ESS requires that $f(\mathbf{s},\mathbf{x})>f(\mathbf{x},\mathbf{x})$, but in fact $f(\mathbf{s},\mathbf{x})=\frac{1}{3}(1,1,1)\mathbf{w}\cdot\mathbf{e_1}=\frac{1}{3}(-4,9,1)\cdot(1,0,0)=\frac{-4}{3}<0=\mathbf{e_1}\mathbf{w}\cdot\mathbf{e_1}=f(\mathbf{x},\mathbf{x})$.

The general intuition is that ESS is essentially a negative definiteness condition, which requires that all trajectories in its neighborhood to come closer and closer together, monotonically. Panel B of the figure suggests that trajectories actually move farther apart as they approach their NNE (or SSW) turning points, although they ultimately do converge.

Lizard games

To further illustrate static and dynamic equilibria, let us recall estimates of two games played by male side-blotched lizards. Section 7 of Chapter 2 used the matrix

$$\mathbf{w} = \begin{pmatrix} 0 & -.49 & .80 \\ .93 & 0 & -.42 \\ -.09 & .78 & 0 \end{pmatrix}. \tag{4.16}$$

as the continuous time estimate for siring success by morphs $o,y,b = 1,2,3$. Familiar calculations yield the fitness difference functions

$$\begin{aligned} \Delta_{12} &= -.93s_1 - .49s_2 + 1.22s_3 \\ \Delta_{13} &= .09s_1 - 1.27s_2 + .80s_3 \\ \Delta_{23} &= 1.02s_1 - .78s_2 - .42s_3. \end{aligned} \tag{4.17}$$

The usual sectoring exercise confirms that none of the intersections of the edges with equal fitness lines $\Delta_{ij} = 0$ are steady states, so neither are they NE. The three-way intersection of the equal fitness lines at the interior state $\mathbf{s}^* \approx (0.358, 0.264, 0.378)$ is, of course, a NE because all three strategies have equal (hence maximal) fitness. The directional cones defined by monotonicity (specifically, version b, rank ordering of rates of change) show that the interior equilibrium is a center, with trajectories spiraling counterclockwise. Centers are not the best candidates for ESS, because generically ESS has the negative definite property, which in turn implies direct convergence and no repetitive cycles. Direct calculation of the eigenvalues for continuous replicator dynamics using Dynamo yields $-0.087 \pm 0.329\sqrt{-1}$, confirming cycles. Thus the interior NE is a stable steady state for replicator dynamics.

Chapter 3 (and its appendix) noted that the game estimated in (4.16) used a possibly biased pairwise estimator of siring success and does not fully account for survival of progeny. A more encompassing fitness estimate uses census counts to measure the how effective each male morph in one generation is in producing adult males in the next generation. We take the maximum likelihood/Dirichlet estimates of discrete dynamics reported there, replace the constrained estimates of 0 fitness by 0.001, and take logs to get the matrix $\begin{pmatrix} -6.9 & -1.4 & -3.9 \\ -2.0 & -2.8 & -1.6 \\ -1.8 & -1.7 & -6.9 \end{pmatrix}$.

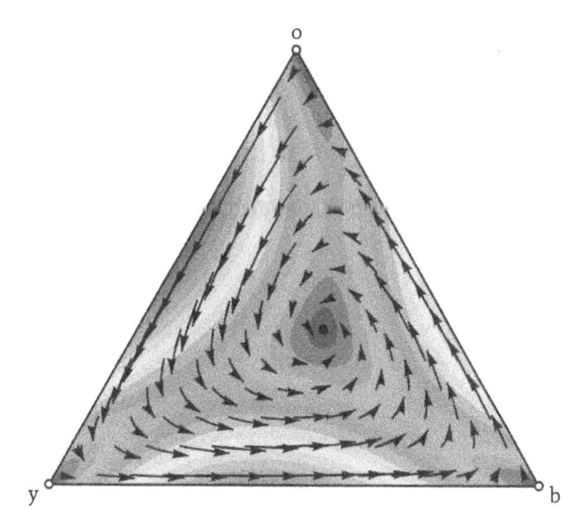

Figure 4.3 Replicator dynamics for game (4.16)

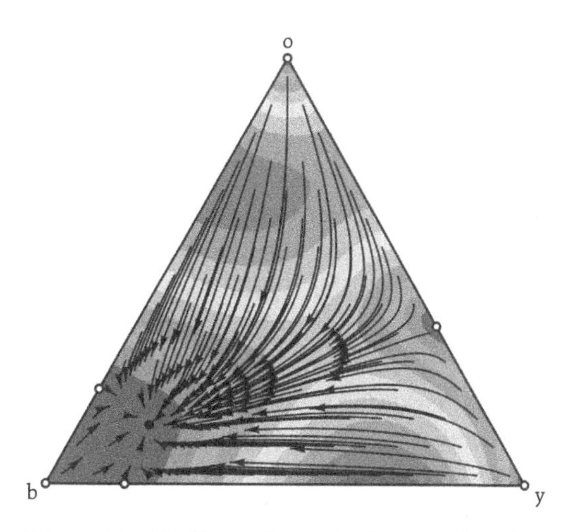

Figure 4.4 Replicator dynamics for game (4.18)

(See the notes for discussion of more accurate methods for obtaining the empirical continuous time fitness matrix.) Imposing the usual normalization to get 0's on the diagonal, we have

$$\mathbf{w} = \begin{pmatrix} 0 & 1.4 & 3.0 \\ 4.9 & 0 & 5.3 \\ 5.1 & 1.1 & 0 \end{pmatrix}. \qquad \textbf{(4.18)}$$

The usual sectoring exercise produces a surprisingly different figure. It is still true that o is the best response (b.r.) to b which is the b.r to y which is the b.r. to o, so the game is a generalized RPS. Nevertheless, trajectories converge directly to the interior NE $\mathbf{s}^* \approx (.14, .71, .15)$ with no spiraling, as seen in Figure 4.4. You can confirm that the three vertices are all saddle points of replicator dynamics, and that three of the six intersections of the edges with equal fitness lines are saddle points but none of them are NE. You can also confirm that the interior NE is indeed an ESS, by verifying part (ii) of the definition.

4.4 EQUILIBRIUM IN MULTI-POPULATION GAMES

Equilibrium concepts and constructions generalize nicely to games with two or more distinct populations. To illustrate, recall from the previous chapter the basic Buyer-Seller game. We saw that it has a unique interior steady state at the solution (p^*, q^*) of the system $\Delta u(\mathbf{q}) = 0 = \Delta v(\mathbf{p})$.

Is (p^*, q^*) a NE of this two-population game? The definition of NE offered in Section 4.3 generalizes directly. Let $(\mathbf{s^1}, \ldots, \mathbf{s^J}) \in \mathcal{S}^1 \times \ldots \times \mathcal{S}^J$ denote a state of a J-population game, and let $f^j(\mathbf{x}; \mathbf{s^1}, \ldots, \mathbf{s^J})$ denote the fitness of strategy $\mathbf{x} \in \mathcal{S}^j$ at that state. Then we have the multi-population

Definition: A state $(\mathbf{s^1}, \ldots, \mathbf{s^J})$ is an NE if, for all $j = 1, \ldots, J$ and all $\mathbf{x} \in \mathcal{S}^j$, (i″) $f^j(\mathbf{s^j}; \mathbf{s^1}, \ldots, \mathbf{s^J}) \geq f(\mathbf{x}; \mathbf{s^1}, \ldots, \mathbf{s^J})$.

As before, the intuition is simply that each player (in each population j) chooses a strategy that gives the highest fitness at the current state.

It is straightforward to check that Proposition 3 generalizes directly. For example, the interior steady state (p^*, q^*) of the basic Buyer-Seller game of Chapter 3 is clearly a NE.

We are now prepared to tackle the question deferred from the previous chapter: How can we tell whether the interior steady state (p^*, q^*) is stable under replicator dynamics? The first approach is to form the Jacobian matrix for the system, and evaluate it at $(p_1^*, q_1^*) = (\frac{1}{9}, \frac{3}{8})$. This yields

$$\mathbf{J} = \begin{pmatrix} (1-2p_1)\Delta u(\mathbf{q}) & p_1(1-p_1)\frac{d\Delta u(\mathbf{q})}{dq_1} \\ q_1(1-q_1)\frac{\Delta v(\mathbf{p})}{dp_1} & (1-2q_1)\Delta v(\mathbf{p}) \end{pmatrix} = \begin{pmatrix} 0 & 8p_1(1-p_1) \\ -9q_1(1-q_1) & 0 \end{pmatrix}$$

$$= \begin{pmatrix} 0 & 64/81 \\ -135/64 & 0 \end{pmatrix},$$

where the second equality uses the fact that we are evaluating at a point where $\Delta u(\mathbf{q}) = 0 = \Delta v(\mathbf{p})$. (No projection matrix is necessary here since we are working in the 2-dimensional state space $(p_1, q_1) \in [0,1]^2$ and not in the 4-d space $(p,q) = ((p_1,p_2),(q_1,q_2))$.) The eigenvalues of $\mathbf{J}$ are $\pm\sqrt{-15}/3$, which are purely imaginary. Hence the steady state appears to be neutrally stable. This is consistent with the figure in Chapter 3, which shows trajectories that move in closed counterclockwise loops (squashed circles), and do not spiral out or spiral in.

Since (p^*,q^*) is not asymptotically stable under replicator dynamics, we anticipate that it is not an ESS. Of course, this begs the question of how to generalize the definition of ESS to multi-population games, and it turns out that there is more than one way to do so. Sandholm (2010, Section 8.3.2) distinguishes between a Cressman ESS (essentially that at least one population component of any small invasion will be less fit), and a Taylor ESS (all components of any invasion will be less fit). He advocates using Taylor ESS with a regularity condition that includes a strong multi-population version of negative definite condition. By any of the ESS definitions, (p^*,q^*) is not an ESS.

The Appendix points out that conclusively establishing neutral stability requires different techniques, and mentions two that work in this case, Lyapunov functions and Liouville's theorem.

4.5 FISHERIAN RUNAWAY EQUILIBRIUM

Convergence to a new equilibrium is especially dramatic when it is far away from, and qualitatively different from, the old equilibrium. This can happen when changes in one population catalyze complementary changes in another population, and vice versa, a situation sometimes called "runaway." For example, about 100 years ago, Jazz emerged as a new musical genre when talented players cultivated a growing audience, which encouraged new talented players and further enlarged the audience. We can get a cultural runaway when the best teachers get the biggest audiences and in so doing generate a faster learning process in their cultural descendants.

Technological "revolutions" follow a similar process. A 100-year-old example here is the internal combustion engine. Its emergence catalyzed new techniques of car manufacture and better engines, which spurred demand for cars and gasoline production and distribution, and so on. The new equilibrium involved sprawling suburbs, vast interstate and local road networks, and huge car, rubber, and gasoline industries.

The basic modeling idea is that, due to changes in the underlying environment, the basin of attraction shrinks for the initial equilibrium, and at some point the state escapes that basin. Then the state begins to accelerate away from that initial equilibrium, picks up speed, and slows again only when it approaches a once-distant new equilibrium.

In this section we will consider a biological example of runaway due to sexual selection. Fisher (1930) verbally formulated a process by which the joint evolution of a male trait, and a female preference for males with that trait, might produce ever more exaggerated levels of the male trait, even if that trait has significant viability costs. The classic example is the male peacock's amazing tail, which seems to captivate female peacocks as well as human feather collectors and other predators.

Lande (1981) solved a continuous trait game, but here we will develop variants of a discrete trait model due to Kirkpatrick (1982), and use equilibrium techniques to spotlight the main issues. The model maintains the following simplifying assumptions.

1. males have two alternative strategies: either plain, or ornamented with mortality cost $m \in (0,1)$; and
2. females have two alternative strategies: either indifferent, or ornament-preferring by factor $a > 1$.

The question is when the boring equilibrium (plain, indifferent) can be escaped, and how far the dynamics will run towards the (ornament, o-preferring) equilibrium, despite its mortality disadvantage.

To get started, we make two further assumptions that we will relax later: (i) female fecundity and survival are independent of female strategy, and (ii) males and females belong to separate populations. Let t = fraction of males with the ornament trait and p = fraction of ornament-preferring females. For this two-population game, the payoff bimatrix is $[\mathbf{U},\mathbf{W}^T] = \begin{pmatrix} 1,1 & 1,1-m \\ 1,1 & 1,a(1-m) \end{pmatrix}$, where the right column refers to ornamented males and the left column to plain males, and the bottom row refers to ornament-preferring females while the top row refers to indifferent females.

Clearly $\Delta U(t) = 1-1 = 0$ independent of t, and $\Delta W(p) = 1 - [(1-m)(1-p) + a(1-m)p] = m - (a-1)(1-m)p$, with root $p^* = \frac{m}{(a-1)(1-m)}$. If $p^* > 1$, then the plain strategy is dominant for males. Otherwise, as shown in Panel A of Figure 4.5, we have a range of equilibria with all plain males ($t = 0$ and $p \in [0,p^*]$) and another range of equilibria with all males ornamented ($t = 1$ and $p \in [p^*,1]$.) All of the equilibria with $p \neq p^*$ are neutrally stable: following a small shift in initial conditions from an initial equilibrium, the evolutionary path under replicator dynamics converges to a nearby equilibrium but not generally to the original equilibrium. Thus one scenario is that, with finite populations, random genetic drift might take the population to an equilibrium $(t,p) = (0,p^* - \epsilon)$ from which a small shift in initial conditions might push p above p^*. The state would then run away to the upper equilibrium with $t = 1$.

Assumption (i) says that the female payoff matrix $\begin{pmatrix} 1 & 1 \\ 1 & 1 \end{pmatrix}$ is degenerate, and that accounts for the continuous sets of neutrally stable equilibria. Now slightly modify that assumption, say to

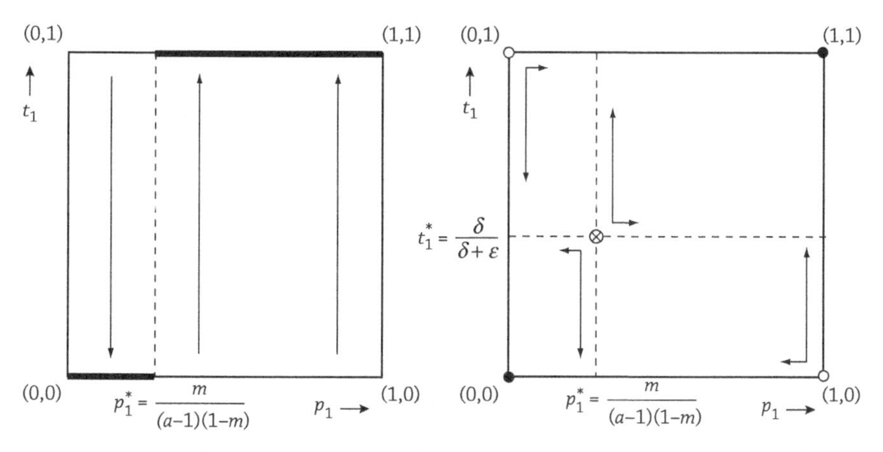

Figure 4.5 Equilibrium in the two-population game. Panel A uses assumption (i) and Panel B uses (i').

i'. Each female morph incurs a small frequency-dependent cost: $\delta > 0$ for indifferent and $\epsilon > 0$ for ornament preferring.

The modified payoff matrix is

$$\mathbf{U} = \begin{pmatrix} 1-\delta & 1 \\ 1 & 1-\epsilon \end{pmatrix}, \tag{4.19}$$

and the modified fitness difference is $\Delta U(t) = 1 - \delta(1-t) - [1 - \epsilon t] = -\delta + (\delta+\epsilon)t$, with root $t^* = \frac{\delta}{\delta+\epsilon}$. This modification yields a generic equilibrium set, with interior saddle-point equilibrium at (p^*, t^*), as shown in Panel B of Figure 4.5. It takes a real push (with force on the order of δ) to escape the basin of attraction of the boring equilibrium $(t,p) = (0,0)$ by reaching the far side of the saddle path (not shown) through (p^*, t^*). Once past the saddle path, the state will then run away to the interesting equilibrium $(t,p) = (1,1)$ where only ornamented males and ornament-preferring females survive.

Assumption (ii) is even more questionable, but relaxing it will introduce new considerations that we defer until the next chapter. We will continue to get runaway dynamics but on a more complex state space.

4.6 DISCUSSION

The main theme in this chapter is the connection between dynamic notions of equilibrium (familiar to engineers) and static concepts (more familiar to economists). It turns out that NE, the central solution concept in classical game theory, is a sufficient condition for steady state in all monotone dynamics, and is a necessary condition for dynamic stability. Another well-known static equilibrium concept, ESS, is a sufficient condition for dynamic stability under the most popular (but not all) dynamic specifications.

In developing its main theme, this chapter offers several examples and some geometric intuition. Chapter 7 will develop the examples further, and cover the gamut of RPS-like games. The present chapter pointed out crucial qualifications, but it skirted by some of the more technical points. Readers seeking a rigorous theoretical treatment should consult Sandholm (2010) and the works cited therein.

Most importantly, the present chapter paves the way to applied work. Chapter 5 will introduce new techniques for dealing with haploid sexual dynamics (and other dual transmission processes), and take a deeper look at the runaway model analyzed earlier. Chapter 14 will begin to extend the analysis to continuous traits (or continuous strategy spaces). Many applications involve socially mediated interactions that are not captured in the random matchings contemplated so far. Chapter 5 will begin to examine the impact, and Chapter 6 will continue the quest by introducing a very useful class of spatial interaction models.

APPENDIX: TECHNIQUES TO ASSESS STABILITY

Here we provide a quick tour of three analytical techniques for assessing dynamic stability. The tour offers intuition and guidance, but to fully master the techniques you will have to work through examples and to consult textbooks such as Hirsch and Smale (1974) and Sandholm (2010).

Assume that dynamics are specified on the n-simplex $\mathcal{S} \subset \mathbb{R}^n$ by an ordinary differential equation (ODE)

$$\dot{s} = V(s), \tag{4.20}$$

where $V : \mathcal{S} \to \mathbb{R}^n$ is a smooth vector-valued function that specifies the tangent vector at each point. For example, with replicator dynamics, V has i^{th} component $V_i(s) = s_i(f(e^i, s) - f(s,s))$. Specification (4.20) also covers most ODE variants of replicator dynamics discussed in Chapter 3. It does not directly cover non-ODE variants—discrete time dynamics, stochastic dynamics, and differential inclusions—but, as explained in Sandholm (2010) and other textbooks, similar techniques turn out to play a central role in their stability analysis.

A dynamic equilibrium is a root $\mathbf{s}^*$ of the equation $V(s) = 0$. Recall that $\mathbf{s}^*$ is Lyapunov stable if solution trajectories $\{s(t) : t \geq 0\}$ to (4.20) remain close to $\mathbf{s}^*$ provided that the initial condition $s(0) = s_o$ was sufficiently close to $\mathbf{s}^*$, and that

$\mathbf{s}^*$ is locally asymptotically stable if, under the same proviso, $s(t) \to \mathbf{s}^*$ as $t \to \infty$. If both conditions hold, we say that $\mathbf{s}^*$ is a stable equilibrium of the dynamics V.

Eigenvalue technique

The first technique is straightforward but often tedious. It reduces the stability question to checking whether a linear algebra property holds.

Take a first-order Taylor expansion of (4.20) around an equilibrium $\mathbf{s}^*$:

$$\begin{aligned} \dot{s} &= V(s) = V(\mathbf{s}^* + y) \\ &= V(\mathbf{s}^*) + DV(\mathbf{s}^*) \cdot y + o(|y|) \\ &= 0 + J \cdot y + o(|y|) \\ &\approx J \cdot y, \end{aligned} \tag{4.21}$$

where $J = DV(\mathbf{s}^*) = ((\frac{\partial V_i(s)}{\partial s_j}))|_{s=\mathbf{s}^*}$ is the Jacobian matrix evaluated at the equilibrium $\mathbf{s}^*$ and $y = s - \mathbf{s}^*$ is the deviation of s from that point. Note that $\dot{y} = \dot{s}$ since $\mathbf{s}^*$ is constant over time, so (4.21) shows that near the equilibrium $\mathbf{s}^*$, the *linear* ODE

$$\dot{y} = J \cdot y \tag{4.22}$$

approximates the true ODE (4.20). The approximation is locally quite good, in that error $o(|y|)$ divided by the Euclidean size $|y|$ of the deviation goes to zero as $|y| \to 0$.

The technique leverages what we know about linear ODEs. The solution to (4.22) with arbitrary initial value $y(0)$ can be written $y(t) = e^{Jt}y(0)$, where the exponentiated matrix is understood as the matrix power series that represents the exp function. For us, the key point is that for a diagonal matrix, with diagonal entries a_i, the solution has i^{th} component $y_i(t) = e^{a_i t}y_i(0)$. Since $e^{a_i t}$ remains bounded iff $a_i \leq 0$ and converges to zero iff $a_i < 0$, we see that in the special case that J is diagonal, the equilibrium $\mathbf{s}^*$ is stable if all entries are negative, is a source if all of them are positive, and is a saddle point if some entries are positive and some are negative.

The Jordan decomposition theorem, a fundamental result in linear algebra, allows us to extend this reasoning to an arbitrary matrix J. The decomposition theorem says that we can change the coordinate system from the usual basis $\{(1,0,\ldots,0),(0,1,0,\ldots,0),\ldots,(0,\ldots,0,1)\}$ to a special new set of coordinates (called unit eigenvectors) so that J is almost diagonal in those coordinates. More specifically, consider the determinant of the matrix $J - \lambda I$, where I is the $n \times n$ identity matrix. That determinant is a degree n polynomial in λ, and so by the fundamental theorem of algebra, it has exactly n complex roots $\lambda_1, \ldots, \lambda_n$. Each root (called an eigenvalue of J) is either a real number $\lambda_j = a_j$ or else is part of a complex conjugate pair $\lambda_j, \lambda_{j+1} = a_j \pm \sqrt{-1}b_j$. Written in eigenvector coordinates, the matrix J has diagonal elements consisting of the real eigenvalues and 2×2 blocks corresponding to the complex conjugate pairs. Using those coordinates, the solution to (4.22) takes the form $y_j(t) = e^{a_j t}e^{\sqrt{-1}b_j t}y_j(0)$. The factor $e^{\sqrt{-1}b_j t}$ represents pure rotation (in the 2-d eigenspace corresponding to the 2×2 block), and its modulus (or Euclidean size) is always 1. Thus once again stability is determined by the factor $e^{a_j t}$, that is, by the sign of the real parts a_j of the eigenvalues. By the Hartman-Grobman theorem, the local stability of $\mathbf{s}^*$ in the original system (4.20) is the same as its stability of the origin in (4.22) as long as none of the $a_j = 0$.

One last subtlety arises from the fact that we are working on the simplex. We are concerned with deviations $y = s - \mathbf{s}^*$ whose components sum to $1 - 1 = 0$, and thus that lie in the $n-1$ dimensional subspace $\mathbb{R}^n_0$. There is no guarantee that $DV(\mathbf{s}^*)\cdot y \in \mathbb{R}^n_0$ even when $y \in \mathbb{R}^n_0$. To ensure that, we set $J = P_0 DV(\mathbf{s}^*)$, where the matrix $P_0 = I - \frac{1}{n}\mathbf{1}_{n\times n}$ projects orthogonally onto $\mathbb{R}^n_0$. For example, for the usual 3-simplex, $P_0 = \frac{1}{3}\begin{pmatrix} 2 & -1 & -1 \\ -1 & 2 & -1 \\ -1 & -1 & 2 \end{pmatrix}$.

To summarize, you implement the eigenvalue technique as follows.

1. At a steady state $\mathbf{s}^* \in \mathcal{S}$ of interest for dynamics V, compute the $n\times n$ matrix $DV(\mathbf{s}^*) = ((\frac{\partial V_i(s)}{\partial s_j}))|_{s=\mathbf{s}^*}$ and the projection matrix $P_0 = I - \frac{1}{n}\mathbf{1}_{n\times n}$, and set $J = P_0 DV(\mathbf{s}^*)$.
2. Compute the eigenvalues $\lambda_1, \ldots, \lambda_n$ of J, by hand or using the Mathematica program or similar software. One of the the eigenvalues should be zero; it corresponds to the eigenvector $(1, 1, \ldots, 1)$ normal to the simplex and to $\mathbb{R}^n_0$.
3. Sort the remaining $n-1$ eigenvalues by their real parts from largest to smallest. By a slight extension of the Hartman-Grobman theorem, we know that $\mathbf{s}^*$ is

   - locally stable if the largest remaining real part is negative,
   - a source (i.e., completely unstable) if the smallest remaining real part is positive, and
   - a saddle if some real parts are positive and some are negative.

Two questions might occur to you at this point. First, what does the imaginary part b_j of eigenvalue j tell us? The quick answer is that $|b_j|$ is proportional to the frequency of the spiral or loop around the equilibrium point, and the sign of b_j tells us whether the spirals go clockwise or counterclockwise. Second, what if the largest remaining eigenvalue (after discarding a zero eigenvalue) also has zero real part? The answer is that in this case, the eigenvalue method is inconclusive. The linear approximation is a center, and each trajectory follows an ellipse with frequency proportional to the imaginary part. Here the higher order terms represented by $o(|y|)$ in (4.21) determine the stability of the original system (4.20).

Jean-Paul Rabanal generously provided the following Mathematica code for finding eigenvalues symbolically for replicator dynamics with an arbitrary 3×3 fitness matrix w. The first few lines input the fitness matrix and define the replicator. The next several lines compute the derivatives for $DV(s)$. The last few lines evaluate the symbolic expressions Jac and (after projection) M at a particular equilibrium point $s = \mathbf{s}^* = (0,3,2)/5$.

```
Clear[c, k, p, q, pe, qe, re, Jac, s, wbar, w, v1, v2, v3];
 w = {{-(1/2), 1, 0}, {-1, 0, 1}, {1, 1, -(1/2)}};
s = {p, q, r}; v1 = {1, 0, 0}; v2 = {0, 1, 0};
v3 = {0, 0, 1};
wbar = s.w.s;
pe = v1.w.s*p - wbar*p;
qe = v2.w.s*q -  wbar*q;
re = v3.w.s*r -  wbar*r;
Jac = Simplify[{{D[pe, p], D[pe, q], D[pe, r]}, {D[qe, p],
    D[qe, q], D[qe, r]}, {D[re, p], D[re, q], D[re, r]}}]
Simplify[Eigenvalues[Jac]]
Jac = Jac /. {p -> (0/5), q -> (3/5), r -> (2/5)}
P = {{2/3, -1/3, -1/3}, {-1/3, 2/3, -1/3}, {-1/3, -1/3,
  2/3}};
M = Jac.P;
Simplify[Eigenvalues[Jac]]
Simplify[Eigenvalues[M]]
TeXForm[\%]
```

Lyapunov technique

The other two techniques can sometimes resolve the case when the largest remaining eigenvalue has 0 real part. That case can be important. In a two-population game with no own-population effects, such as the Buyer-Seller game in the previous chapter, the diagonal entries of the Jacobian matrix typically have a Δw-type factor. At an interior equilibrium these entries therefore are zero, that is, the trace is zero, which means that the eigenvalues sum to 0. Zero real parts thus are generic in such cases.

The Russian mathematician Alexander Lyapunov (1857–1918) developed a technique that allows direct assessment of stability if you can find a special sort of real-valued function.

Definition 1. *A continuously differentiable function $L : \mathcal{S} \to \mathbb{R}$ is a (strict) Lyapunov function for V if $L(s(t))$ is (strictly) monotone increasing in t on every solution trajectory $\{s(t) : t \geq 0\}$ for V under (4.20).*

Proposition 7. *Let $\mathbf{s}^*$ be an isolated strict maximum of a strict Lyapunov function for V. Then $\mathbf{s}^*$ is a locally stable equilibrium point for V under (4.20).*

The logic behind this result is simple and elegant. Say that the strict local maximum value is $L(\mathbf{s}^*) = 10$, and that $N \subset \mathcal{S}$ is a neighborhood of $\mathbf{s}^*$ that excludes other maxima. Then the local v-upper contour sets $U(v) = \{s \in N : L(s) \geq v\}$ shrink down to the single point $\{\mathbf{s}^*\}$ as $v < 10$ approaches 10. By hypothesis $0 < L(\dot{s}(t)) = \nabla L \cdot \dot{s}(t) = \nabla L \cdot V(s(t))$, where $\nabla L = (\frac{\partial L}{\partial s_1}, \ldots, \frac{\partial L}{\partial s_1})$. The positive dot product means that there is an acute angle between the tangent $\dot{s}(t) = V(s(t))$ to the trajectory and the boundary of the upper contour set. That is (or as can be seen from the definition), the trajectory enters and can't leave smaller and smaller upper contour sets as t increases. With a little care for detail, one can show that "can't leave" implies Lyapunov stability and "smaller and smaller" implies asymptotic stability.

Sometimes Lyapunov functions are defined as monotone decreasing rather than monotone increasing; to see the equivalence, simply change the sign of L. Non-strict Lyapunov functions can be used to verify Lyapunov stability but, of course, not asymptotic stability. One can show that there are closed orbits (i.e., cycles around an equilibrium point) if there is a weak Lyapunov function that is actually constant on each trajectory.

Variations of relative entropy (discussed in the appendix to Chapter 1) often provide Lyapunov functions for replicator dynamics. Chapter 7 of Sandholm (2010) shows how to construct Lyapunov functions for several special subclasses of games and dynamics. Unfortunately, however, there is no general algorithm for constructing these wonderful functions.

Liouville technique

The French mathematician Joseph Liouville (1809–1882), among many other accomplishments, developed a technique that can help resolve the case where remaining eigenvalues have zero real part. It again considers subsets of $\mathcal{S}$ that may or may not shrink to a single point $\mathbf{s}^*$, but the sets are not defined using a Lyapunov function. They instead use the general solution to the ODE.

For given V, the general solution to the ODE (4.20) maps an arbitrary initial condition $s_o \in \mathcal{S}$ and an arbitrary time $t \in [0,\infty]$ to the time-t value $s(t) = \phi(t,s_o)$ of the solution emanating from the given initial condition. Liouville's theorem begins with an arbitrary open set $A_0 \subset \mathcal{S}$ containing the equilibrium point $\mathbf{s}^*$ of interest. The theorem compares the volume (or measure) $\mu(A_0) > 0$ of that set of initial conditions with the volume $\mu(A_t)$ of the set of points $A_t = \phi(t,A_o)$ that, at a given later time $t > 0$, are on solution trajectories emanating from A_o.

Liouville's theorem says that the volume changes at rate

$$\frac{d\mu(A_t)}{dt} = \int_{A_t} tr[J(s)]d\mu(s), \qquad (4.23)$$

where $J(s)$ is the projection of the Jacobian matrix (defined in step 1 of the previous subsection) evaluated at a point $s = s(t) \in A_t = \phi(t,A_o)$, and $d\mu$ indicates the volume integral. Recall that the trace $tr[M]$ of any $n \times n$ matrix M is the sum of its diagonal elements, a real number which (according to a standard result in linear algebra) coincides with the sum of its eigenvalues.

The theorem is especially helpful when the Jacobian has trace zero, as in games like Buyer-Seller. Then the integrand in (4.23) is zero, so the theorem tells us that the volume of A_t remains constant. But this means that the sets can not shrink to an interior equilibrium point $\mathbf{s}^*$, as they must if that point were asymptotically stable. Since the volume remains constant, the interior equilibrium must either be a saddle (with expansion along the directions of eigenvectors for eigenvalues with positive real parts just balancing contraction in the other directions), or else a center, with trajectories as closed loops. It follows that in games like Buyer-Seller with zero real parts for all eigenvalues, the interior equilibrium must be a center.

For completeness, we note that edge or corner equilibria may be stable even if the Jacobian has trace zero, since the direction of expansion may point outside the simplex. A negative trace indicates that the sets A_t shrink, a necessary condition for stability of an interior equilibrium. Of course, a positive trace always implies instability, but that already follows from the eigenvalue analysis.

When is a NE an ESS?

Below is a cookbook for determining whether a Nash equilibrium $\mathbf{s}^*$ is a Regular ESS (RESS). Sandholm (2010) and other texts explain the mild additional conditions that make an ESS a RESS. We are grateful to professor Sandholm for suggesting this method to us in personal correspondence.

Assume for the moment that the game is linear and for one population, with $n \times n$ payoff matrix W.

Step 1. Check that $\mathbf{s}^*$ is a quasi-strict NE, that is, that every pure best reply i is active, i.e., that $i \in B(\mathbf{s}^*) \implies i \in A(\mathbf{s}^*)$. If not, then $\mathbf{s}^*$ is not an RESS, and you are done.

Step 2. Take the submatrix of W restricted to the simplex face $A(\mathbf{s}^*)$, that is, ignoring all strategies extinct at $\mathbf{s}^*$. Symmetrize that submatrix $\tilde{W}$ by adding it to its transpose, to obtain $\hat{W} = \tilde{W} + \tilde{W}^T$.

Step 3. Use the Eigenvalue Technique described earlier in this appendix (with the projection matrix and discarding a 0 eigenvalue) to compute the relevant eigenvalues of $\hat{W}$. If and only if they all are negative, the original NE is a RESS.

If the game is nonlinear with fitness function $f(a,s)$, then set $W = ((\frac{\partial f}{\partial s_j}(e^i, \mathbf{s}^*)))$ and proceed. Multi-population versions are left to the reader, with the reminder to distinguish between Cressman ESS and Taylor ESS.

EXERCISES

1. Recall from the previous chapter the 3×3 RPS matrix for the European common lizard with free parameters c, k. Use comparative statics techniques to see how the location of the interior equilibrium changes as the parameter c changes.
2. Verify the assertions made in the text regarding the payoff matrix (4.18) for male side-blotched lizards: that the three vertices are all saddle points of replicator dynamics; that three of the six intersections of the edges with equal fitness lines are saddle points but none of them are NE; and that the interior NE is indeed an ESS. For the last verification, you may use part (ii) of the classic definition.
3. From Bomze (1983), pick an example with an interior equilibrium (or, alternatively, take the interior ESS in the previous exercise) and assess its stability using eigenvalue techniques.
4. Show that every payoff matrix for continuous dynamics can be normalized so that zeros appear on the diagonals. Hint: first prove the following *Lemma: The* Δw *functions of any payoff matrix are unchanged when any real number* c *is added to each entry in a single column of the matrix.*

NOTES

Nash's (1950) dissertation foreshadows much of the discussion in the present chapter. It justifies what is now called NE in two different ways. The first emphasizes the rationality of players, who assume that other players are rational. The second, called "mass action," introduces the concept of population games, and explains NE as a state where further adaptation has no scope to improve payoffs.

There is now a vast technical literature on ESS, its relation to NE, and its ability to identify stable steady states of various dynamics. To clarify one point in the text, the negative definite property characterizes what is called a Regular ESS, but it can fail in a knife-edge case of an ESS that is not Regular. Sandholm (2010, Chapter 8) includes a high-level and thorough recent treatment of such matters, and Weibull (1995) includes a slightly less abstract treatment. Early classic treatments include Maynard Smith and Price (1973), Maynard Smith (1982), Zeeman (1980), and Taylor and Jonker (1978).

The ODE expression V in equation (4.20) need not be differentiable. Standard textbooks, such as Hirsch and Smale, show that a slightly stronger continuity property (Lipshitz continuous) suffices to guarantee local existence and uniqueness of a solution $s(t)$ to the ODE in $\mathbb{R}^n$ given an arbitrary initial condition $s(0) = s_o \in \mathbb{R}^n$.

BIBLIOGRAPHY

Bomze, Immanuel M. "Lotka-Volterra equation and replicator dynamics: A two-dimensional classification." *Biological Cybernetics* 48, no. 3 (1983): 201–211.

Fisher, Ronald Aylmer. *The genetical theory of natural selection*. Clarendon Press, 1930.

Friedman, Daniel. "Evolutionary games in economics." *Econometrica: Journal of the Econometric Society* 59, no. 3 (1991): 637–666.

Hirsch, Morris, W., and Stephen Smale. *Differential equations, dynamical systems, and linear algebra*. Academic Press, 1974.

Kirkpatrick, Mark. "Sexual selection and the evolution of female choice." *Evolution* 36, no. 1 (1982): 1–12.

Lande, Russell. "Models of speciation by sexual selection on polygenic traits." *Proceedings of the National Academy of Sciences* 78, no. 6 (1981): 3721–3725.

Maynard Smith, John. and G. R. Price. "The logic of animal conflict." *Nature* 246 (1973): 15–18.

Maynard Smith, John. *Evolution and the theory of games*. Cambridge University Press, 1982.

Nash, John F. "Non-cooperative games." PhD Dissertation, Mathematics Department, Princeton University (1950).

Oprea, Ryan, Keith Henwood, and Daniel Friedman. "Separating the Hawks from the Doves: Evidence from continuous time laboratory games." *Journal of Economic Theory* 146, no. 6 (2011): 2206–2225.

Sandholm, William H. *Population games and evolutionary dynamics.* MIT Press, 2010.

Taylor, Peter D., and Leo B. Jonker. "Evolutionary stable strategies and game dynamics." *Mathematical Biosciences* 40, no. 1 (1978): 145–156.

Weibull, Jörgen W. *Evolutionary game theory.* MIT Press, 1995.

Zeeman, E. Christopher. "Population dynamics from game theory." In *Global Theory of Dynamical Systems,* Zbigniew Nitecki and Clark Robinson, eds., pp. 471–497. Springer Berlin Heidelberg, 1980.

5

Social Games

In a sense, we have been considering social games since Chapter 2—there is a social aspect to any interaction for which a player's payoff depends on what the other players are doing. In this chapter we will go further, and consider several kinds of social structures that modulate strategic interactions.

What do we mean by "modulate"? So far we have assumed that the intensity of interaction is directly proportional to prevalence. Hawks, for example, encounter other hawks with frequency s_H and encounter doves with frequency $s_D = 1 - s_H$. But what if hawks had their own hangouts, and so encountered other hawks more often? Or (more profitably) if they tended to avoid other hawks and could find more doves?

Clearly there are many biological games in which such departures from random encounters affect payoffs and equilibrium. Virtual worlds have their own rules of encounter, most of them not uniformly random. In human society, major examples of nonrandom encounters include marriage, jobs, schools, and housing—the overall population distribution of demographic traits (age, sex, income, etc.) will not predict well a particular individual's marriage partner, nor who fills a given job, attends a given school, and lives in a given neighborhood.

Useful models should be able to take into account such departures from uniform random encounters, and this chapter presents several helpful modeling tools. The first is called assortativity. It specifies directly how often one sort of strategist encounters another, and in 2×2 symmetric games it can be summarized in an single index (or function). We'll work out the algebra carefully for the Hawk-Dove game, and sketch the impact on other types of 2×2 symmetric games.

Different tools are needed to deal with more complicated games. Section 5.2 introduces a new tool that we call an adjustment matrix, which works for any finite game. Such matrices are combined with the usual fitness or payoff matrix using a linear algebra operation known as the Hadamard product. We illustrate with some simple examples.

The next technique is also new, but more specialized to deal with transmission from two parents. In biological applications, one parent is female and the other male, but the technique also works for some forms of vertical meme transmission. Either way, the technique allows us to distinguish between selection via fertility versus selection via survival, and to model their combined effects. Fertility selection distorts random associations of gametes combining sexually, and is captured by multi-matrices. Survival selection distorts the genotype shares produced from gametes to reflect experience in later life and maturation, and its impact is captured via a Hadamard product.

The last few sections work with the Price equation, originally proposed by George Price in 1970. Now widely used but still somewhat controversial, the Price equation enables us to model group-mediated social interactions. We offer a streamlined version of its well-known analysis of the evolution of cooperation in a social dilemma. A later section proposes a new extension of the Price equation to continuous-time settings.

Perhaps the most explicit way to model social interaction is to make it spatial, so an individual directly interacts only with her neighbors rather than with the population as a whole. Because spatial interaction involves very different techniques, we defer its discussion until Chapter 6. Most of the more interesting applications of the tools introduced here are also deferred to later chapters, especially to Chapters 9, 13, and 14.

5.1 ASSORTATIVE MATCHING

To illustrate assortativity in 2×2 games, consider the basic HD payoff matrix $\begin{pmatrix} \frac{v-c}{2} & v \\ 0 & \frac{v}{2} \end{pmatrix}$. Let $x = s_H$ be the current fraction of H play in the population, and let $p(x) \in [0,1]$ be the probability that a given H player will be matched with another H player. With random matching in a large population, of course, we'd have $p(x) = x$. A larger p would arise if H players tend to hang out together, and $p(x) < x$ indicates that H players avoid each other. Likewise, let $q(x) \in [0,1]$ be the probability that a given D player will be matched with a H player. Here $q(x) < x$ would arise if doves are able to avoid hawks, and $q(x) > x$ if they encounter hawks relatively often.

The number N_{HD} of H-D pairings in a population of size N is the probability $1 - p(x)$ that any hawk matches with a dove times the number xN of hawks, so $N_{HD} = Nx(1 - p(x))$. Likewise, the number of D-H pairings is the number of doves $(1 - x)N$ times the per-player matching probability $q(x)$, so $N_{DH} = N(1 - x)q(x)$. Of course, by the nature of a matching, it must be the case that $N_{HD} = N_{DH}$, so we have

$$x(1-p(x)) = (1-x)q(x). \tag{5.1}$$

Another way to write the relation is

$$\frac{q(x)}{x} = \frac{1-p(x)}{1-x}, \tag{5.2}$$

which shows that the distortion ratio defined by q can be expressed in terms of p by reflection in the interval $[0,1]$. Thus p and q are different sides of the same coin that describes matchings.

Bergstrom describes that coin more symmetrically by defining an index of assortativity, $a(x) = p(x) - q(x) \in [-1,1]$. As written, $a(x)$ captures the difference in encounter rates that hawks experience between own strategy and other strategy. However, the corresponding difference for doves is $(1-q(x)) - (1-p(x)) = p(x) - q(x) = a(x)$. Thus the index captures the difference in encounter rates between own strategy and other strategy that *any* player experiences.

Positive values of $a(x)$ indicate that players are disproportionately likely to meet their fellow strategists, and in the extreme case $a(x) = 1$ they never meet other strategists. Negative values indicate that (relative to unbiased random meetings) they are less likely to meet their fellow strategists, and indeed never do in the other extreme case $a(x) = -1$.

Of course, $p(x)$ and $q(x)$ must be nondecreasing functions bounded between 0 and 1. One plausible parametric specification, illustrated in Panel A of Figure 5.1, is that $p(x)$ takes the form x^{α}. If $\alpha \in (0,1)$ we have a bias towards meeting H, and if $\alpha > 1$ then we have avoidance of H. Of course, given any p, we can recover q from (5.1) or (5.2). A nice example is when α is a positive integer n. In this

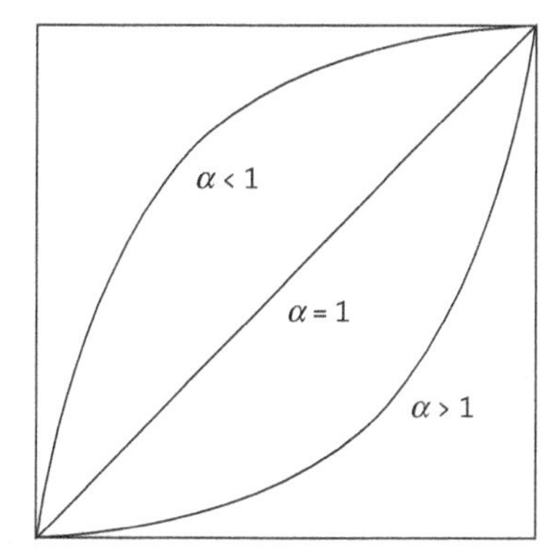

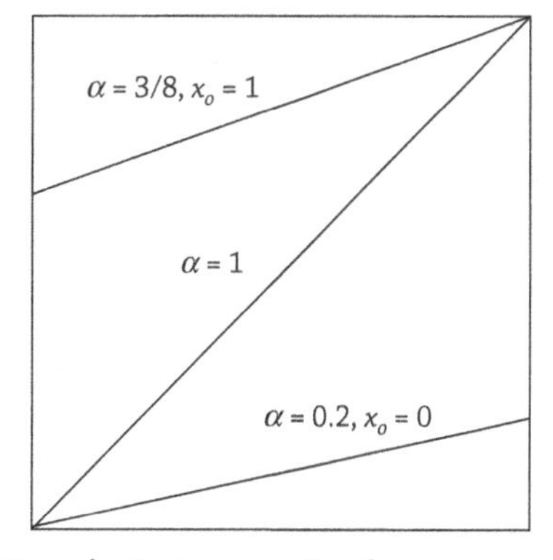

Figure 5.1 Probability distortion functions for 2×2 games. Panel A: Multiplicative functions $p(x) = x^{\alpha}$. Panel B: Linear functions $p(x) = \alpha x + (1-\alpha)x_o$.

case, $q(x)=\sum_{k=1}^{n} x^k$, as can be easily verified from inserting $p(x)=x^n$ into (5.1), and recalling the telescope identity $(1-x)\sum_{k=1}^{n} x^k = x - x^{n+1}$. It follows that $a(x) = -\sum_{k=1}^{n-1} x^k$, which is strictly negative for $n>1$. Similar but slightly more complicated algebra establishes that we have strictly positive assortativity when $\alpha = 1/n$ for integer $n>1$. A chapter-end exercise invites the reader to verify that in this case $q(x) = \frac{x}{\sum_{k=0}^{n-1} x^{k/n}}$ and $a(x)=p(x)r(x,n)>0$, where $r(x,n)\in(0,1)$.

Sometimes it is more convenient to use a linear parametrization, as in Panel B of Figure 5.1. The matching function takes the form $p(x)=\alpha x+(1-\alpha)x_o$ with slope $\alpha \in (0,1]$; the intercept $(1-\alpha)x_o$ uses a second parameter $x_o \in [0,1]$ that ensures that the function stays within its proper bounds. For example, take $p(x) = \alpha x + 1 - \alpha$ (using $x_o = 1$). Then equation (5.1) tells us that $q(x) = \frac{x}{1-x}(1 - p(x)) = \frac{x}{1-x}(\alpha - \alpha x) = \alpha x$, so $a(x) = p(x) - q(x) = 1 - \alpha$. Thus the assortativity index is constant in such cases.

Distortions in matching probabilities have fitness consequences. For the general HD payoff matrix above, we have $W_H = \frac{(v-c)}{2}p(x) + v(1-p(x)) = v - \frac{(v+c)}{2}p(x)$ and $W_D = 0q(x) + \frac{v}{2}(1-q(x))$. The fitness difference is

$$\Delta W(x) = \frac{1}{2}[v(1-a(x)) - cp(x)]. \tag{5.3}$$

The linear distortion example yields $\Delta W(x) = \frac{1}{2}[(v\alpha - c(\alpha x + 1 - \alpha)]$. Setting $\Delta W(x) = 0$, we see that the equilibrium H share is

$$x^* = \frac{v}{c} + 1 - \frac{1}{\alpha} \tag{5.4}$$

whenever $\alpha \in [\frac{c}{v+c}, 1]$. In the undistorted case $\alpha = 1$, the expression (5.4) reduces to the usual $x^* = v/c$. When $\alpha < \frac{c}{v+c}$, the probability $p(x) = \alpha x + 1 - \alpha$ that hawks meet other hawks is so high that $\Delta W(x) < 0$ for all x, and the only NE is pure dove.

Bergstrom (2002, 2003) emphasizes prisoner's dilemma games, especially the additive PD game $\begin{pmatrix} b-c & -c \\ b & 0 \end{pmatrix}$, in which $c>0$ is the personal cost of cooperation and $b>c$ is the per-capita benefit. Applying straightforward algebra to the additive PD, he finds that $\Delta W_{C-D}(x) = a(x)b - c$. Thus, if assortativity $a(x)$ is zero or negative, the second strategy D is dominant, but if $a(x)>0$ exceeds the cost/benefit ratio $\frac{c}{b}$, then C will gain share. The intuition, that cooperation can evolve in the PD if its benefits are sufficiently focused on cooperators, is developed further in Chapter 13.

Notice that if a is a strictly increasing function of x such that $a(0) < \frac{c}{b} < a(1)$, then the PD game just described becomes a coordination game, with its two basins of attraction separated by $x^* = a^{-1}(\frac{c}{b})$. In the chapter-end exercises, you are invited to further explore assortativity in 2×2 coordination games.

5.2 SOCIAL TWISTS

The assortativity index $a(x)$, and its predecessors mentioned in the notes, are very handy for symmetric 2×2 games, which have 1-dimensional state spaces. However, as far as we know, nobody has found a good way to generalize $a(x)$ to higher dimensions. In this section we take a different tack, and show how to use an $n\times n$ adjustment matrix A to capture twists in n-strategy symmetric game payoffs due to various sorts of social mediation.

We begin in familiar territory with a general 2×2 symmetric game $W = \begin{pmatrix} w_{11} & w_{12} \\ w_{21} & w_{22} \end{pmatrix}$ and again let $x = s_1$ be the population share employing the first strategy. When matching probabilities are uniformly random, so $p(x) = q(x) = x$ and $a(x) = 0$, we know how to represent the fitness difference by matrix multiplication, $\Delta W = (1,-1)W\cdot(x, 1-x)$. Can we use matrix multiplication to represent the overall fitness difference when there are given matching probabilities $p(x), q(x) \neq x$? That is, is there some way to adjust the underlying payoff matrix W and construct a payoff matrix $U = ((U_{ij}))$ that accounts for assortative matching so that $\Delta W = (1,-1)U\cdot(x, 1-x)$?

A little scratchwork gives a positive answer. When $p(x) = q(x) = x$, the fitness difference is the sum of four terms of the form $\pm w_{ij}z$, where $z = x$ or $1-x$; while for general encounter distortions $p(x), q(x) \neq x$, the fitness difference is again the sum of four terms of the same form except that in the top row $i = 1$, we have $z = p(x)$ or $1-p(x)$, and in the bottom row $i = 2$ we have $z = q(x)$ or $1-q(x)$. Hence the matrix we desire is $U = \begin{pmatrix} w_{11}\frac{p(x)}{x} & w_{12}\frac{1-p(x)}{1-x} \\ w_{21}\frac{q(x)}{x} & w_{22}\frac{1-q(x)}{1-x} \end{pmatrix}$. Note for future reference that equation (5.2) tells us that the two off-diagonal adjustments are the same; we will write them as $\frac{q(x)}{x}$.

Hadamard product

To represent U as a matrix operation involving W we use the Hadamard product. In general, if $A = ((a_{ij}))$ and $B = ((b_{ij}))$ are both $n\times n$ matrices, then their Hadamard product $A\circ B = C$ is defined as entry-by-entry multiplication, so $c_{ij} = a_{ij}b_{ij}$ for each i, j. In the present instance, you can see directly that $U = W\circ A$ where $A = \begin{pmatrix} \frac{p(x)}{x} & \frac{q(x)}{x} \\ \frac{q(x)}{x} & \frac{1-q(x)}{1-x} \end{pmatrix}$.

Due to the dependence of the matrix coefficients on x, we see that this Hadamard product trick doesn't generally simplify the assortativity approach of the previous section. However, the special case of a constant adjustment matrix A is useful even in the 2×2 case. For example, return to the Hawk-Dove matrix W noted earlier, and let $A=\begin{pmatrix}1 & 1\\ 1 & a\end{pmatrix}$, where the parameter $a\geq 1$ captures the degree to which doves are able to interact more frequently with other doves. The socially modulated fitness matrix then is $W\circ A=\begin{pmatrix}\frac{v-c}{2} & v\\ 0 & \frac{av}{2}\end{pmatrix}$ with fitness difference

$$\Delta(W\circ A)=(1,-1)[W\circ A]\cdot s\left(\frac{v-c}{2},v-\frac{av}{2}\right)\cdot(x,1-x)$$
$$=\left(\frac{v-c}{2}-(v-\frac{av}{2})\right)x+v-\frac{av}{2}=\frac{1}{2}[(av-v-c)x-(av-2v)].$$

Solving $\Delta(W\circ A)=0$, we obtain the equilibrium $x^{*}=\frac{2-a}{1+\frac{c}{v}-a}$ when $a\in[0,2]$. Note that the expression for x^{*} again reduces to the usual $\frac{v}{c}$ at the neutral value $a=1$, that the equilibrium fraction of hawks decreases (and of doves increases) as the discrimination parameter a increases from its neutral value towards 2, and that only doves survive in equilibrium when $a\geq 2$.

What if hawks can also discriminate? Of course, they benefit from dispersal, i.e., reducing the frequency of encounters with other hawks. Normalizing the D-H encounter rate at 1, we have the combined encounter matrix

$$A=\begin{pmatrix}b & 1\\ 1 & a\end{pmatrix},\quad 0<b\leq 1\leq a<2. \tag{5.5}$$

Using (5.5) in the calculations just performed yields

$$x^{*}=\frac{2-a}{2-a+b(\frac{c}{v}-1)}. \tag{5.6}$$

Since $\frac{c}{v}>1$ in any Hawk-Dove game, it is easy to see from (5.6) that the better hawks are at dispersal (i.e., the smaller is $b>0$), the larger is their equilibrium share. Of course, it is still also true that doves increase their equilibrium share when $a<2$ increases, that is, when they are better able to find each other.

The real advantage of the Hadamard product approach is that it generalizes to matrix games with $n>2$ pure strategies. Consider a general $n\times n$ encounter matrix $A=((A_{ij}))$ and fitness matrix $W=((W_{ij}))$. Then the fitness of strategy i is

$$W_i=e^{i}[W\circ A]\cdot s=\sum_{j=1}^{n}W_{ij}A_{ij}s_j, \tag{5.7}$$

while the corresponding mean fitness is

$$\bar{W} = s[W \circ A] \cdot s = \sum_{i=1}^{n} \sum_{j=1}^{n} W_{ij} A_{ij} s_i s_j. \tag{5.8}$$

With these modulations, discrete time replicator dynamics are still written $s_i' = \frac{W_i}{\bar{W}} s_i$.

Adjustment matrix properties

Are there any reasonable restrictions to impose on an $n \times n$ adjustment matrix A? Typically, when considering discrete time dynamics, we want the underlying fitness matrix W to have non-negative entries. For example, in the standard Hawk-Dove game we would include a baseline fitness $\beta > 0$ sufficiently large to ensure non-negative fitness for H-H encounters. With parameters $\beta = 2, c = 6$, $v = 4$ the H-D game matrix is $W = \begin{pmatrix} \beta + \frac{v-c}{2} & \beta + v \\ \beta + 0 & \beta + \frac{v}{2} \end{pmatrix} = \begin{pmatrix} 1 & 6 \\ 2 & 4 \end{pmatrix}$. Note that the fitness difference ΔW is independent of β.

The point is that we want the adjusted fitness matrix $W \circ A$ also to have non-negative entries. This will happen for general underlying matrices W if and only if

Property 1. The adjustment matrix $A = ((A_{ij}))$ has non-negative entries (i.e., $A_{ij} \geq 0 \; \forall i, j$).

If A captures a pairwise matching process then there is a second property. In this case, A_{ij} represents the ratio of actual $i - j$ pairings N_{ij} relative to the expected number $N s_i s_j$ of such pairings under random pairwise matching, so $N_{ij} = A_{ij} N s_i s_j$. Likewise, the number of actual $j - i$ pairings is $N_{ji} = A_{ji} N s_i s_j$. But (by definition of pairings) these numbers must be equal. Hence

Property 2. The adjustment matrix $A = ((A_{ij}))$ is symmetric (i.e., $A_{ij} = A_{ji} \; \forall i, j$).

However, payoffs can be twisted by social interactions beyond pairwise matching. Three-way interactions, or "playing the field" more generally, can have asymmetric impacts. For example, suppose that doves are more social in a manner that doesn't affect the frequency of interactions but that makes it more difficult for hawks to take the resource away from them. In this case the encounter rates remain at $p(x) = x = q(x)$, but the probability that a hawk can take the resource from a dove drops from $P = 1$ to some value $P < 1$. In that case, Property 2 would be violated and the adjustment matrix would have entries $A_{HD} < 1 < A_{DH}$.

RPS twists

Consider a simple example with $n = 3$ and, for variety, in continuous time. The most basic RPS game in this setting has payoff matrix

$$\mathbf{w} = \begin{pmatrix} 0 & -1 & 1 \\ 1 & 0 & -1 \\ -1 & 1 & 0 \end{pmatrix}. \tag{5.9}$$

Each invading strategy is advantaged if it can target (i.e., achieve efficiency $a \geq 1$ in encountering) the strategy that it does best against. For example, the invading rock type $(i = 1 = R)$ targets the common scissors type $(j = 3 = S)$, and scissors targets paper $(j = 2 = P)$. A reasonable conjecture is that this sort of targeting will speed up the cycle (Sinervo et al. 2007). Conversely, each type may, to some extent, be able to evade its nemesis. For example, the invading rock type may be able to shrink losses to paper by factor $b < 1$. This twist might slow down the cycle.

To explore these conjectures, we apply the adjustment matrix

$$\mathbf{A} = \begin{pmatrix} 1 & b & a \\ a & 1 & b \\ b & a & 1 \end{pmatrix} \tag{5.10}$$

with $0 < b \leq 1 \leq a$. That is, we replace the basic payoff matrix $\mathbf{w}$ with the modulated matrix $\mathbf{u} = \mathbf{w} \circ \mathbf{A} = \begin{pmatrix} 0 & -b & a \\ a & 0 & -b \\ -b & a & 0 \end{pmatrix}$ when calculating mean fitness $\bar{u} = \mathbf{su} \cdot \mathbf{s}$ and continuous replicator dynamics $\dot{s}_i = [\mathbf{e_i u} \cdot \mathbf{s} - \bar{u}]s_i$, $i = 1, 2, 3$.

Calculating the Jacobian matrix (with the appropriate projection, as in the Appendix of Chapter 4), and evaluating at the equilibrium $(1, 1, 1)/3$, we obtain the eigenvalues 0 and $\frac{1}{6}(b - a \pm \sqrt{-3}(a + b))$. The zero eigenvalue corresponds to the eigenvector $(1, 1, 1)$ and simply reflects the fact that replicator dynamics remain in the hyperplane of vectors whose components sum to a constant (namely 1.0, since we are in the simplex). The other eigenvalues are complex conjugates. The imaginary part (involving the square root of a negative number) determines the cycle frequency, which is proportional to $(a + b)$. Thus we see that better targeting $(1 < a \uparrow)$ and lesser evasion ability $(1 > b \uparrow)$ both tend to speed up the RPS cycle. The real part of the eigenvalues is proportional to the parameter difference $(b - a)$. Since $b \leq 1 \leq a$, the real part is negative and the equilibrium is stable under replicator dynamics; better targeting $(a \uparrow)$ and better evasion ability $(b \downarrow)$ imply tighter spirals and more rapid convergence to equilibrium.

Three comments seem appropriate at this juncture. First, it is reasonable to suppose that targeting behaviors will evolve rapidly in a population because they most enhance fitness of the strategy gaining share. The genes for targeting thus can "hitchhike" and increase share in lockstep with the strategy genes that are currently gaining share. By contrast, evasion behaviors "swim against the current" since they disproportionately support strategies whose shares are declining. Thus the most realistic case would seem to be $a/b > 1$. To our knowledge, this intuitive asymmetry has not yet been explored formally or empirically.

Second, Hadamard products are convenient to work with because they commute and can be decomposed. In the previous example, $\mathbf{u} = \mathbf{w} \circ \mathbf{A} = \mathbf{A} \circ \mathbf{w}$ by commutativity, and for $\mathbf{A} = \begin{pmatrix} 1 & b & a \\ a & 1 & b \\ b & a & 1 \end{pmatrix}$, with $\mathbf{B} = \begin{pmatrix} 1 & b & 1 \\ 1 & 1 & b \\ b & 1 & 1 \end{pmatrix}$ and $\mathbf{C} = \begin{pmatrix} 1 & 1 & a \\ a & 1 & 1 \\ 1 & a & 1 \end{pmatrix}$ we have the decomposition $\mathbf{A} = \mathbf{B} \circ \mathbf{C}$. These convenient properties are inherited from the commutativity of ordinary multiplication of real numbers ($xy = yx$), the fact that 1 is the multiplicative identity ($1x = x$), and the definition of the Hadamard product as element-by-element multiplication.

Last, it is not always obvious how to estimate adjustment matrices from field data. The maximum likelihood techniques described in an appendix to Chapter 3 are designed to recover the entries of a modulated matrix $\mathbf{u} = \mathbf{w} \circ \mathbf{A}$. Of course, if the payoff matrix $\mathbf{w}$ is known a priori, then the estimate of encounter matrix component a_{ij} is simply the estimated composite matrix entry u_{ij} divided by the known payoff matrix entry w_{ij}. To avoid dividing by 0, it may be best to work with discrete time matrices. If (as usually is the case in biology applications) $\mathbf{w}$ is unknown, but (as may be the case) a one- or two-parameter adjustment matrix seems plausible, then it may still be possible to get useful estimates of those adjustment parameters, such as $0 < b < 1 < a$ in the previous example. Identifying these parameters is much easier if the data include subsets across which either the basic payoffs differ or the adjustment parameters differ, but not both. Fitze et al. (2014) provide a laboratory experiment that estimates the elements of the adjustment matrix for female European common lizards who prefer male RSP genotypes that provide high fitness by virtue of the production of rare genotypes in the next generation, a type of adjustment arising from male and female games.

5.3 INHERITANCE FROM TWO PARENTS

As promised in the introduction, we now present techniques for describing evolutionary dynamics that arise from fertility selection combined with survival selection, and apply them to the Kirkpatrick model of runaway sexual selection introduced in the previous chapter. To preview, we will derive the general equation (5.11) below to express in compact matrix notation the complexity of sexual reproduction. At the center of the equation is a multi-matrix **Q** that summarizes the transmission rules from parental genotypes (or memotypes) to progeny. After specifying the other elements of the equation—survival selection matrices **F** and **M** for female and male parents, their respective sexual preferences **A** and **B**, and normalization factor $c > 0$—we will obtain

$$\mathbf{s}' = c\mathbf{s}^F \mathbf{F}\mathbf{A} \circ [\mathbf{Q}^{\mathbf{1}}, \ldots, \mathbf{Q}^{\mathbf{n}}] \circ \mathbf{B}\mathbf{M} \cdot \mathbf{s}^M. \tag{5.11}$$

There is a lot to explain, so we work with a natural example, the simplified Kirkpatrick model. Recall that in this model males are either plain (allele T1) or ornamented (T2), while females are either indifferent to male ornaments (allele P1) or prefer them (P2). The analysis in the previous chapter kept things simple by imposing an assumption labeled (ii), that males and females belong to genetically separate populations, each with a single locus. Of course, assumption (ii) is biological nonsense, so we now more realistically assume

(ii′) males and females belong to the same population, in which all individuals possess both T and P loci.

Thus the (haploid) genotypes of either sex are T1P1, T1P2, T2P1, T2P2, and the T allele controls the phenotype exhibited by males while the P allele controls the phenotype exhibited by females. We interpret (ii') to imply that the genotype shares are the same for both sexes, so $\mathbf{s}^F = \mathbf{s}^M = \mathbf{s} = (s_{T1P1}, s_{T1P2}, s_{T2P1}, s_{T2P2})$. The evolving state $\mathbf{s}$ therefore is a point in the tetrahedron S^4, and we can recover the phenotype shares $t = s_{T2P1} + s_{T2P2}$ and $p = s_{T1P2} + s_{T2P2}$ whenever convenient.

A dynamical process describes the transition from the current state $\mathbf{s} \in S^4$ to the successor state $\mathbf{s}' \in S^4$ describing the genotype distribution in the next generation. We break down the transition into four steps. The first step is to compute the proportions of adult males, $\mathbf{M} \cdot \mathbf{s}$, where the 4×4 selection matrix $\mathbf{M}$ captures fitness costs, the effects of male-male competition, and so on. In the current application the only relevant aspect is the ornament's fitness cost $m \in [0, 1)$, so

$$\mathbf{M} = \begin{pmatrix} 1 & 0 & 0 & 0 \\ 0 & 1 & 0 & 0 \\ 0 & 0 & 1-m & 0 \\ 0 & 0 & 0 & 1-m \end{pmatrix}.$$

Similarly, the potential moms have proportions obtained by multiplying by some 4×4 matrix $\mathbf{F}$. For simplicity we return here to assumption (i) and drop the small frequency-dependent costs, so $\mathbf{F}$ is the identity matrix and the shares of potential moms simply reflect the current state $\mathbf{s}$ of the gene pool.

An attraction matrix $\mathbf{A}$ captures female departures from random mating. In the present example, $\mathbf{A} = \begin{pmatrix} 1 & 1 & 1 & 1 \\ 1 & 1 & a & a \\ 1 & 1 & 1 & 1 \\ 1 & 1 & a & a \end{pmatrix}$ for a given attraction parameter $a > 1$. In general, males also have an attraction matrix $\mathbf{B}$ but in the present application (as in many other settings) we assume males are equally likely to mate with any available female, so the $\mathbf{B}$ matrix is trivial (i.e., every entry is 1) and we can ignore it in the calculations below.

The most interesting and complicated step is to specify the distribution of progeny genotypes from each possible pairing of moms and dads. We need a multi-matrix with as many components as there are alternative genotypes (or memotypes). Thus in the present application we use a quad-matrix, and Table 5.1 gives three possible examples. For now we focus on the simple case shown in Panel A, where the male trait is inherited only from the father, and the female trait is inherited only from the mother. For example, the quad-matrix entry $[0,0,0,1]$ in row (female genotype) T1P2 and column (male genotype) T2P1 indicates that all resulting progeny will be of the fourth phenotype, T2P2, and none will be of the other three phenotypes.

We now combine the four steps as in equation (5.11) to obtain the distribution $\mathbf{s}'$ in the next generation. The unnormalized fraction z_i of genotype i is obtained by multiplying the adult female vector $\mathbf{z}^F$ into the mating matrix $A \circ Q^i = ((a_{ij}q^i_{ij}))$ for that genotype, and then taking the dot product with the vector $\mathbf{z}^M$ of adult males. The state $\mathbf{s}'$ we seek is obtained by normalizing $\mathbf{z}$ by the sum $c^{-1} = z_T = (1,1,1,1)\mathbf{z} = \sum_{i=1}^{4} z_i$. In the present example we have $\mathbf{z}^M = (s_{T1P1}, s_{T1P2}, s_{T2P1}(1-m), s_{T2P2}(1-m))$ and $\mathbf{z}^F = (s_{T1P1}, s_{T1P2}, s_{T2P1}, s_{T2P2})$, while (for example) for the fourth genotype T2P2 the mating matrix is $A \circ Q^4 = \begin{pmatrix} 0 & 0 & 0 & 0 \\ 0 & 0 & a & a \\ 0 & 0 & 0 & 0 \\ 0 & 0 & a & a \end{pmatrix}$. Hence

$$
\begin{aligned}
z_4 &= (0, 0, as_{T1P2} + as_{T2P2}(1-m), as_{T1P2} + as_{T2P2}(1-m)) \\
&\quad \cdot (s_{T1P1}, s_{T1P2}, s_{T2P1}, s_{T2P2}) \\
&= a(s_{T2P1} + s_{T2P2})(s_{T1P2} + s_{T2P2}(1-m)). \qquad \textbf{(5.12)}
\end{aligned}
$$

Table 5.1. PARENTAL TRANSMISSION QUAD-MATRICES (Q)

Panel A. Same-sex transmission.

	Males:			
Females	T1P1	T1P2	T2P1	T2P2
T1P1	[1,0,0,0]	[1,0,0,0]	[0,0,1,0]	[0,0,1,0]
T1P2	[0,1,0,0]	[0,1,0,0]	[0,0,0,1]	[0,0,0,1]
T2P1	[1,0,0,0]	[1,0,0,0]	[0,0,1,0]	[0,0,1,0]
T2P2	[0,1,0,0]	[0,1,0,0]	[0,0,0,1]	[0,0,0,1]

Panel B. Memetic transmission strictly from father to all progeny.

	Males:			
Females:	T1P1	T1P2	T2P1	T2P2
T1P1	[1,0,0,0]	[0,1,0,0]	[0,0,1,0]	[0,0,0,1]
T1P2	[1,0,0,0]	[0,1,0,0]	[0,0,1,0]	[0,0,0,1]
T2P1	[1,0,0,0]	[0,1,0,0]	[0,0,1,0]	[0,0,0,1]
T2P2	[1,0,0,0]	[0,1,0,0]	[0,0,1,0]	[0,0,0,1]

Panel C. Transmission with recombination rate $r \in [0, \frac{1}{2}]$.

	Males:			
Females:	T1P1	T1P2	T2P1	T2P2
T1P1	[1,0,0,0]	$[\frac{1}{2},\frac{1}{2},0,0]$	$[\frac{1}{2},0,\frac{1}{2},0]$	$[\frac{1-r}{2},\frac{r}{2},\frac{r}{2},\frac{1-r}{2}]$
T1P2	$[\frac{1}{2},\frac{1}{2},0,0]$	[0,1,0,0]	$[\frac{r}{2},\frac{1-r}{2},\frac{1-r}{2},\frac{r}{2}]$	$[0,\frac{1}{2},0,\frac{1}{2}]$
T2P1	$[\frac{1}{2},0,\frac{1}{2},0]$	$[\frac{r}{2},\frac{1-r}{2},\frac{1-r}{2},\frac{r}{2}]$	[0,0,1,0]	$[0,0,\frac{1}{2},\frac{1}{2}]$
T2P2	$[\frac{1-r}{2},\frac{r}{2},\frac{r}{2},\frac{1-r}{2}]$	$[0,\frac{1}{2},0,\frac{1}{2}]$	$[0,0,\frac{1}{2},\frac{1}{2}]$	[0,0,0,1]

Similar calculations using the mating matrices for the other three genotypes yield

$$z_1 = (s_{T1P1} + s_{T1P2})(s_{T1P1} + s_{T2P1}(1-m)), \tag{5.13}$$

$$z_2 = (s_{T1P1} + s_{T1P2})(s_{T1P2} + s_{T2P2}(1-m)), \text{ and} \tag{5.14}$$

$$z_3 = (s_{T2P1} + s_{T2P2})(s_{T1P1} + s_{T2P1}(1-m)). \tag{5.15}$$

Their sum is

$$z_T = \sum_{i=1}^{4} z_i = (s_{T1P1} + s_{T1P2})(s_{T1P1} + s_{T1P2} + (1-m)(s_{T2P1} + s_{T2P2})) \tag{5.16}$$

$$+ (s_{T2P1} + s_{T2P2})(s_{T1P1} + as_{T1P2} + s_{T2P1}(1-m) + as_{T2P2}(1-m)),$$

so the successor state is

$$\mathbf{s}' = (z_1, z_2, z_3, z_4)/z_T. \tag{5.17}$$

Readers curious about the connection to the two-population game discussed in the previous chapter should consider the initial state $\mathbf{s} = (1-p, p, 0, 0)$ for $p \in [0,1]$. It corresponds to the $(t,p) = (0,p)$ neutral equilibrium of that game (for $0 \le p \le p^* = \frac{m}{(a-1)(1-m)}$). From (5.12)–(5.17) we have $z_1 = 1-p$ and $z_2 = p$ while $z_3 = z_4 = 0$, so $z_T = 1$ and $\mathbf{s}' = (1-p, p, 0, 0) = \mathbf{s}$. That is, these are all equilibrium points. Similar calculations for $\mathbf{s} = (0, 0, 1-p, p)$ yield $z_1 = z_2 = 0$ and $z_3 = (1-p)(1-m), z_4 = ap(1-m)$ so $z_T = (1-p+ap)(1-m)$. We have fixed points in this case only when $p = 0$ or $p = 1$. It follows that the runaway dynamics are only slightly different than in our first two-population game, in that they take us all the way to the interesting corner equilibrium $(t,p) = (1,1)$ once we escape the boring case $t = 0$.

The connection to the two-population game is not surprising for this simple quad-matrix, since it allows no mixing of male and female lines of inheritance. Male traits whose locus is on the Y chromosome work like this, and the same quad-matrix describes same-sex memetic transmission, for example, the father songbird transmits his song to sons, and the mother transmits her song preference to daughters. Panel B of Table 5.1 shows the slightly different (but still very simple) quad-matrix for the case that daughters also get their song preferences from their dads, which is typical for most songbirds.

The more complicated quad-matrix in Panel C deals with recombination, a biologically important process for genes that reside on autosomal chromosomes (non-sex chromosomes) that are shared freely between the sexes. As shown in Figure 5.2, recombination allows offspring with new genotypes that combine elements of parental genotypes in new ways. When both parents share the same allele at both loci (i.e., for the diagonal elements of Q), then there is nothing to recombine. Recombination therefore is relevant only on the off-diagonal elements of the quad-matrix, where the parents differ at one or both loci. When parents differ at both loci, recombination can even create two new genotypes that are not seen in either parent.

Consider, for example, a pairing of a T2P1 mom with a T1P2 dad, and first consider the case where the two loci reside on different chromosomes. In this case, after parental gametes combine and resegregate, the gene for female preference and male trait appear independently in progeny, so all four genotypes are equally likely.

Now consider the case where both loci reside on the same chromosome. Then one strand of that chromosome may cross over its counterpart from the other parent at a point between the two loci, as shown in Figure 5.2. Such a crossover produces genotypes not seen in either parent, in this example T1P1 and T2P2.

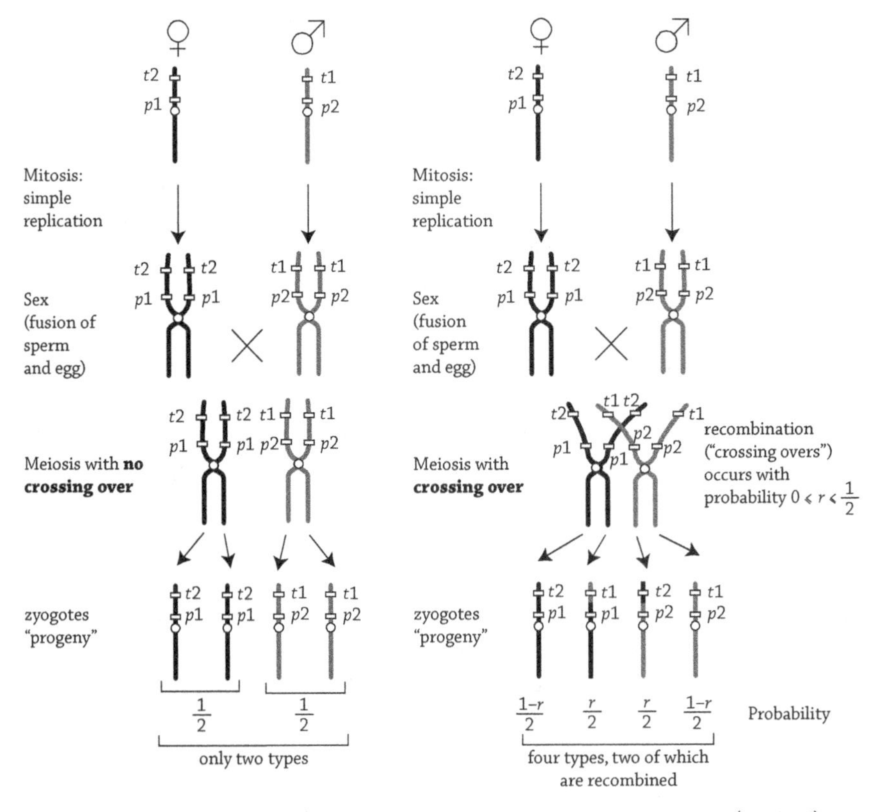

Figure 5.2 Recombination when two loci reside on the same chromosome (haploid). With no crossing over in meiosis, as shown on the left side of the figure, the offspring genotypes are simply the same as one of the parent genotypes, and both are equally likely. With recombination, as on the right side of the figure, offspring genotypes can differ from parent genotypes if a crossover occurs between the two loci. The probability $r \in [0, \frac{1}{2}]$ of such a crossover is an increasing function of the distance between the two loci. The process is only slightly different for diploid organisms, in that meiosis occurs between maternal and paternal copies in the diploid zyogote that then produces haploid gametes.

The probability r of such a crossover is larger the farther away the two loci are on the chromosome; at most it is $1/2$ (as when the loci are on different chromosomes), and it approaches $r = 0$ when the loci are right next to each other. Thus for general r we get the quad-matrix entry $[\frac{r}{2}, \frac{1-r}{2}, \frac{1-r}{2}, \frac{r}{2}]$ shown in the T2P1 row and T1P2 column of Table 5.1C.

The equilibrium calculation in this case is the same as for the simpler cases considered earlier, but of course the algebra is messier when $r > 0$. As shown in the appendix at the end of the chapter, the present model has a continuum of equilibria that lie along the slanted line segment shown in Figure 5.3. It turns out that the slope of the segment reflects the strength of selection for survival

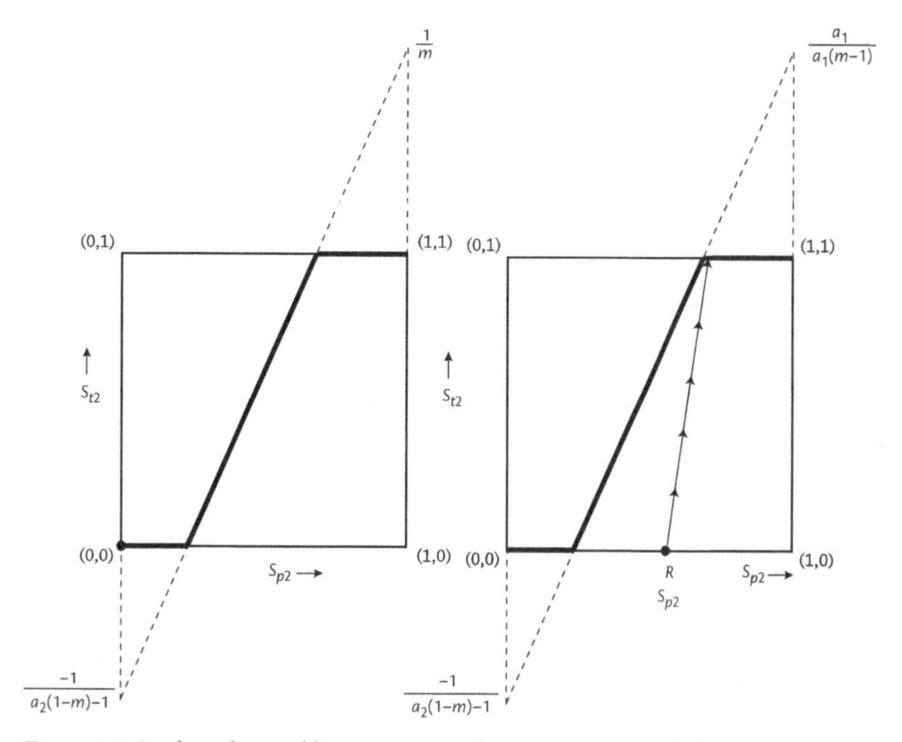

Figure 5.3 On the right, equilibrium correspondence runaway sexual selection, with a preference a_2 of P2 females for T2 males, for the case of $m = 0.4$ and $a_2 = 3$. On the left, a runaway model that also considers a preference of P1 females for T1 males given by parameter a_1. The heavy line represents the equilibrium line or steady state line. In the right panel, runaway is possible for frequencies of female preference $s_{p2} > R$ will bypass the equilibrium line and rapidly fix on T2, if T2 mutates from T1 at some low frequency (from Kirkpatrick 1982).

relative to the strength of sexual selection. All points along this line are of equal fitness, so selection will sweep the population to the line, but once on the line, the population will simply drift slowly under the influence of stochastic factors, especially in finite populations.

5.4 THE STANDARD PRICE EQUATION

Sometimes strategic interactions are modulated by groups—each individual belongs to some group, and interactions within the group are different (and typically much more frequent) than interactions outside the group. A standard analytical tool for dealing with that sort of social structure is known as the Price equation. That equation actually describes the evolution of *any* phenotypic trait, as we shall now explain. The next section shows how to use the Price equation to analyze interactions mediated by distinct social groups.

The full Price equation (PE) is:

$$\Delta\bar{z} = \frac{Cov[W,z]}{\bar{W}} + \frac{E[W\Delta z]}{\bar{W}}, \tag{5.18}$$

where $\Delta\bar{z}$ represents the change from one discrete time period to the next in the mean value of a phenotypic trait (or, more generally, of any scalar value associated with individuals), E is the expectation operator, $\bar{W} = EW$ is as usual mean fitness and $\bar{z} = Ez$ is the mean trait value, and $Cov(W,z) = E[(W - \bar{W})(z - \bar{z})] = E(Wz) - \bar{W}\bar{z}$ is the covariance of fitness with the trait values.

Equation (5.18) partitions the overall change in the mean trait into two terms. The first term is usually referred to as *selection*. We will see that it captures the change in the mean trait value due to changes in frequencies of different morphs, as fitter morphs have more progeny than less fit, all else held constant. Intuitively, the mean trait value increases to the extent that the trait is positively correlated with fitness. The second term captures "all else" and is often referred to as *inheritance*. In some applications, progeny can carry traits that systematically differ from traits carried by their parents, and this term will capture such an inheritance bias.

The Price equation is a mathematical identity that can be proved by unwinding the definitions. To see this, let there be a finite or infinite number of disjoint groups indexed by i, let n_i denote the current population of group i, and let $n = \sum_i n_i$ denote the total current population. Let n'_i denote the number of progeny produced by members of the ith group, so the total number of progeny is $n' = \sum_i n'_i$. The fitness of group i is by definition

$$W_i = \frac{n'_i}{n_i} \tag{5.19}$$

and average fitness is

$$\bar{W} \equiv E[W] = \frac{\sum_i W_i n_i}{n} = \frac{\sum_i \frac{n'_i}{n_i} n_i}{n} = \frac{n'}{n}. \tag{5.20}$$

The average trait value among progeny is

$$\bar{z}' = \frac{\sum_i z'_i n'_i}{n'}. \tag{5.21}$$

Recall for the record that

$$Cov(W,z) = E[Wz] - \bar{W}\bar{z}. \tag{5.22}$$

The change in progeny's trait relative to parent is denoted $\Delta z_i \equiv z_i' - z_i$, so

$$E[W\Delta z] = E[W(z' - z)] = E[Wz'] - E[Wz]. \quad (5.23)$$

Adding equation (5.22) to (5.23), we get

$$Cov(W,z) + E[W\Delta z] = E[Wz'] - \bar{W}\bar{z}. \quad (5.24)$$

Use the definition of the expectation operator and then (5.19) to unpack the first term on equation (5.24)'s right-hand side:

$$E[Wz'] = \frac{\sum_i W_i z_i' n_i}{n} = \frac{\sum_i \frac{n_i'}{n_i} z_i' n_i}{n} = \frac{\sum_i n_i' z_i'}{n} = \frac{n'}{n} \frac{\sum_i n_i' z_i'}{n'} = \bar{W}\bar{z}', \quad (5.25)$$

where the last expression uses (5.20) and (5.21). Insert that last expression into (5.24) to obtain

$$Cov(W,z) + E[W\Delta z] = \bar{W}\bar{z}' - \bar{W}\bar{z} = \bar{W}\Delta\bar{z}, \quad (5.26)$$

which is a version of the full Price equation with $\bar{W}$ cleared from the denominators.

The transmission term $E[W\Delta z]$ is often hard to work with, and makes recursion problematic. In some applications, one can argue that that term is small and can be neglected. In that case, (5.26) reduces to the so-called simple Price equation

$$\bar{W}\Delta\bar{z} = Cov(W,z). \quad (5.27)$$

5.5 GROUP-STRUCTURED PRICE EQUATION AND COOPERATION

The previous derivation is a good warmup exercise for our main goal, constructing a group-structured version of the Price equation. In particular, we will use it to show how cooperation can evolve in a social dilemma game (linear public goods or multiplayer prisoner's dilemma) when strategic interaction is modulated by a group structure (Cooper and Wallace 2005).

Pick a particular group i as above and look inside. Members $j = 1, \ldots, n_i$ of that group may differ in trait level $z \geq 0$ representing their contribution (or degree of cooperation or altruism). In the game of interest, an individual's fitness W_j is decreasing in this contribution but increasing in the group average contribution $\bar{z}_i$. For now, we focus on a particular group and streamline notation by dropping the i subscript, so we will write $\bar{z}_i$ as just $\bar{z}$. In this streamlined notation, the fitness function in our linear social dilemma game is

$$W_j = \frac{n'_j}{n_j} = k - az_j + b\bar{z}. \tag{5.28}$$

The cost parameter $a > 0$ reflects the personal sacrifice that contributions entail, and the benefit parameter $b > a$ reflects the shared value of group members' total contribution (that total being $n_i\bar{z}_i$ in the unstreamlined notation). The baseline fitness parameter $k > 0$ is chosen to satisfy the technical condition that (in this discrete time setting) fitness W_j is never negative.

For the moment we assume perfect transmission of z from parent to progeny so that we can use the simple Price equation. It says

$$\begin{aligned} \bar{W}\Delta\bar{z} &= Cov(W,z) = Cov[k - az + b\bar{z}, z] \\ &= Cov[k,z] - Cov[az,z] + Cov[b\bar{z},z] = 0 - aCov[z,z] + 0 \\ &= -aVar[z] \le 0. \end{aligned} \tag{5.29}$$

The second line used the facts that covariance is bilinear and that the covariance of any variable with any constant is zero, while the last line used the fact that the covariance of any variable with itself is its variance. Thus the average contribution or degree of altruism will decay within the group ($\Delta\bar{z} < 0$) unless variance is zero, that is, unless all members maintain exactly the same contribution level $z_i = \bar{z}$. This implies that cooperation within the group will be destabilized by any invasion of cheaters with a lower contribution level.

Things get more interesting when we remember that there may be many groups, not just one. Unstreamlining the notation, the fitness function is

$$W_{ij} = \frac{n'_{ij}}{n_{ij}} = k - az_{ij} + b\bar{z}_i. \tag{5.30}$$

This function again says that costs of contribution (or altruism) are borne personally, while the benefits are shared within the group but not across groups. The parameter restriction $0 < a < b$ again ensures that the game is a social dilemma. In this setting, we drop the assumption that transmission is perfect and use the full Price equation:

$$\bar{W}\Delta z = Cov[k - az + b\bar{z}_i, z] + E[(k - az + b\bar{z}_i)\Delta z]. \tag{5.31}$$

The inheritance (second) term is the expectation across all groups i of the non-negative factor $(k - az + b\bar{z}_i)$ times the change Δz_i of the mean contribution in that group from one generation to the next. We have just seen that that change is negative to the extent that the contribution cost a is substantial and that there is variance in the contribution rate z_{ij} within group i. Hence the second term is negative, or possibly zero. The selection (first) term reduces to

$$Cov[k - az + b\bar{z}_i, z] = (b - a)Var(z) > 0, \qquad (5.32)$$

as you can see by reusing the key facts in the derivation of equation (5.29), together with the fact that $E\bar{z}_i$ and Ez are both just the overall mean trait value $\sum_i \sum_j n_{ij} z_{ij} / \sum_i \sum_j n_{ij}$.

We conclude that cooperation can be sustained in social dilemmas with the proper group structure. The net social benefit $(b - a)$ and the variance across groups $Var(z_i)$ both must be positive and large enough to offset the negative transmission term. Moreover, in light of equation (5.29), the within-group variance must be small in order to prevent rapid erosion of cooperation within the group. The idea is that more cooperative groups achieve higher fitness, and thus grow sufficiently rapidly to offset erosion within each group as well as imperfect transmission. See the notes for pointers to the extensive literature on such matters.

5.6 GROUP STRUCTURE AND ASSORTATIVITY IN LIZARDS

We can make concepts of group structure, attraction, and impacts on game dynamics less abstract with a concrete example drawn from the now familiar side-blotched lizards. This species is one of the few where complete fitness accounting is available for all males and the group structure is fully described. Sinervo and Clobert (2003) and Sinervo et al. (2006) considered the effects of social networks of nearest neighbors on the fitness of the three male strategies. The complete set of equations for each type of pairwise encounter are supplied in Sinervo et al. (2006), and here we only consider a few of these relationships salient to the impact of group structure on male fitness.

A bit of background is in order. Male morphs with pure blue throats find blue males at a higher rate than other morph pairs and compared to random, and orange males are repulsed by orange males at a higher rate than other morphs, and compared to random. Yellow males are neutral with respect to type. Thus, the adjustment matrix for the lizards has the form (for $i = o, b, y$):

$$\mathbf{A} = \begin{pmatrix} b & 1 & 1 \\ 1 & a & 1 \\ 1 & 1 & 1 \end{pmatrix} \qquad (5.33)$$

with $0 < b \leq 1 \leq a$. Data suggest approximate values of $a = 3$ for blue and $b = 0.26$ for orange (Sinervo and Clobert 2003).

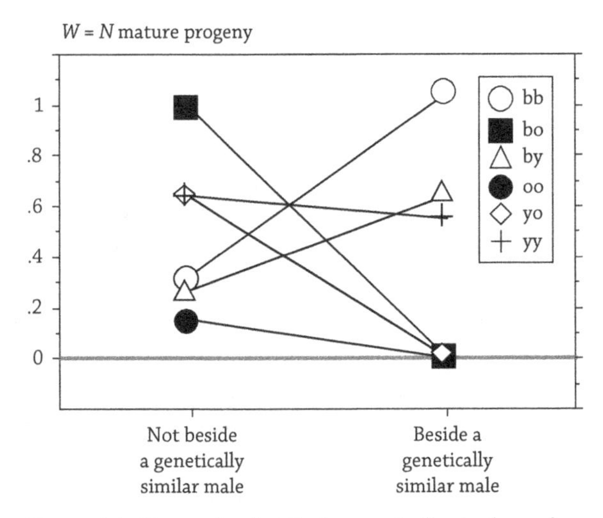

Figure 5.4 Blue males that find a genetically similar male neighbor have higher fitness than those that do not. For other morphs, the fitness increment from genetically similar neighbors is either nil (yellow) or negative (orange).

Blue males share remarkably many genes but pedigrees established by DNA testing show clearly that neighboring blue males typically are completely unrelated. Gene mapping studies show that blue males have many loci for *finding* and *signaling* to blue males and the best blue males at this task, those with the most shared genes for mutual attraction and signaling, obtain the highest fitness (Figure 5.4). Genetic similarity (or lack of variation) in a group of blue males stabilizes cooperation and these males obtain very high fitness compared to other blue males that lack genetically similar neighbors, or compared to the other color genotypes.

The situation is much more asymmetric if we consider the impact of orange neighbors on the fitness of each of the two blue males. Typically, only one blue male experiences the impact of an orange male and thus this blue male's territory buffers his partner from the aggressive orange males. Sinervo et al. (2006) computed the ΔW for blue males as a function of the number of O males in their neighborhood, which also depends on the variation in orange neighbors: $\Delta W_{b,b}(t) = \beta_5 \Delta N_O$ (equation [4] therein). A consequence of the RPS cycles in orange alleles is that the males experience high numbers of O and then nearly no O males, which causes oscillations in the difference in fitness between two genetically similar blue neighbors.

The side-blotched lizard system exhibits conditions that are best described as altruism where the male nearest the orange males experiences a fitness cost that benefits his partner, when orange neighbors are common. Conditions for

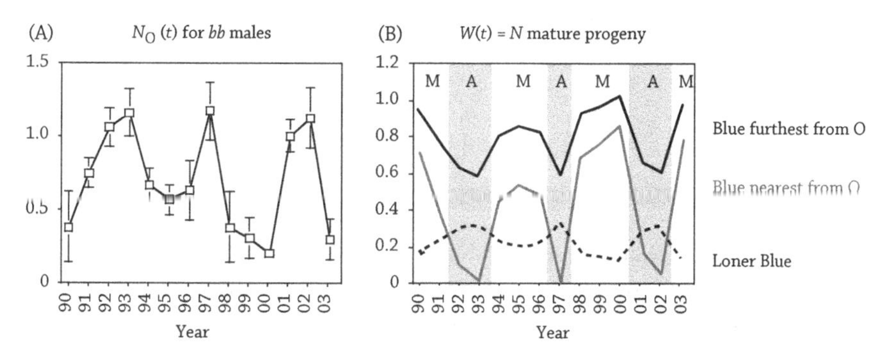

Figure 5.5 (A) Cycles in the number of O neighbors for blue, due to the RPS game, (B) generate a cycle of altruism (labeled A) and mutualism (M) in Blue males.

altruism will be discussed further in Chapter 13, but altruism is usually contrasted to a strategy that does not engage in the social behavior, such as "loner blue males" that do not have genetically similar male neighbors. When orange is rare, both cooperative males experience high fitness, compared to the loner strategy and thus their relationship is mutualistic. Salient with respect to the group-structured Price equation is the finding that similarity (low variance) between blue males enhances their fitness, while variation in neighboring groups, such as the orange males, destabilizes mutualism and enforces altruism between blue males.

5.7 PRICE EQUATION IN CONTINUOUS TIME

We've not seen it mentioned in the literature, but the full Price equation has a continuous time analogue. The idea is to decompose the rate of change $\dot{\bar{z}}$ of the mean trait value, instead of $\Delta\bar{z}$, the change over one period. Again, there will be two components, one representing selection and the other, the residual, that can be thought of as departures from faithful transmission.

In continuous time, we want to think of shares as well as traits varying continuously, so we will write $s_i = \frac{n_i}{n}$ and treat s_i as a differentiable function of time. Likewise we shall regard z_i, the mean trait value in group i, as a differentiable function of time. As usual, we will use Newton's dot notation to indicate time derivatives (e.g., $\dot{s}_i = \frac{ds_i}{dt}$).

Under these conventions, the mean trait value is

$$\bar{z} = \sum_{i=1}^{n} s_i z_i. \tag{5.34}$$

The product rule from calculus tells us that its time derivative is

$$\dot{\bar{z}} = \sum_{i=1}^{n} [\dot{s}_i z_i + s_i \dot{z}_i] = \sum_{i=1}^{n} \dot{s}_i z_i + \sum_{i=1}^{n} s_i \dot{z}_i. \tag{5.35}$$

Remarkably, we are already almost done. The first term represents selection—holding trait values constant for each group i, it captures the change in the mean trait value due to changing shares. To see the covariance connection, substitute the standard continuous replicator equation $\dot{s}_i = (w_i - \bar{w})s_i$ into that term to obtain

$$\begin{aligned}\sum_{i=1}^{n} \dot{s}_i z_i &= \sum_{i=1}^{n} s_i (w_i - \bar{w}) z_i = \sum_{i=1}^{n} s_i (w_i - \bar{w})(z_i - \bar{z} + \bar{z}) \\ &= \sum_{i=1}^{n} s_i (w_i - \bar{w})(z_i - \bar{z}) + \sum_{i=1}^{n} s_i (w_i - \bar{w}) \bar{z} \\ &= \mathrm{Cov}(w,z) + (\bar{w} - \bar{w})\bar{z} = \mathrm{Cov}(w,z).\end{aligned} \tag{5.36}$$

Thus the selection term is precisely the covariance of fitness with trait values!

The second term in (5.36) captures the shift in the mean trait due to changes in the underlying trait values, and thus can be interpreted as transmission. It can be written as $E(\dot{\bar{z}}|s = \text{constant})$. It captures shifts over time in the location of the discrete trait set $\{z_i : i = 1, \ldots, n\}$ or alternatively, if the trait set is constant over time, it captures non-fitness-related migrations across trait values. For example, with Mendelian genetic transmission, $\dot{z}_i = a(z_i' - z_i)$, where z_i' is the mean trait of progeny of individuals with trait z_i, and the rate constant $a > 0$ accounts for the time to maturation.

To summarize, the continuous time Price equation for a discrete set of traits is

$$\dot{\bar{z}} = \mathrm{Cov}(w,z) + E(\dot{\bar{z}}|s = \text{constant}). \tag{5.37}$$

As in the standard discrete time version, the first term represents selection, and the second term represents transmission (in)fidelity.

What if the trait set itself is continuous? To analyze that situation properly, we need to be more specific about how the set of traits changes over time. There are several different possibilities, and some of the more interesting ones require techniques beyond the scope of this book. The notes here (and in Chapter 14) include pointers to existing literature.

5.8 DISCUSSION

In this chapter we have assembled and compared several distinct ways to represent social interactions that twist fitness or payoffs away from the baseline defined by random pairwise matching. The assortativity index $a(x)$ is a nice way to summarize distortions in pairwise matching for the lowest dimensional evolutionary games, but doesn't seem to generalize beyond that. Symmetric adjustment matrices can do that job, but only for state-independent distortion rates. General $n \times n$ adjustment matrices, possibly asymmetric, can capture arbitrary state-independent distortions that arise from social processes beyond pairwise matching.

Sexual (or fertility) selection is an important form of social transaction, but it is complicated to trace its interaction with survival selection. Building on linear algebra introduced earlier, we present new techniques that cope with the complications, and illustrate them for a variant of the runaway model from the previous chapter. The techniques seem suitable for meme transmission from two parents as well as gene transmission.

There are yet other ways to think about social structures that modulate strategic interaction. Group structures are natural in many contexts, especially in analyzing cooperation and altruism. We briefly summarize known results based on the Price equation, and offer an extension to continuous time.

Group structures are an important special case of network structures, which explicitly specify the other individuals with whom any agent interacts. Roughly speaking, groups partition the network into components with complete connections within each component and sparse connections across components. Cellular automata are an especially useful network structure that is explicitly spatial with tidy neighborhoods. We will study them in detail in Chapter 6. Finally, group structure is intimately related to the evolution of cooperation, subjects we explore in Chapters 6 and 13.

APPENDIX: EQUILIBRIUM IN THE KIRKPATRICK (1982) MODEL

Here we draw on the material in the present chapter to provide more details on the Kirkpatrick (1982) model introduced in Chapter 4. Recall that the model dealt with runaway sexual selection, with two alternative alleles (1=ornamented, 2=plain) on a single locus (T) determining male phenotypes, and two alleles (2=ornament preferring, 1=indifferent) on another locus determining female phenotypes. The underlying genotypes (T1P1, T1P2, T2P1, T2P2) are the same for both sexes. Thus the evolving state variable is the shares vector for the four genotypes.

An equilibrium is given by the steady state condition that the new state $\mathbf{s}' \in \mathbf{S}^4$ is the same as the old state $\mathbf{s}$, that is, it is a root of the equation $\mathbf{s}' - \mathbf{s} = \mathbf{0}$. This condition is expressed by three equations in three unknowns, along with the usual simplex condition that the four components of the vectors sum to 1.0.

Kirkpatrick finds it convenient to re-express the three equilibrium equations in terms of another three variables: $s_{T2} = s_{T2P1} + s_{T2P2}, s_{P2} = s_{T1P2} + s_{T2P2}$, and $D = s_{T1P1}s_{T2P2} - s_{T1P2}s_{T2P1}$, where the new quantity D refers to the degree to which multilocus genotypes are not freely combining in ratios expected by chance alone. Due to the combination of survival selection and sexual selection, some multilocus genotypes are over-represented in both sexes (e.g., T2P2) and some are under-represented (e.g., T1P2).

The equations have a relatively simple form once we collect messy terms into expressions abbreviated as A, B, and R:

$$A = \frac{1}{(1 - ms_{T2})}\left[1 + s_{P2}\left(\frac{1}{\bar{W}} - 1\right)\right], \tag{5.38}$$

$$B = \frac{1-m}{(1 - ms_{T2})}\left[1 + s_{P2}\left(\frac{a_2}{\bar{W}} - 1\right)\right], \tag{5.39}$$

and

$$\begin{aligned} R = \frac{1}{(1 - ms_{T2})}&\left[D^2\left[\frac{1 - a_2(1-m)}{\bar{W}} - m\right]\right. \\ &+ D\left[1 + ((1 - s_{T2})(1 - s_{P2}) + s_{T2}s_{P2})\left(\frac{1}{\bar{W}} - m\right)\right. \\ &\left.+ \frac{a_2}{\bar{W}}(1-m)(s_{P2}(1 - s_{T2}) + (1 - s_{P2})s_{T2})\right] + s_{T2}(1 - s_{T2})s_{P2}(1 - s_{P2}) \\ &\left.\times\left[\frac{1 - a_2(1-m)}{\bar{W}} - m\right]\right]. \end{aligned} \tag{5.40}$$

Equilibrium now can be characterized by roots of the following equations:

$$\Delta s_{T2} = \frac{1}{2}(B - 1) \tag{5.41}$$

$$\Delta s_{P2} = \Delta s_{T2}\frac{D}{s_{T2}(1 - s_{T2})} \tag{5.42}$$

$$\Delta D = \frac{1}{4}D[A + B + AB - 3] - \frac{1}{4}R[1 + A(1 - s_{T2}) + Bs_{T2}] \tag{5.43}$$

In particular, setting the expressions in (5.41) and in (5.42) equal to 0, we obtain a single equation for the equilibrium shares of the male trait and of the female preference:

$$s^*_{T2} = \left[\frac{1}{m} + \frac{1}{a_2(1-m)-1}\right] s^*_{P2} - \frac{1}{a_2(1-m)-1}, \quad \text{for } \frac{m}{(a_2-1)}(1-m) < s^*_{P2} < \frac{a_2 m}{a_2-1} \tag{5.44}$$

The equilibrium for D^* can be obtained by noticing that when genotypes are in equilibrium, the term in the first set of brackets in equation (5.43) equals 0, which implies $R=0$, and thus the resulting quadratic for D can be easily solved for its two roots:

$$D^* = \frac{s^*_{P2}(1-ms^*_{T2})}{m} \pm \sqrt{\frac{s^*_{P2}(1-ms^*_{T2})}{m} - s^*_{T2}(1-s^*_{T2})s^*_{P2}(1-s^*_{P2})} \tag{5.45}$$

of which the larger root is always feasible, while the smaller root is not. Notice that D^* is completely independent the recombination rate, which means that the runaway will happen regardless of whether the male and female traits sit on different chromosomes. Non-random mating and the relative success of T2 over T1, due the attraction of P2 for T2, will couple the fate of the male and female genotypes generating a standing level of D^* when s_{t2} and s_{p2} arrive at the line of equilibria shown in Figure 5.3.

Now consider a slight extension of the model in which the parameter $a_1 > 1$ describes the preference of P1 females for T1 males, while the preference parameter $a > 1$ of P2 females for T1 males is now relabeled as a_2. The attraction matrix is now $\mathbf{A} = \begin{pmatrix} a_1 & a_1 & 1 & 1 \\ 1 & 1 & a_2 & a_2 \\ a_1 & a_1 & 1 & 1 \\ 1 & 1 & a_2 & a_2 \end{pmatrix}$.

In this modification of the model, if the frequency of P2 drifts up to high frequency (e.g., s^R_{p2} in Figure 5.3) in the absence of any genetic variation in the male trait (e.g., along the x-axis, then above a certain value of female preference), the stage is set for a runaway. If T2 mutates from T1 at any point above s^R_{p2} (i.e., invades from low frequency), then the frequency of T2 will shoot up and ultimately go to 100%. This is the sense in which positive frequency dependent selection due to the interaction of exaggerated male traits and female preferences for them can drive a runaway. As this happens average fitness in the population is driven downwards as more and more males express T2, and thus a viability cost given by m.

EXERCISES

1. Section 5.1 introduces distortions of matching probabilities. Suppose these take the form $p(x) = x^{\alpha}$. Find the NE of the otherwise basic HD game (say with $c = 6$ and $v = 4$) as a function of the common $\alpha > 0$, starting with $\alpha = 2$ and $\alpha = 1/2$.
2. Using results from Chapter 2, show that any 2×2 coordination game can be written without loss of generality (in terms of evolutionary dynamics driven by fitness differences) as $\begin{pmatrix} w_1 & 0 \\ 0 & w_2 \end{pmatrix}$, where $w_1, w_2 > 0$. Write down the expression for the equal payoff locus $\Delta w = 0$ when interactions are governed by assortativity index $a(x)$. Under what conditions does that expression define an interior NE which separates the basins of attraction of the two pure NE? When can that expression give a non-interior value, outside of $[0, 1]$? In such cases, which subtype (PD or LF) of a DS game do we have?
3. The text claims that $a(x) > 0$ when $p(x) = x^{\alpha}$ and $\alpha = 1/n$ for any integer $n > 1$. Verify that claim by deriving the expressions $q(x) = \frac{x}{\sum_{k=0}^{n-1} x^{k/n}}$ and $a(x) = p(x)r(x,n) > 0$, where $r(x,n) = \frac{\sum_{k=1}^{n-1} x^{k/n}}{\sum_{k=0}^{n-1} x^{k/n}} \in (0,1)$. Hint: For the negative assortativity case $\alpha = n$, the text used algebraic tricks using powers of x and telescoping sums. Try similar tricks using powers of $x^{1/n}$.
4. Section 5.2 used the basic HD game to introduce adjustment matrices. Analyze the basic PD game in a similar fashion, emphasizing the possibility that cooperators may be able to interact with fellow cooperators at a rate $a > 1$. Show that the pure equilibrium $s_D = 1$ is unaffected by values of the parameter a near 1, but that there is some a^* such that a different equilibrium appears for $a > a^*$.
5. Verify that the RPS twist eigenvalues have real part $(b - a)/6$ as asserted in Section 5.2. The verification can either be analytically or using the Mathematica code in the appendix to Chapter 4, or (preferably) both. Hint: if proceeding analytically, it may be helpful first to solve for particular values such as $b = 0.5$ and $a = b + 1$.

NOTES

The idea of assortative mating goes back at least to Sewall Wright (1921), who used a correlation measure. Bergstrom (2002) notes that his index $a(x)$ reduces to Wright's correlation coefficient m when matching is pairwise and $a(x)$ is constant. Previously, Cavalli-Sforza and Feldman (1981) showed that one way to obtain such an m (hence a constant $a(x)$) in a symmetric 2×2 game is to match each player to her own type with probability m and match at random with probability $1-m$.

Regarding the limitation to symmetric 2×2 games, Bergstrom writes (private communication, May 2011): "One can make a case that if you are considering monomorphic equilibria, then the only (or at least main) interesting case is that where there are only 2 strategies to be considered at a time—the reigning equilibrium strategy and rare mutant strategies. Because of assortativity, the rare mutants are not [necessarily] rare to each other, but the only other type they encounter with significant probability are the equilibrium types. The equilibrium types almost always meet only their own type." Of course, this book is mainly concerned with polymorphic equilibria.

To our knowledge, the general idea of adjustment matrices in Section 5.2 is new. Precursors can be found in Sinervo et al. (2007). A private communication from Bergstrom (October 2014) alerted us to the need to distinguish such matrices that represent pairwise matching processes from those that represent more general social mediation.

Gardner et al. (2007) show the formal connections between multilocus models of the sort described in the previous chapter (for runaway) and the equations of Hamilton and Price.

The derivations in Sections 5.4 and 5.5 are standard. There is a huge literature on the Price equation, some of it quite critical (e.g., van Veelen et al. 2012). We have found Steven Frank a particularly lucid expositor and defender of the Price equation (e.g., Frank 2012).

BIBLIOGRAPHY

Bergstrom, Theodore C. "Evolution of social behavior: Individual and group selection." *Journal of Economic Perspectives* (2002): 67–88.

Bergstrom, Theodore C. "The algebra of assortative encounters and the evolution of cooperation." *International Game Theory Review* 5, no. 3 (2003): 211–228.

Cavalli-Sforza, Luigi Luca, and Marcus W. Feldman. *Cultural transmission and evolution: A quantitative approach.* Princeton University Press, 1981.

Cooper, Ben, and Chris Wallace. "Group selection and the evolution of altruism." *Oxford Economic Papers* 56, no. 2 (2004): 307–330.

Fitze, Patrick S., V. Gozalez-Jimena, L. M. San-Jose, B. Heulin, and B. Sinervo. "Frequency-dependent sexual selection with respect to progeny survival is consistent with predictions from rock-paper-scissors dynamics in the European common lizard." *Frontiers in Ecology and Evolution* 2 (2014): 1–11.

Frank, Steven A. "Natural selection. IV. The Price equation." *Journal of Evolutionary Biology* 25, no. 6 (2012): 1002–1019.

Gardner, Andy, Stuart A. West, and Nicholas H. Barton. "The relation between multilocus population genetics and social evolution theory." *The American Naturalist* 169, no. 2 (2007): 207–226.

Kirkpatrick, Mark. "Sexual selection and the evolution of female choice." *Evolution* 36, no. 1 (1982): 1–12.

Price, George R. "Selection and covariance." *Nature* 227 (1970): 520–521.

Sinervo, Barry, and Jean Clobert. "Morphs, dispersal behavior, genetic similarity, and the evolution of cooperation." *Science* 300, no. 5627 (2003): 1949–1951.

Sinervo, Barry, Alexis Chaine, Jean Clobert, Ryan Calsbeek, Lisa Hazard, Lesley Lancaster, Andrew G. McAdam, Suzanne Alonzo, Gwynne Corrigan, and Michael E. Hochberg. "Self-recognition, color signals, and cycles of greenbeard mutualism and altruism." *Proceedings of the National Academy of Sciences* 103, no. 19 (2006): 7372–7377.

Sinervo, Barry, Benoit Heulin, Yann Surget-Groba, Jean Clobert, Donald B. Miles, Ammon Corl, Alexis Chaine, and Alison Davis. "Models of density-dependent genic selection and a new rock-paper-scissors social system." *The American Naturalist* 170, no. 5 (2007): 663–680.

van Veelen, Matthijs, Julián García, Maurice W. Sabelis, and Martijn Egas. "Group selection and inclusive fitness are not equivalent; the Price equation vs. models and statistics." *Journal of Theoretical Biology* 299 (2012): 64–80.

Wright, Sewall. "Systems of mating." *Genetics* 6, no. 2 (1921): 111–178.

6

Cellular Automaton Games

A cellular automaton (CA) is a discrete spatial simulation. Each location (or "cell") has a current state and set of nearest neighbors. Each cell transitions to a (possibly) new state each period according to well-defined rules involving the nearest neighbors. As explained in the notes, CAs are a mid-20th century invention that has captured the imagination of some eminent mathematicians and applied researchers in the physical sciences as well as in biology, social science, and computer science.

This chapter will review recent CA research applied to spatial evolutionary games, show how it generalizes, and suggest some new applications. The basic idea is simple. Instead of interacting directly with an entire player population (or a random sample), each player interacts directly only with a set of nearest neighbors. But those interactions have ripple effects, as her neighbors interact with their neighbors, and they interact with theirs. Behaviors thus can spread throughout entire populations, and we will see that surprising spatial patterns can emerge.

For our exposition of basic evolutionary game techniques, as for the TV show *Star Trek*, space is the final frontier. This chapter includes new material showing how CAs connect with other techniques for modeling spatial interaction, including density dependence and assortative interactions. Subsequent chapters are application-oriented, and will show how the techniques shed light on a variety of research questions.

6.1 SPECIFYING A CA

Neighborhood structure

CAs are played over regular networks, usually simple grids in the 2-dimensional plane or the 1-dimensional line. In one dimension, most cells have two neighbors, one on the left and one on the right. If the grid is finite, its endpoints need special treatment because they have only one neighbor, and that difference could affect the direct neighbor and perhaps indirect neighbors as well. Perhaps you are interested in such boundary effects, but if not, you can decree that the two endpoints are each other's neighbors. In that case, the CA is played on a circle, and all cells have two neighbors. (Occasionally, as noted in the notes, one sees CA models that are essentially 1-dimensional yet have larger neighborhoods.)

Two-dimensional grids come in many varieties. If you chop the plane up into proper triangles, then each cell is a triangle with three neighbors, those with which it shares an edge. If you chop up the plane with the usual rectangular grid, then each cell is a rectangle with four neighbors, those with which it shares an edge; this is the most widely used structure, and its neighborhoods are called Von Neumann. Alternatively, you can consider as neighbors also those rectangles for which the cell shares a corner, yielding eight neighbors in what is called the Moore neighborhood. A honeycomb (hexagonal) structure yields six neighbors for each cell. Another possibility (which we have not seen used in the literature) is the Moore analogue of the triangular structure, in which each cell has up to 12 neighbors.

To avoid boundary effects in finite grids in two dimensions, just decree the right edge cells to be neighbors of corresponding left edge cells to make a cylinder, and then decree the bottom cells to be neighbors of corresponding top edge cells to make a torus, like the surface of a bagel. The torus is widely used in CA simulations.

If you prefer the sphere to the torus, you are out of luck if you want to keep regular Von Neumann, Moore, or hexagonal neighborhoods. The topology just doesn't work. But you can put your CA on the sphere if you use the three-neighbor arrangement described in the previous paragraph, or a Moore analogue. Topologists have shown that any orientable surface can be properly triangulated, so these neighborhoods work on pretty much any surface you are likely to encounter.

Games and transition rules

Any CA has a transition rule specifying when and how a cell's state (or strategy) changes. A fairly general rule is that the update probability is a specified decreasing function of the cell's current fitness, or payoff. As a special case, the probability can be zero above a specified payoff threshold and 1.0 below the threshold. Throughout this chapter, we will use another special case: each period each cell's state is updated with a fixed probability q, independent of current payoff and system history.

When a cell is due for an update, how is the new state determined? In most CA specifications, the update depends on the current state of the cell and its neighbors, independent of previous history. This is called a Markov or stationarity property, and we will assume it for the rest of the chapter. So the question becomes, how does the current state of the neighborhood determine the new state?

The CA is an evolutionary game if the rule involves payoffs determined by the neighbors' states (i.e., fitness determined by local interaction) and/or involves the neighbors' current payoffs w_k (i.e., imitation or transmission). In this chapter, we will usually use the following logit update rule: the updated state of cell i is the same as the current state of neighbor $k \in N_i$ with probability $p_k = ce^{\beta w_k}$, where N_i is the set of all neighbors of cell i and as usual the normalizing constant c satisfies $1/c = \sum_{j \in N_i} e^{\beta w_j}$. In the limit as the payoff sensitivity $\beta \to \infty$, this stochastic rule becomes deterministic: the update is always to the state of the neighbor with highest current payoff. In the limit $\beta \to 0$, the updated state is independent of fitness, and imitates a random neighbor.

It can make a difference whether updates are done simultaneously (as in a discrete-time replicator), or one at a time (asynchronously, as in the continuous-time replicator). Except when otherwise specified, the updates in this chapter are simultaneous.

Simulation package

We have created an [R] package *CAgames* that can simulate the main examples covered in the present chapter, and a wide variety of other cellular automaton games. For comparative purposes, the package can also can simulate discrete time replicator dynamics for the same games with no spatial structure. Connecting these two extremes, the user can specify a fraction $\alpha \in [0,1]$ of the interactions that determine fitness that come from neighbors, with the remaining fraction $1 - \alpha$ coming from the population at large. As $\alpha \to 0$ the fitness calculation involves only the delta functions featured in earlier chapters, and is location-independent.

The parameters (α, β, q) and the payoff matrix parameters of all CA games introduced in this chapter can be manipulated via a graphical user interface. The simulation package also covers the two-strategy, two-population games such as Buyer-Seller or the two-population HD games of Chapter 4. If you don't mind a little programming, you can download the code for the [R] workspace as an [R] package *CAgames* from the CRAN site nearest you but also available on the software repository Friedman and Sinervo (2015). This will allow you to manipulate the code directly and develop spatial games beyond those covered in this chapter.

It should now be clear that there are a vast number of variations on CA evolutionary games. To our knowledge there has been little systematic investigation of which variations matter in which circumstances. Evidently this is a fertile area for a master's thesis and perhaps even for a doctoral dissertation!

6.2 PRISONER'S DILEMMA

We begin our discussion of published work with the familiar PD, as usual labeling the first of the two alternative strategies as C or Cooperate and labeling the second as D or Defect. The payoff matrix

$$\mathbf{w} = \begin{pmatrix} b-c & -c \\ b & 0 \end{pmatrix}. \tag{6.1}$$

has two parameters, with $c > 0$ representing the costs of cooperation and $b > c$ representing the benefits. Such matrices automatically satisfy the PD inequalities $w_{DC} > w_{CC} > w_{DD} > w_{CD}$ and $2w_{CC} > w_{DC} + w_{CD} > 2w_{DD}$ noted in Chapter 2. The cost-to-net-benefit ratio $\rho = c/(b-c) \in (0, \infty)$ quantifies the strategic tension between self-interest and group interest.

Hauert and Doebeli (2004) introduced a cellular automaton using this parametrization of the PD game. They worked with a 50×50 Von Neumann torus with unspecified update and transition parameters. Figure 6.1 uses exactly the same neighborhood structure with death (update) rate $q = 0.125$, fitness entirely from local interaction ($\alpha = 1$), and payoff sensitivity $\beta = 0.5$. The three panels of the figure illustrate the spatial patterns typical after hundreds of periods when $\rho = 1, \frac{1}{2}$, and $\frac{1}{3}$. For $\rho > 1$ cooperators disappear, and for $\rho < \frac{1}{4}$ defectors go extinct.

The long run survival of Cooperate (and the long-run extinction of Defect) for favorable values of ρ are surprising at first, but can be understood intuitively. For sufficiently low values of ρ, islands of cooperation emerge, as in the left panel of Figure 6.1. These congealed social neighborhoods protect internal members of the cooperating group from attack by an egoist strategy, thereby enhancing Cooperate's fitness. Egoist strategies like Defect can only prey on the edge

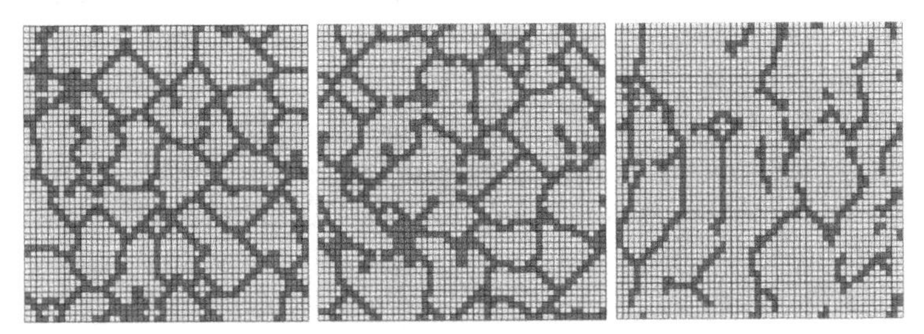

Figure 6.1 Typical snapshots of prisoner's dilemma CAs for matrix parameters $b=2, c=1$ (left panel); $b=3, c=1$ (center panel); and $b=4, c=1$ (right panel). Lighter cells are currently playing C and darker cells are currently playing D.

members of the cooperative clusters, reducing the number of individuals that they can exploit and hence reducing their fitness.

When the cost of cooperation is sufficiently low relative to its net benefit, only an isolated Defect cell can achieve higher fitness than Cooperate, but such a pattern is not stable in the long run. It will either (a) randomly die and necessarily be replaced by Cooperate (since that is the only alternative seen in the neighborhood), or (b) flip a neighboring cell to Defect, in which case it loses fitness since it has fewer exploitable neighbors. Hence, despite the fact that Defect is a dominant strategy in a well-mixed (or non-spatial) environment, it loses its dominance in a spatial setting where Cooperate cells on average have more Cooperate neighbors than do Defect cells.

Thus the explicitly spatial CA structures allow cooperation to persist for reasons similar to assortativity or group structure, the implicitly spatial structures featured in Chapter 5. We will take a broader look at such matters in Chapter 14.

6.3 SNOWDRIFT

The game Snowdrift (SD) illustrates a slightly different tension between self- and group interest, where cooperation by either player benefits both players. The story is that a snowdrift blocks the road and two drivers arrive from opposite directions. The cost $c > 0$ of digging out can be borne by either or shared equally, and it brings benefit $b > c$ to each. (The high-cost case $2b > c > b$ is sometimes considered, but it leads to a payoff matrix of type PD.) Thus the cost-to-net benefit ratio is now $\rho = c/(2b - c)$, and the payoff matrix is

$$\mathbf{w} = \begin{pmatrix} b - \frac{c}{2} & b - c \\ b & 0 \end{pmatrix}. \tag{6.2}$$

Recall from Chapter 2 that this payoff matrix is the same game type as HD, so there is a unique symmetric mixed strategy NE with fraction s^* cooperators satisfying the equation $w_C(s) = w_D(s)$. Straightforward algebra yields $s^* = \frac{2b-2c}{2b-c} \in (0,1)$ and thus $s^* + \rho = 1$. In a well-mixed population, we would expect to see this negative linear relationship between the equilibrium fraction of cooperators and the cost/benefit ratio.

Things are a bit different when the game is played on a CA, since your neighbor faces a different strategy distribution than you do. Building a CA analogous to that of the PD game, Hauert and Doebeli (2004) show that once again we obtain connected regions of cooperation and defection. However, as illustrated in the center panel of Figure 6.2, the cooperative regions look very different. They are not clumps, but instead they tend to be branched noodles, quite long and tangled

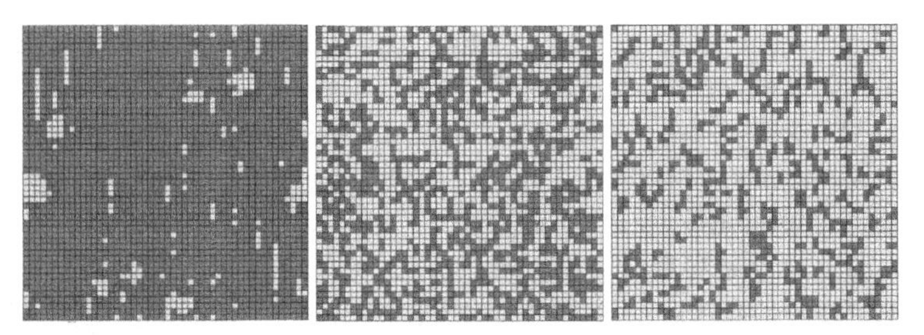

Figure 6.2 Typical snapshots for Snowdrift game played on a CA. The parameter values from left to right are $b=2$, $b=3$, and $b=4$, holding constant $c=2$, death rate $d=0.125$, and global dispersion $\alpha=0.0$. Below $b=2$ cooperators go extinct, and above $b=5$ defectors go extinct (not shown).

but only 1 to 3 cells thick. Intuitively, the reason is that it pays to have neighbors that adopt a different strategy, leading to the formation of long filaments of cooperation alternating with fields of defectors.

6.4 PUBLIC GOODS GAMES WITH TWO STRATEGIES

Recall from earlier chapters public goods (PG) games played in well-mixed groups of fixed size $N \geq 2$. Here we specify a PG game with two alternative strategies: C = contribute and D = free ride. Any individual contributing incurs an individual cost $k > 0$ and generates public return rk, where $r > 1$ is the gross rate of return on the public good. With n_C contributors in a given group, the total return is rkn_C, and this public good is distributed equally to all N group members. Thus free riders get payoff

$$w_D = rk\frac{n_C}{N} \tag{6.3}$$

and contributors get the same payoff less incurred cost,

$$w_C = w_D - k. \tag{6.4}$$

Since $w_D > w_C$ in any given group, we need spatial or other considerations to maintain the cooperative strategy C. Before considering CAs in the next subsection, we build intuition here by considering random assortment of players into moderate-sized groups.

Let s_C be the share of contributors in the overall population. Then the probability that the $N-1$ other members of your group contain exactly n_C contributors is

$$\binom{N-1}{n_C} s_C^{n_C}(1-s_C)^{N-1-n_C}, \tag{6.5}$$

whose expected value is $s_C(N-1)$. Thus your expected payoff if you do not contribute is

$$w_D = \frac{r}{N} k s_C (N-1). \tag{6.6}$$

Your expected payoff if you do contribute is

$$w_C = \frac{r}{N} k s_C (N-1) - k + \frac{rk}{N}, \tag{6.7}$$

so the payoff difference is

$$\Delta w_{CD} = (\frac{r}{N} - 1)k. \tag{6.8}$$

If $r > N$ then $\Delta w_{CD} > 0$, so strategy C gains share and goes to fixation. If $r < N$ then D wins, but the population will still go extinct, since fitness is zero when nobody contributes.

This model with fixed group size thus is unstable and seems artificial. To deal with the problem, Hauert et al. (2006) proposed density-dependent competition, for the moment still in a non-spatial (or well-mixed) model. There is a fixed death rate d, and fitness is as before plus a constant b, a term for the per capita birth rate. The net per capita growth rate of the population is therefore $b-d$. The maximum population size is N but the actual number of players $S \leq N$ can be less. We have just seen that D is favored at high densities (i.e., for $S \geq r$). However, this will lower the share of cooperators, resulting in a decline in their population size. That in turn lowers the fitness of defectors, thus eventually lowering effective group size to $S < r$. At these lower population densities Cooperate has the higher payoff, and can reinvade. This drives up population density, re-establishing conditions for the invasion of defect. Thus the density-dependent PG game can generate oscillatory density dynamics.

That's the intuition. Here's the math. Write continuous replicator equations for the shares s_C and s_D of the maximum population size, with the residual share $s_V = 1 - \frac{S}{N}$ representing vacant space. In this game, births only occur in vacant spaces, so the growth rate of Cooperate is $s_V(W_C + b) - d$, and similarly for Defect. Thus the replicator ODEs are:

$$\dot{s_C} = s_C(s_V(W_C + b) - d) \tag{6.9}$$

$$\dot{s_D} = s_D(s_V(W_D + b) - d) \tag{6.10}$$

$$\dot{s_V} = -\dot{s_C} - \dot{s_D}. \tag{6.11}$$

The authors find it easier to rewrite the equations after a change in variables to density = S/N and relative share of cooperators, $\frac{s_C}{s_C + s_D}$. Figure 6.3 shows phase portraits for the ODE system using the new variables. As expected, we see

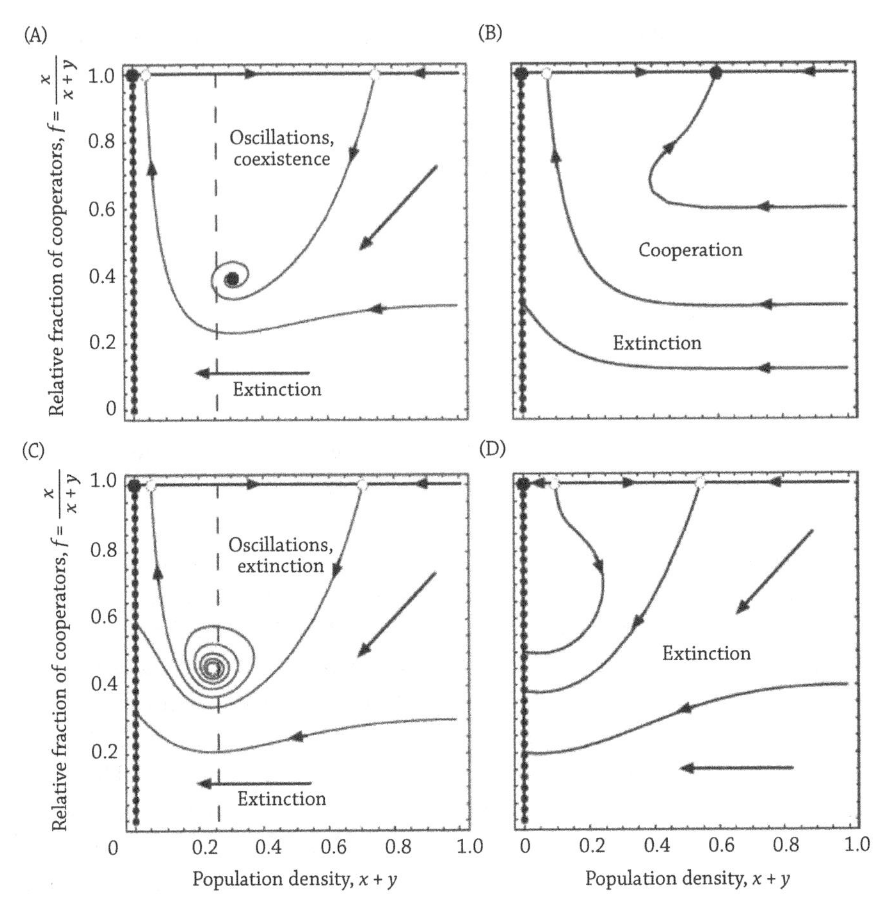

Figure 6.3 Phase portraits for PG game. Here $x = s_C$ is the share of cooperators, $y = s_D$ is the share of free riders, and $s_V = 1 - x - y$ is the share of vacancies into which they are born. For unspecified parameter values, C and D coexist in Panel A or fix on C as death rate goes up in Panel B Extinction of both C and D is possible in Panel C with low return r, and is inevitable in Panel D with high death rate d. Source: Hauert et al. (2006)

oscillations and coexistence for values of $r > Sk$. The system fixes on Cooperate if d is too high, or it goes extinct (both C and D vanish) if r is too low or d is very high.

A CA version of the density-dependent PG game is available in the [R] package *CAgame* and output from this game can be compared with a ODE version of the game also implemented in [R]. Whereas simple 2x2 games like HD and CO will not oscillate, density-dependent games have the more interesting possibility of oscillations. This is yet another interesting property of spatial games, be they implicit formulations of spatial interactions (see equations 6.9–6.11) or explicit formulations involving CA; vacancies can alter the dynamics.

6.5 SPATIAL ROCK-PAPER-SCISSORS DYNAMIC

Density dependence takes us outside the domain of evolutionary games per se, but we can return to that domain by introducing a third strategy. Retain the public goods game strategies C and D, but instead of vacant cells opened up by death, include a loner (L) strategy, which receives a low fixed payoff, but is completely resistant to the free-rider effects of D because L plays with no neighbors. However, the low payoff of L is susceptible to the higher payoff strategy of C, generating a three-strategy RPS game (recall that C is invadable by D). Instead of joining the cooperative, loners prefer to take a smaller fixed payoff $k\sigma$, where $r-1>\sigma>0$. This yields the payoff equations

$$w_D = rk\frac{n_C}{N} \tag{6.12}$$

$$w_C = w_D - k \tag{6.13}$$

$$w_L = k\sigma \tag{6.14}$$

As above, this game has a rich analytical solution provided by Hauert et al. (2002), which assumes probabilities of group formation follow random assortment. One derives the probability that S players of the sampled group members N are actually willing to join a public goods game. In the case $S=1$, where no other players engage in PG, this player is assumed to play L and obtain payoff σ. This event occurs with probability s_L^{N-1}. For players willing to join in a public goods game (be they cooperators or defectors), the probability of finding $S-1$ co-players to join the group $(S>1)$, among $N-1$ other players in the sample is

$$\binom{N-1}{S-1}(1-s_L)^{S-1}(s_L)^{N-S}. \tag{6.15}$$

The probability that n_C of these players are cooperators and $S-1-n_C$ are defectors is

$$\binom{S-1}{n_C}\left(\frac{s_C}{s_C+s_D}\right)^{n_C}\left(\frac{s_D}{s_C+s_D}\right)^{S-1-n_C}. \tag{6.16}$$

Because the game is impacted primarily by requirements on the magnitude of r relative to N, without any loss of dynamics k, the costs of cooperation can be assumed to be 1. After first deriving equations for W_i and some algebra, this spatially implicit game yields a simple fitness difference equation (plotted in Figure 6.4)

$$\Delta W_{DC} = 1+(r-1)s_L^{N-1} - \frac{r}{N}\frac{1-s_L^N}{1-s_L}. \tag{6.17}$$

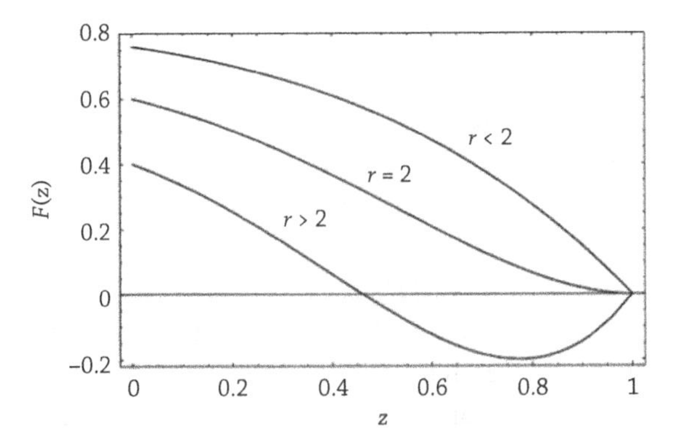

Figure 6.4 The ΔW_{DC} function for *Defect – Cooperate* for the public goods game with the additional strategy Loner, for various values of the gross return, r. When $r > 2$ it pays to cooperate, but both Loner and Defect are also retained in a version of the RPS game (from Hauert et al. (2002)).

The fitness advantage of Defect over Cooperate depends on the shares of strategies engaging in public goods $(1 - s_L)$ and the shares of Loners (s_L), but is independent of the Loner's payoff. In this sense, Loners resemble the vacant spaces in the simple two-player game, which in this game derive positive fitness, albeit at a constant rate. The function for ΔW_{DC} has a simple form (Figure 6.4) and conditions for coexistence of Cooperate-Defect depend on r.

When $r > 2$ it pays to cooperate (Figure 6.4), and in this case all three strategies can be retained in a rock-paper-scissors public goods game with closed loop orbits shown in Figure 6.5B. If $r < 2$ all orbits converge on the vertex e_L, as in Panel A of Figure 6.5. Panel C shows best-reply dynamics, for which the NE is attractor. Finally, in the case of individual-based models with finite population size, the orbits are quite robust (Figure 6.5D).

The CA for the public goods rps can be found in the [R] package *CAgames*, and one can use it to explore the impact of varying the parameters such as r as well as dispersal d on the spatial configurations for Cooperate, Defect, and Loner. In the case of very low r values, such as $r = 2.2$, traveling waves of cooperate are chased by Defect, which are in turn chased by loner, and cooperate chases loner to complete the spatial dynamic (Figure 6.6AC). For very high r values, such as $r = 3.8$, loner is eliminated, but Cooperate can still persist in the face of defect (Figure 6.6B). However, in the case of some random movement across the lattice (d), even with the high value of $r = 3.8$, all three strategies can be retained (Figure 6.6C,D). This effect occurs because randomly drawn s_L get mixed within fields of Defect, allowing them to transiently rise in frequency before engaging

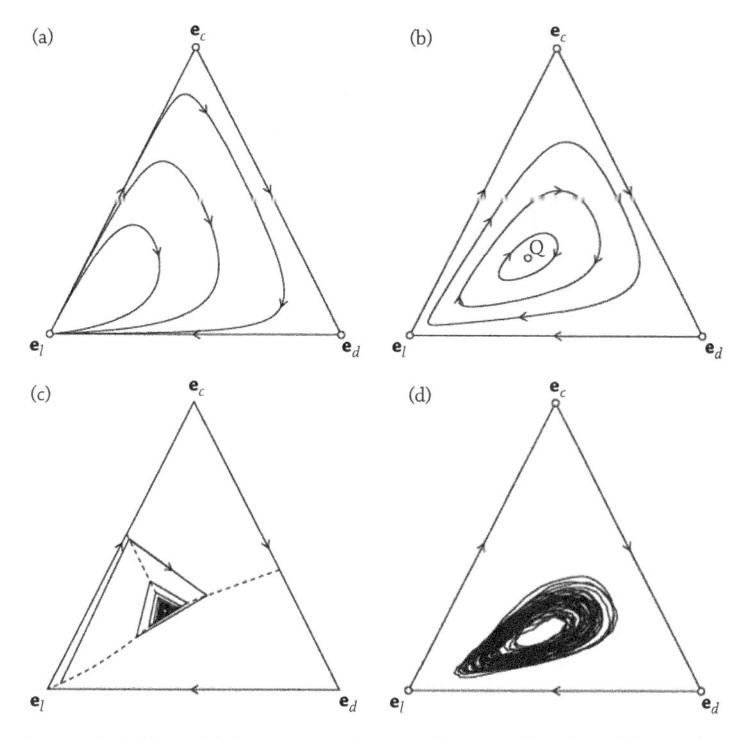

Figure 6.5 The ΔW function for *Defect* – *Cooperate* for the public goods game with the additional strategy Loner, for various values of the interest rate, r. When $r > 2$ it pays to cooperate, but both Loner and Defect are also retained in a version of the RPS game.

with Cooperate, which has very high fitness and thus will eliminate Loner. The dispersal escape for Loner, with successful Cooperators remaining clustered, is another emergent feature of this game.

The spatial clusters of Cooperate formed on the lattice are entirely analogous to the clusters formed in the PD game. Thus, spatial structure enhances the maintenance of the strategy Cooperate, which persist for all values of $r > \sigma > 1$ (instead of the more stringent condition of $r > N$, derived above). In the two-player game, Cooperate can only persist for very large values of r, and in the absence of density-dependent competition, defect would be eliminated (Figure 6.3). In the three-player PG, loners protect cooperation by exactly the same mechanism as vacant space protects the r-strategy of cooperate in the two-player PG game.

The simulation package *CAgames* includes a generic 3×3 symmetric game as well as the parameterized three-strategy PG game. Using a generic rock-paper-scissors payoff matrix, one can observe traveling waves of strategies—as

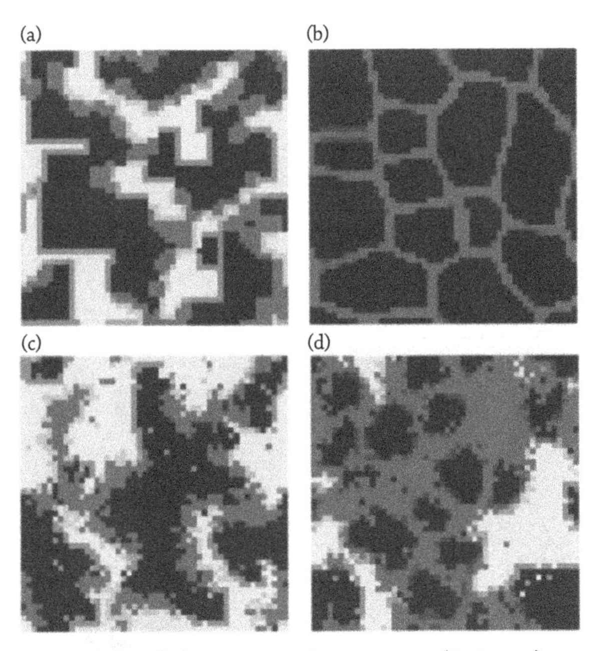

Figure 6.6 Cellular automata for Cooperate (dark grey), Defect (medium grey), and Loners (light grey) for the public goods game for various values of r, the interest rate allowing for (a) persistence of Cooperate ($r > 2$), or (b) extinction of Cooperate ($r <= 2$). In (C) and (D) the same values of r are used as in (A) and (B), but the game has 80% drawn from the local region, as in a typical CA, while 20% is drawn from the across the CA, simulating the impact of dispersal on persistence of Cooperate.

one strategy beats an adversary on its advancing edge, it is displaced on its trailing edge by its nemesis. In Figure 6.7, waves of dark grey (scissors) wash over light grey (paper), followed by waves of medium grey (rock) that wash over dark grey, and finally light grey washes back over medium grey to complete the cycle.

6.6 APPLICATION TO BACTERIAL STRAINS

The importance of spatial dynamics is illustrated with an example from nature in the form of the rps game found in the bacteria *Eschericia coli*. Kerr et al. (2002) carried out an experiment in which they competed 3 genotypes of bacteria: sensitive to colicin (S), colicin producing or colicinogenic (C), and resistant to colicin (R).

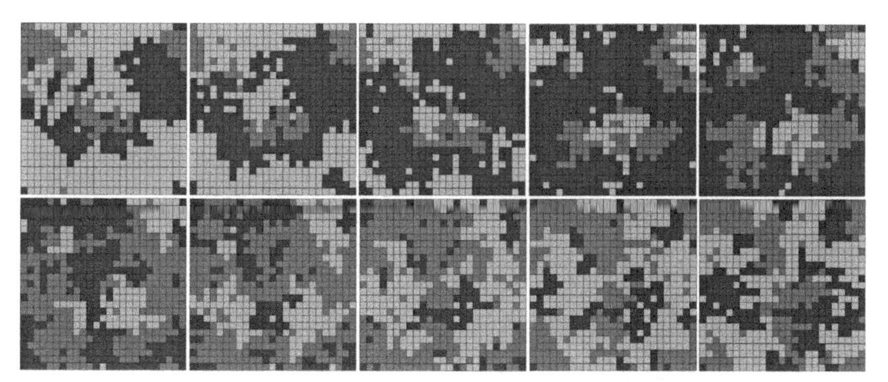

Figure 6.7 A generic RPS (medium grey, light grey, dark grey) game played out on a 25 × 25 von Neumann torus. Every 10th period is shown from period 250 to 340.

The C strategy produces a toxin that kills off any strain that is undefended, such as the strain S. The R strategy forgoes production of colicin, the toxin, but does retain the resistance gene which detoxifies colicin. The S strategy lacks this gene and thus grows faster than R in the absence of colicin, because it does not have the production costs of resistance. These interactions generate an rps game.

Kerr et al. (2002) used a CA model to illustrate that the payoffs in the bacterial RPS require space to retain all strategies (Figure 6.8). If strategies are randomized across the CA (e.g., by random dispersal), the bacterial RPS cannot sustain diversity and two of the strategies will be eliminated (Figure 6.8). Test out the CA models developed by Kerr et al. (2002) with the *CAgames* package.

They also tested the key prediction of CA model, that spatial dynamics are critical to maintenance of all three strategies. By plating out the bacterial strains on petri dishes they considered two treatments during the replication phase where a new plate of bacteria was generated: retain the growing spatial interactions across the plate (e.g., standing waves) or randomize the bacteria during transfer of the plate. In the randomized spatial treatment strains were lost, while strains were retained if the spatial dynamics of colony growth was preserved at the time of transfer from a crowded plate to a fresh uncrowded plate. The inference is that strains chase each other (according to rps dynamics) in traveling waves.

Following up on the results from the agar plate experiments of Kerr et al. (2002), Kirkup and Riley (2004) created a bacterial rps in caged mice. Each mouse can be thought of as an element of the lattice, which interact with a population of mice in a cage—an amorphous CA, but a CA nonetheless with spatial movement of strains occurring slowly from mouse to mouse as they infect each other with new strains. In this real-life CA played out in the mouse colony, C strains displaced S, and R replaced C, and S replaced R (Figure 6.9).

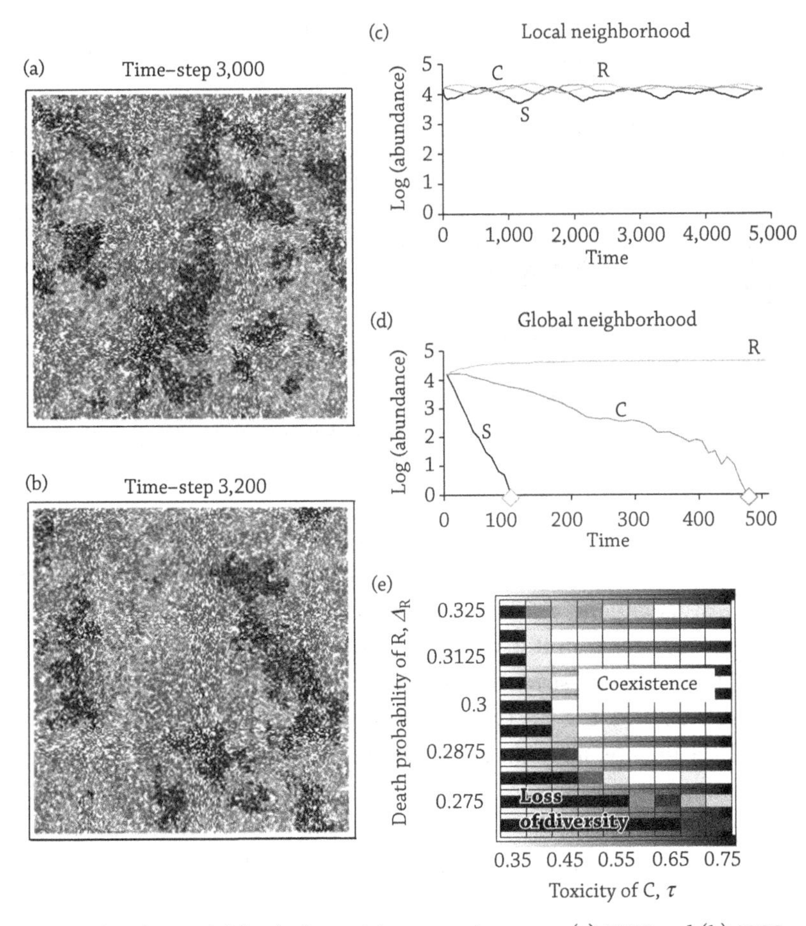

Figure 6.8 CA model for the bacterial game at time-steps (a) 3000 and (b) 3200 where medium grey is C, light grey is S, and dark grey is R, while empty lattice points are white. (c) RPS cycles of CSR are observed over the long-term, but if the game is run on a randomized lattice (d) C and S are lost and R will win out illustrating the importance of spatial interactions. (e) Strain coexistence also depends on values of toxicity of C and death rate of R (from Kerr et al. 2002).

6.7 BUYER-SELLER GAME AS A TWO-POPULATION CA

Recall the basic two-population Buyer-Seller game from Chapter 4 where

$$\mathbf{u} = \begin{pmatrix} 1 & -1 \\ -4 & 2 \end{pmatrix} \quad \text{and} \quad \mathbf{v} = \begin{pmatrix} -8 & 2 \\ 0 & 1 \end{pmatrix}$$

We can use these parameters in *CAgames* to generate the expected cycles of buyers ($i = lazy, testing$) and sellers ($j = honest, cheat$), but in this case it is played out over the CA grid (Figure 6.10).

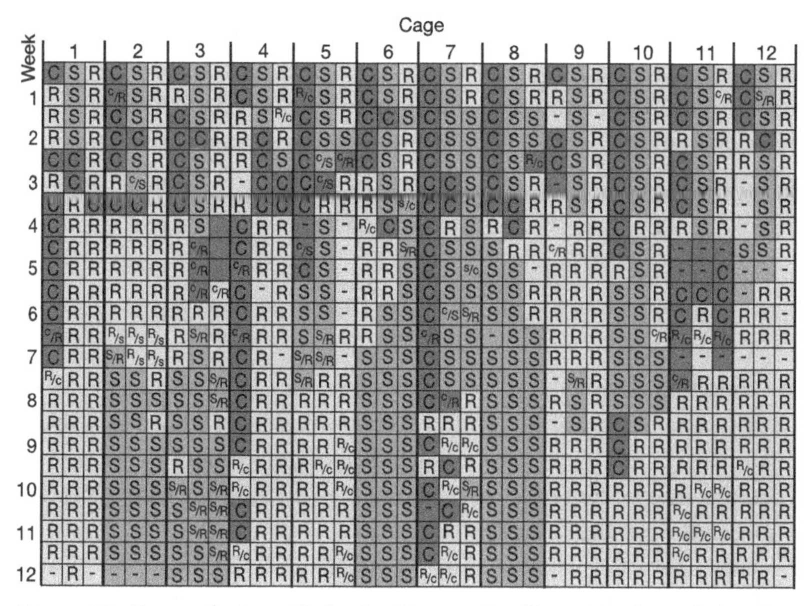

Figure 6.9 Co-caged mice with the dominant strain of bacteria color coded evolve over time: colicinogenic (C), resistant to colicin (R) and sensitive (S) (from Kirkup and Riley 2004).

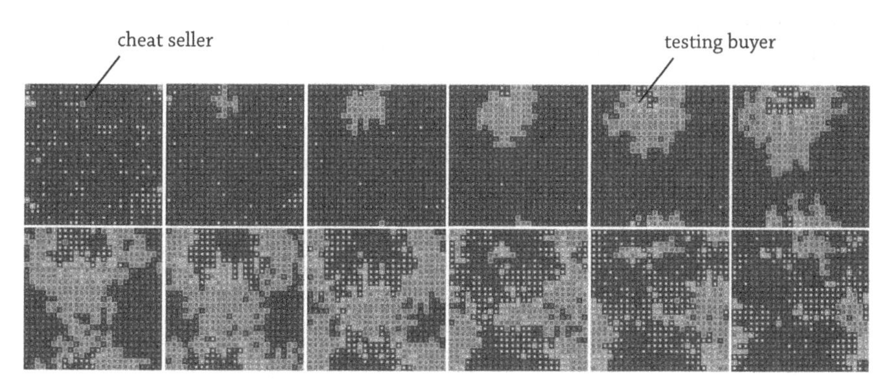

Figure 6.10 The Buyer-Seller game played out on a CA grid for periods 30 to 140 in steps of 10 periods.

Each cell of the two-population CA is coded by two colors; the outer color reflects the seller strategy and the inner color the buyer strategy. Thus, a color light grey (outer) and dark grey (outer) are *cheat* seller and *honest* seller, respectively, while medium grey (inner) and white (inner) are *lazy* buyers and *testing* buyers. Notice the single cell of *cheat* seller in the first panel of Figure 6.10 spreads rapidly into a population of *honest* sellers and *lazy* buyers, but eventually this rapid spread of cheat *sellers* (light grey) is chased by an expanding bubble of *testing* buyers (white), thereby re-establishing *honesty* among the sellers (dark grey). These cycles propagate constantly across the CA.

A more general solution of the Buyer-Seller game on the CA grid requires a slightly different standardized game of buyers and sellers defined by the following payoffs, where all diagonal elements are set to 1 while off diagonals can freely vary subject to the typical constraint on a discrete game (e.g., $a,b,c,d > 0$).

$$\mathbf{w} = \begin{pmatrix} 1 & b \\ a & 1 \end{pmatrix} \quad \text{and} \quad \mathbf{u} = \begin{pmatrix} 1 & d \\ c & 1 \end{pmatrix}$$

The conditions favorable to cyclical dynamics can be determined by varying the four parameters $a, b, c,$ and d to determine conditions where the cycles are extinguished. Extending the analysis slightly in Chapters 3 and 4, it is easy to see that one obtains clockwise cycles when $0 < a,b < 1 < c,d$, and counterclockwise cycles when the inequalities are reversed.

Consider now a CA with the usual Von Neumann neighborhood of 4 cells. Playing around with the simulation package, one soon sees that parameter values for sellers c and d must be more tightly coupled to support spatial cycles. For example, set $a,b = 0.9$ and $c = 10$ in the package, and vary d from 3.0 down to 2.6 in increments of 0.1. You will see that cycles are extinguished for $d < 2.7$.

Now change the neighborhood architecture to a Moore neighborhood of 8. The Buyer-Seller game now cycles at the d values that extinguished cycles in the Von Neumann architecture. Why? In a Von Neumann neighborhood of four, a cell at the leading edge of an advancing wave can spread to two or three new cells, while in a Moore neighborhood of eight, it can spread to three to seven new cells. The extra neighbors thus promote traveling waves.

We urge users to find other regularities of two-population CAs. As far as we know, there are no published articles on the topic, so the field is wide open for new analytical and empirical results.

EXERCISES

1. Using the [R] CA simulator for the SD game, vary the relative value b versus c and verify that the game changes from filaments to clusters of cooperators as the game passes from SD to PD.
2. Develop your own CA specification and try it out on the simulator, in either a 2×2, 3×3, or two-population game.
3. In the Buyer-Seller CA simulations described in the text, we set $a,b = 0.9$ and $c = 10$ in the package, and varied d from 3.0 down to 2.6 in increments of 0.1, then changed from a Von Neumann to Moore neighborhood and

repeated the simulations. Develop a simulation strategy to map out the entire space where c and d support cyclical behavior. You can automate the process in R and record simulation results, which are output to the screen console, or turn on your screen capture software to make movies. Compare the c and d space for cycles in Von Neumann and Moore neighborhoods. Then derive a general explanation of why the parameter supporting cycles space changes between the two neighborhood structures.

4. Repeat the simulation exercise described above for some of the other cyclical games in this chapter.

NOTES

Cellular automata were invented in the middle of the 20th century by some of the greatest mathematicians of the era, including John von Neumann, Stan Ulam, and Norbert Weiner, but first attracted widespread attention (especially among computer scientists) with the presentation of John H. Conway's "game of life" in *Scientific American* (Gardner 1970). In Conway's CA, cells are squares on a Von Neumann grid, and the current state is either vacant ("dead") or occupied ("alive"). A cell stays alive next period if either two or three of the four neighbors (cells with which it shares an edge) are alive this period, and a dead cell comes to life if it has three live neighbors. Otherwise, a cell dies or stays dead, from loneliness or overcrowding. Despite these very simple specifications of states, neighborhoods and transition rule, it turned out that many seemingly innocuous initial patterns give rise to remarkable spatial dynamics—the live clusters could die, grow, cycle, explode, or go chaotic.

About the same time, Nobel Laurate-to-be Thomas Schelling (1971) used simple CAs to illustrate racial segregation. In an early example of "emergent behavior," slightly biased individual state transitions sufficed to produce outcomes that resemble those of abhorrent top-down policies like apartheid.

Other neighborhood structures beyond those discussed in this chapter are occasionally seen in the literature. In CAs on the circle, some authors count two or more cells on each side as neighbors (i.e. those with network path length $\leq n$ for some $n > 1$). For $n = 2$, you can visualize the CA as a strip (or circular ribbon) chopped up into triangles, with the interior edges forming an edge-to-edge zig-zag. For some larger n you can visualize higher dimensional rings properly triangulated.

Wolfram (2002) argues that most scientific models can be reformulated as CA, and that they should play a leading role in physics, chemistry, geology, and other fields as well as in biology, social sciences, and (of course) computer science. Our own range of application is modest by comparison!

Nowak and May (1992) pioneered CA games, which are games where the cell's state is a strategy and the transition rule is based on payoffs arising from strategic interaction with neighbors. Later work (cited in sections 2 and 3) by Dobeli and coauthors sometimes refer to a pairwise approximation that yields ordinary differential equations.

The public goods games specifications assume rival consumption as with a heritage trail with limited capacity, and so the payoff is divided by the number of users. By contrast, for a pure public good such as a TV broadcast or national defense, consumption is nonrival and one does not divide by N.

A reviewer suggests a connection to ideal free settlement and isodar dispersion methods, and recommends Brown (1998) as well as later work extending these methods to continuously varying (as opposed to discrete) resource landscapes. Another reviewer recommends the survey Szabo and Fath (2007),

Readers also may want to check out Hauert's VirtualLabs software, available online.

BIBLIOGRAPHY

Brown, Joel S. "Game theory and habitat selection." *Game Theory and Animal Behavior* (1998): 188–220.

Friedman, Daniel, and Barry Sinervo. *Evolutionary Game Theory: Simulations*. UC Santa Cruz (2015). Software. doi:10.7291/d1mw2x.

Gardner, Martin. "Mathematical games: The fantastic combinations of John Conway's new solitaire game 'Life.'" *Scientific American* 223, no. 4 (1970): 120–123.

Hauert, Christoph, Joe Yuichiro Wakano, and Michael Doebeli. "Ecological public goods games: Cooperation and bifurcation." *Theoretical Population Biology* 73, no. 2 (2008): 257–263.

Hauert, Christoph, and Michael Doebeli. "Spatial structure often inhibits the evolution of cooperation in the snowdrift game." *Nature* 428, no. 6983 (2004): 643–646.

Hauert, Christoph, Miranda Holmes, and Michael Doebeli. "Evolutionary games and population dynamics: Maintenance of cooperation in public goods games." *Proceedings of the Royal Society B: Biological Sciences* 273, no. 1600 (2006): 2565–2571.

Hauert, Christoph, Silvia De Monte, Josef Hofbauer, and Karl Sigmund. "Replicator dynamics for optional public good games." *Journal of Theoretical Biology* 218, no. 2 (2002): 187–194.

Hauert, Christoph, Silvia De Monte, Josef Hofbauer, and Karl Sigmund. "Volunteering as red queen mechanism for cooperation in public goods games." *Science* 296, no. 5570 (2002): 1129–1132.

Kerr, Benjamin, Margaret A. Riley, Marcus W. Feldman, and Brendan JM Bohannan. "Local dispersal promotes biodiversity in a real-life game of rock-paper-scissors." *Nature* 418, no. 6894 (2002): 171–174.

Kirkup, Benjamin C., and Margaret A. Riley. "Antibiotic-mediated antagonism leads to a bacterial game of rock-paper-scissors in vivo." *Nature* 428, no. 6981 (2004): 412–414.

Nowak, Martin A., and Robert M. May. "Evolutionary games and spatial chaos." *Nature* 359, no. 6398 (1992): 826–829.
Schelling, Thomas C. "Dynamic models of segregation." *Journal of Mathematical Sociology* 1, no. 2 (1971): 143–186.
Szabó, György, and Gabor Fath. "Evolutionary games on graphs." *Physics Reports* 446, no. 4 (2007): 97–216.
Wolfram, Stephen. *A new kind of science*. Vol. 5. Wolfram Media, 2002.

PART TWO

APPLICATIONS

7

Rock-Paper-Scissors Everywhere

Rock-Paper-Scissors (RPS) is an ancient game. Humans played it at least since the the Han Dynasty 2000 years ago, and today it is known as ShouShiLing in China, as JanKenPon in Japan, and RoShamBo in much of the world. There are now international tournaments for computerized agents to play RPS, as well as tournaments for humans (see Fisher 2008).

RPS goes back much further in the natural world. A variety of bacteria, plants, and animals interact strategically via RPS games that govern mating behavior as well as predator-prey relationships.

Among all possible strategic interactions, RPS games are distinctive for their intransitive dominance relations: strategy R (rock) beats strategy S (scissors), which beats strategy P (paper), which beats strategy R (rock). It thus seems plausible that evolutionary dynamics would be cyclic in such games. But do we actually see cycles? If so, when are they damped, spiraling into a stable interior (mixed) equilibrium, and when do they increase in amplitude? And what are the economic and biological consequences?

This chapter is a bridge between the foundational material developed in Part I and the more substantial applications in Part II. As such, it builds on the theoretical distinction introduced in Chapter 2 between negative and positive returns to scale (to use economics jargon), or (to use biological jargon) between negative frequency dependence (or apostatic selection) and positive frequency dependence (or antiapostatic selection). Chapter 3 suggested that RPS braids together negative and positive frequency dependence, and Section 7.1 takes that thought a little further. It shows that 3×3 payoff matrices include four different types of RPS games, and shows how moving the $\Delta = 0$ loci can morph one RPS type game into another, or into a non-RPS game.

Section 7.2 draws on those theoretical insights and examines human economic behavior. It notes the difficulties in observing RPS cycles in today's economy, and in early lab experiments. Then it shows how a recent experiment produced beautiful and persistent cycles, even where basic theory predicted they would die out.

Section 7.3 extends the analysis of mating games beyond those of the lizards considered in Chapters 2 and 3. It describes the set of unifying social strategies underlying many RPS games in the natural world: cooperation, aggression (or despotism), and deception.

The rest of the chapter reviews accumulated evidence concerning signaling and mimicry in prey and learning by their predators. It notes some partial models in the literature, and introduces a new co-evolutionary (two-population) model involving RPS-like dynamics.

7.1 SOME RPS THEORY

When does a 3×3 matrix $A = \begin{pmatrix} a_{11} & a_{12} & a_{13} \\ a_{21} & a_{22} & a_{23} \\ a_{31} & a_{32} & a_{33} \end{pmatrix}$ define an RPS game? We say that A is a *generalized RPS* game if it has an intransitive strict best response (BR) in pure strategies. To spell this out, we require that

1. $a_{21} > a_{11}, a_{31}$ (i.e., strategy 2 is the BR to strategy 1). The substantive requirement here is only that 1 is not a BR to itself; if 3 were the BR to 1, then just rewrite the matrix A with labels 2 and 3 switched.
2. $a_{32} > a_{22}, a_{12}$ (i.e., strategy 3 is the BR to strategy 2).
3. $a_{13} > a_{33}, a_{23}$ (i.e., strategy 1 is the BR to strategy 3).

In terms of the simplex triangle, this amounts to saying that in a generalized RPS game, each vertex is the unique best reply to some other vertex.

The definition leaves open for each strategy whether the second best response is that strategy itself or the remaining strategy (e.g., whether (i) $a_{21} > a_{11} > a_{31}$ or (ii) $a_{21} > a_{31} > a_{11}$). Geometrically speaking, the question at at each vertex k (where fitness w_i is simply the matrix coefficient a_{ik}) is whether (i) the common type is the second-best response, or (ii) the common type is its own worst response. We refer to case (ii) as apostatic (i.e., strong negative frequency dependence) because there both rare types do better than the common type.

In the RPS game that children play, case (i) always applies: Rock is a better response to itself than is scissors (though of course paper is the best response), while scissors is better against itself than is paper, and paper is better against itself than is rock. We will refer to this configuration, illustrated in Figure 7.1A, as true RPS. At the opposite extreme, case (ii) applies at each vertex, a configuration illustrated in Figure 7.1D that we will call apostatic RPS. Due to circular symmetry (RPS is the same game as PSR and SRP), there are only two mixed cases, illustrated in panels B and C of the figure, where (i) holds at either one or two of the three vertices.

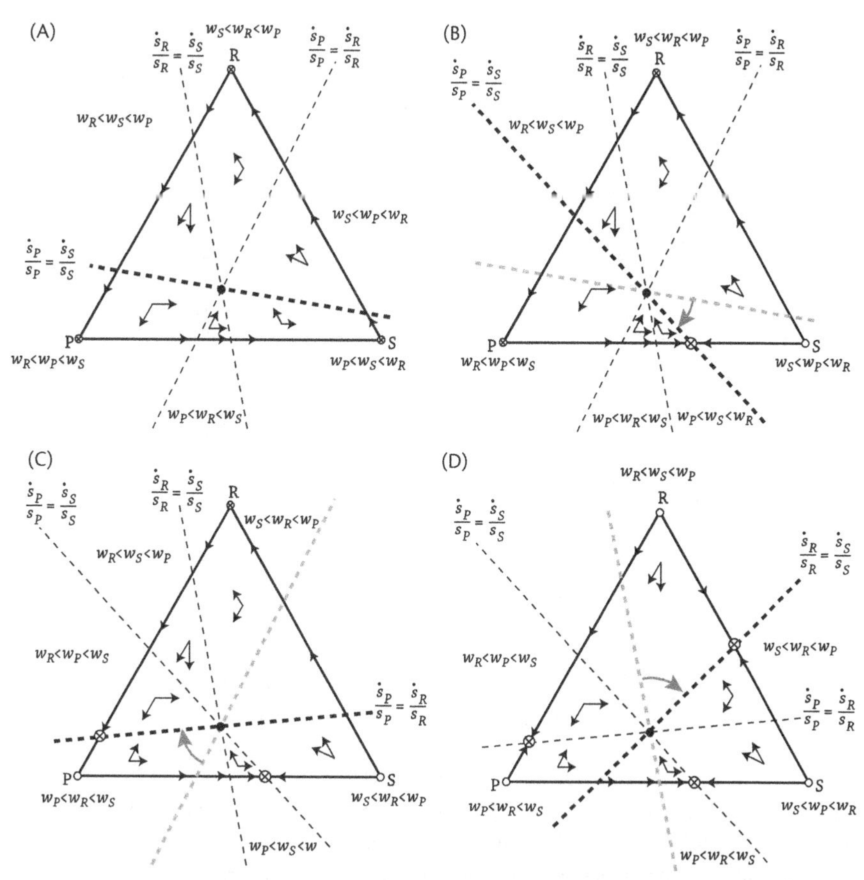

Figure 7.1 Four types of RPS games and their conversions. Beginning with a true RPS game (all three vertices of type i) in Panel A, the figure shows that rotating one of the equal payoff lines (heavy dashed grey line to heavy dashed black line) against the flow coverts the vertex it crosses from type i (true RPS) to type ii (apostatic). Thus Panel B shows a RPS game with one apostatic vertex, Panel C shows a game with two apostatic vertices, and Panel D shows a pure apostatic RPS game. As vertices are converted, new saddle points appear on the edges.

Changing the matrix entries a_{ij} shifts or rotates the equal payoff lines $0 = \Delta_{i-j} = w_i(s) - w_j(s)$. Suppose we have $a_{ji} > a_{ii} > a_{ki}$, where vertex i is case (i). Continuously change the matrix entries a_{ii}, a_{ki} until we have case (ii) where $a_{ji} > a_{ki} > a_{ii}$. Geometrically, this rotates the equal fitness line $0 = \Delta_{i-j}$ (in the direction counter to RPS flow) so that its intersection with an edge of the triangle moves across vertex i to another edge. Figure 7.1 illustrates how such rotations convert each vertex from case (i) to case (ii), thereby moving from true RPS in Panel A to pure apostatic-RPS in Panel D.

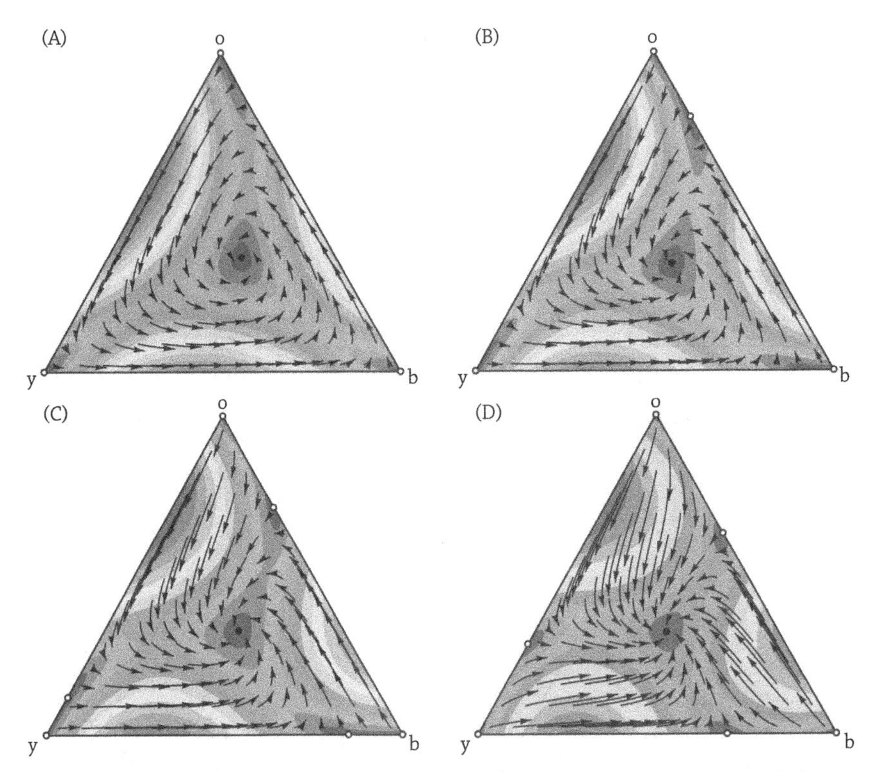

Figure 7.2 Continuous replicator dynamics for $\ln \mathbf{U}$. Modified fitness $\mathbf{U}$ is specified in equations (7.1)–(7.3); panels A-D respectively use parameter values $a = 1, 1.5, 2, 4$. Note that the edge equilibria and saddle paths characteristic of apostatic RPS appear in panels B–D. The lifetime fitness game for male lizards presented in Chapter 3 resembles Panel C.

Why should matrix entries change? As we saw in Chapter 5, a fixed basic fitness matrix can be modulated by taking the Hadamard product with a parametrized adjustment matrix. For side-blotched lizards, choosy females can modulate the male competition game. As explained in Chapter 14, Alonzo and Sinervo (2001) proposed a female preference for avoiding the common type of male, since that type is likely to do poorly in the next generation, and mating with the invading rare type. Here we propose a simpler version of the preference rule, avoid common type. The preference can be implemented either by active choice of copulation partners or by sperm sorting once fertilized by multiple genotypes (Calsbeek and Sinervo 2001). The corresponding adjustment matrix is

$$\mathbf{A} = \begin{pmatrix} 1 & a & a \\ a & 1 & a \\ a & a & 1 \end{pmatrix} \tag{7.1}$$

with $1 \leq a$. Assume the basic male fitness matrix presented in Chapter 2,

$$\mathbf{W} = \begin{pmatrix} 1.00 & 0.62 & 2.23 \\ 2.55 & 1.00 & 0.65 \\ 0.91 & 2.19 & 1.00 \end{pmatrix}. \tag{7.2}$$

Then overall fitness is

$$\mathbf{U} = \mathbf{W} \circ \mathbf{A}. \tag{7.3}$$

Beginning with negligible female preference parameter $a = 1$, we have a true RPS overall fitness matrix $\mathbf{U} = \mathbf{W}$. When the preference parameter $a > 1/.91 \approx 1.1$, however, the R-vertex for $\mathbf{U} = \mathbf{W} \circ \mathbf{A}$ becomes apostatic, as do the remaining two vertices when $a > 1/.65 \approx 1.5$ and when $a > 1/.62 \approx 1.6$. Figure 7.1 illustrates the impact on continuous replicator dynamics for overall fitness $\ln \mathbf{U}$.

What if matrix parameter changes rotate an equal fitness line in the opposite direction? As illustrated in Figure 7.3, rotating with RPS flow (instead of counter to it) converts vertex i into a locally dominant strategy, $a_{ii} > a_{ji}, a_{ki}$, extinguishing the RPS cycle. In the figure, after rotation the strategy paper dominates both scissors and rock in the region containing the paper vertex. Under sign-preserving dynamics, a trajectory beginning at the point marked with an asterisk (*) in the

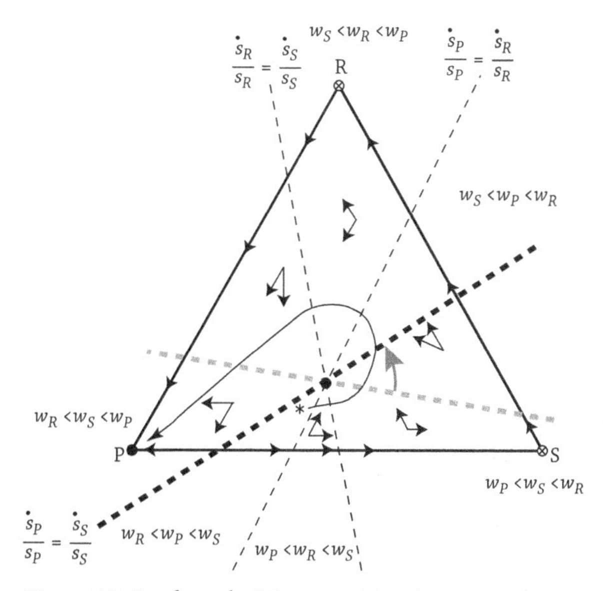

Figure 7.3 Breaking the RPS game. Rotating an equal payoff line in the direction of the flow (instead of counter to it) past the vertex of type (i) makes strategy i its own strict best reply, hence dominant in a neighborhood of that vertex. The new game is not generalized RPS.

figure makes one counterclockwise turn and then converges to the vertex P. The game is no longer even generalized RPS.

As an example, consider again female preference in lizards. Females who chose males with self-similar genotypes will produce progeny with highly integrated phenotypes (Lancaster et al. 2014). As explained in Chapter 14, there can be significant fitness advantages for such preferences. The upshot is an adjustment matrix with larger entries on the main diagonal, regardless of whether males are common or rare:

$$\mathbf{A} = \begin{pmatrix} b & 1 & 1 \\ 1 & b & 1 \\ 1 & 1 & b \end{pmatrix} \tag{7.4}$$

with $1 \leq b$. The overall fitness matrix $\mathbf{U} = \mathbf{W} \circ \mathbf{A}$ remains true RPS as the parameter b starts to increase from the baseline value $b = 1$. However, one vertex after another becomes locally dominant when b moves above 2.19, 2.23, and 2.55. Figure 7.4 uses the same conventions (and the same package, Sandholm's Dynamo) as in the previous figure to show how endless cycles of increasing amplitude with $b = 1.5$ break up when $b = 3$.

Of course there are many sorts of 3×3 matrices besides the four types of generalized RPS. Bomze (1983, 1995) classifies all 3×3 payoff matrices, and shows that only a few others have an equilibrium point attractor in the interior of the simplex. They include Zeeman's game discussed in Chapter 4 and the autocatalytic game illustrated in Figure 7.4B, which we will see again in Chapter 10.

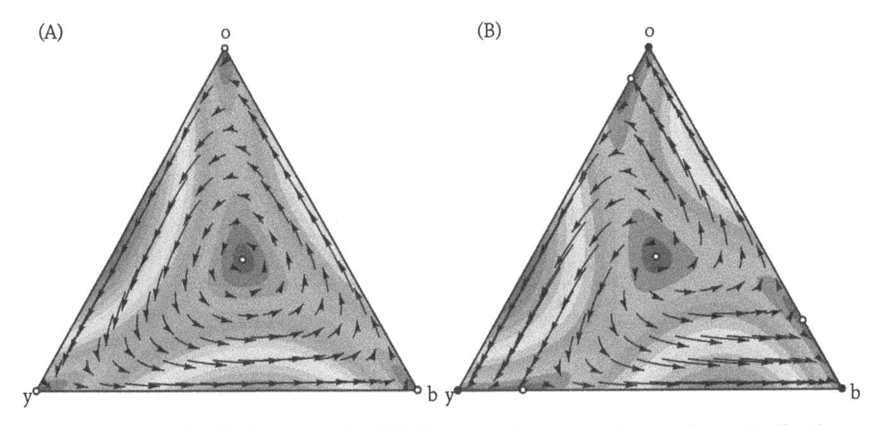

Figure 7.4 Modulated Uta game. Panel A shows continuous replicator dynamics for the true RPS matrix obtained by taking the logarithm of the entries in $[\mathbf{W} \circ \mathbf{A}]$, where the parameter in (7.4) is $b = 1.5$. Panel B is the same except that $b = 3$ and so the matrix is autocatalytic: it has three basins of attraction, one for each vertex, separated by the saddle paths leading to the three edge equilibria.

7.2 HUMANS PLAY RPS IN THE LAB

There is a long but controversial tradition of describing economic events in terms of cycles. Many 19th- and early 20th-century economists (including Clement Juglar, Karl Marx, Joseph Schumpeter, and Arthur Burns) believed that national economies followed semi-regular cycles of expansion, excess, recession, and recovery. Late 20th- and early 21st-century macroeconomists still refer to their main subject as "business cycles," but most of them use equilibrium models that have little role for repetitive cycles beyond simple seasonal effects like stockpiling for Christmas sales. Likewise, the notion that financial markets cycle regularly between "bull" and "bear" fell out of favor among financial economists in the second half of the 20th century. The financial crisis beginning in 2008 seems to have revived interest in irregular financial market cycles in the spirit of Minsky (1982); see for example Friedman and Abraham (2009) or Akerlof and Shiller (2010).

At the micro level of particular goods, some economists (e.g., see Harlow 1960) argued that, due to production lags, many agricultural products have price cycles. For example, high pork prices might encourage farmers to raise lots of young hogs, but when the hogs all mature and are sold to meat packers they create a glut and depress prices. This decreases the number of young hogs raised, eventually leading to shortage and higher prices, which initiates the next cycle. For manufactured goods, Edgeworth (1925) predicted a price cycle in which sellers try to increase market share when there is excess capacity by undercutting each other's price slightly. At a low enough price, however, demand may exceed production capacity, and then it is in someone's interest to jump to a much higher price and capture the unsatisfied demand, until price undercutting resumes.

A logical problem with simple models of price cycles is that speculators make tidy profits if they can buy when prices are low and sell when prices are high. Such speculation adds to demand and boosts prices when they are low, and adds to supply and depresses prices when they are high. Thus speculation should undermine a predictable cycle in storable goods, including financial assets. An empirical problem is that the field data tend to be ambiguous at best. A few studies seem supportive, for example, Lach (2002) finds some evidence for Edgeworth-style cycles in individual seller price data on coffee, flour, chicken, and refrigerators. But most other studies find no cycles.

Some economists have sought sharper evidence in the laboratory. As described in Chapter 8, a researcher can recruit human subjects, create a laboratory environment that strips away complications and ambiguities, and see how profit-motivated humans respond to the chosen environment. Beginning in 1967, several researchers created simplified hog cycle environments in the lab. Plott (2008) summarizes the generally non-cyclical results: human subjects quickly learn to speculate, and predictable price cycles typically do not persist. A slight exception is reported in Cason et al. (2005), who find some evidence for Edgeworth-style cycles in their lab study.

Surprisingly, it is only quite recently that experimenters have tried out the basic cycling rock-paper-scissors game in the lab. Hoffman et al. (2015) compare behavior with three different true RPS payoff matrices of the form $\begin{pmatrix} 0 & -1 & b \\ b & 0 & -1 \\ -1 & b & 0 \end{pmatrix}$, where the treatments are $b = 0.1, 1.0$, and 3.0. The reader can confirm that the unique NE $= (1,1,1)/3$ is an ESS (hence in theory dynamically stable) when $b=3$, but not in the other two treatments. Each of the 30 lab sessions had 8 human subjects matched simultaneously with all others for 100 periods. The authors report that NE predicts quite well the time-averaged observed state, and that the mean distance from NE is similar to that of binomial sampling error, except in the $b=0.1$ (unstable) treatment, when the mean distance is larger.

Most importantly for our purposes, Hoffman et al. (2015) report no evidence of cycles. A possible explanation is that higher order speculation is at work. For example, suppose that in period 53 the most prevalent strategy is rock. If players had noticed in the previous 52 periods that most players chose paper in periods following rock, then they would tend to choose scissors. But if scissors had tended to prevail following rock, then they would choose rock instead, and so on. Thus speculation would tend to undermine any predictable pattern that appeared in this simultaneous choice/discrete time setup.

In a different lab study initiated about the same time, Cason et al. (2014) ran RPS games using another three true RPS matrices:
$S = \begin{pmatrix} 36 & 24 & 66 \\ 96 & 36 & 30 \\ 24 & 96 & 36 \end{pmatrix}$, $U_a = \begin{pmatrix} 60 & 0 & 66 \\ 72 & 60 & 30 \\ 0 & 72 & 60 \end{pmatrix}$, and $U_b = \begin{pmatrix} 60 & 72 & 30 \\ 0 & 60 & 66 \\ 72 & 0 & 60 \end{pmatrix}$. Readers can confirm (see chapter-end exercises) that the interior NE in these games is not at the centroid, but rather at $(1,1,2)/4$, and that it is dynamically stable for matrix S but is unstable for U_a and U_b. Trajectories are predicted to run counterclockwise for S and U_a, but run clockwise for U_b.

Probably the most important difference from previous lab work on cycles is that the games were played in continuous time, with asynchronous choices. Figure 7.5 shows an example of a player's decision screen. The upper-left corner shows the (practice) payoff matrix B, and subjects chose actions by clicking on locations on the "heat map" simplex in the lower left. Clicking a vertex gave a pure strategy, while clicking an interior point gave the corresponding mixture. In one treatment called CI, the player's mixture changed instantly (actually, within 50 milliseconds) and in another treatment called CS, the mixture changed steadily, moving from the previous mixture to the target (clicked) mixture over a couple of seconds.

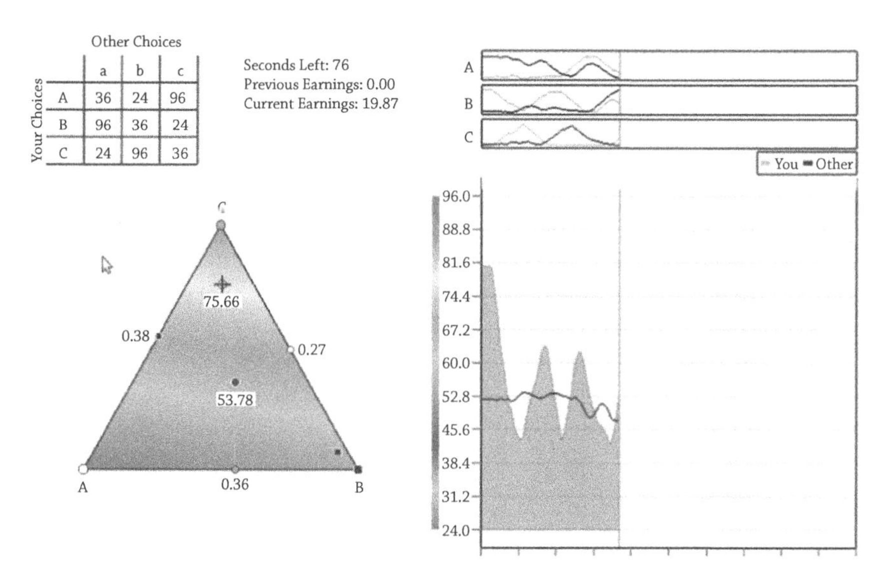

Figure 7.5 ConG Software: CS treatment (10 sec transit)

The thermometer to the right of the simplex shows how heat map colors correspond to current payoff flow rates, given the current average mixture $x(t)$ chosen by other players. This visual display of the payoff drastically reduces the player's cognitive load, since otherwise she would continually have to approximate, for each mixture y she might choose, the inner product $yB \cdot x(t)$ that gives the payoff flow.

The upper-right side of the screen presents in real time the dynamic time path of strategies selected by the player and the population average. The lower-right panel shows the payoff flow received by the player and the population average; the gray area represents the player's accumulated earnings so far in the current period.

Over time, $x(t)$ changes as players adjust their strategy choices, and the heat map morphs accordingly, like an animated weather map. Players (most of them undergraduates in economics, biology, and engineering) treated the experiment as a nicely paid multiplayer video game; typically they went home with more than \$20 after less than 2 hours in the lab.

So what happened? Figures 7.6 and 7.7 display the evolving population mixtures during some sample periods in the CS (continuous slow adjustment) treatment. They show a projection of the mixture triangle into the (x,y) plane, so mixture probability for the third strategy ("Scissors") is suppressed. The vertical axis represents time remaining in the 180-second period, so moving downward in the figure corresponds to moving forward in time. The NE appears as a vertical red line at $(x,y) = (.25,.25)$. Figure 7.6 shows 10 counterclockwise cycles around the

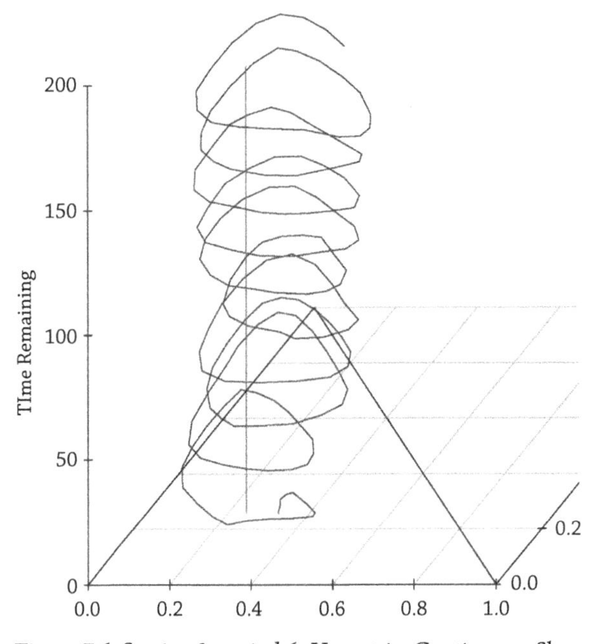

Figure 7.6 Session 2, period 6: U_a matrix, Continuous-Slow.

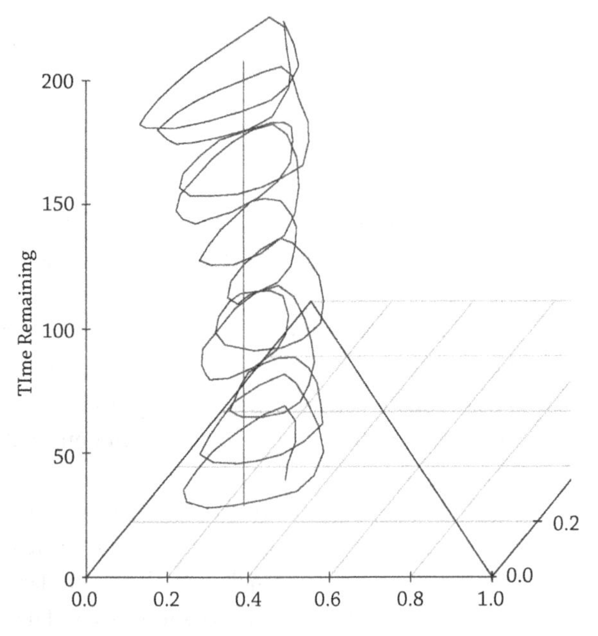

Figure 7.7 Session 11, period 20: U_b matrix, Continuous-Slow.

NE for a sample period using the unstable matrix U_a. The first few cycles (at the top) seem centered on the centroid $(x,y) = (0.33, 0.33)$ but last few cycles center closer to the NE. The amplitude falls only slightly by the end of the period. Figure 7.7 shows 11 cycles for the reverse unstable matrix U_b. They are similar to those for U_a, with one major exception: as predicted, the cycles are clockwise.

Cason et al. (2014) document that these figures are fairly typical for the CS treatment with the U matrices, and the paper argues that those cycles are consistent with evolutionary game models that assume some inertia. No inertia is imposed in the CI treatment, and the paper reports considerably faster and more erratic cycles under CI.

Classic game theory predicts that, on average, play will approximate NE. Indeed, time averages over the three-minute continuous time periods (and over 20 subperiods in discrete time treatments) fairly closely approximated NE for sessions using the stable (S) matrix. However, for the U matrices, evolutionary game theory provides an alternative prediction of central tendency called TASP (for Time Average over the Shapley Polygon; see the chapter-end exercises for the definition and computations), and it consistently outperformed NE. Other results from this experiment will be discussed in Chapter 8.

The paper concludes that the results "...are quite supportive of evolutionary game theory. EGT offers short-run predictions, where classic game theory has little to say, and those predictions for the most part explained our data quite well. EGT's long-run predictions either agreed with those of classic game theory or else were more accurate in explaining our data."

7.3 RPS MATING SYSTEMS

In nature, many mating systems have converged on the RPS game, and here we summarize examples from marine crustaceans (isopods), fish, birds, and damselflies. Many of these systems have alternative strategies of aggression, cooperation, and deception, while some combine aggression and cooperation into a single strategy that competes with two different deceptive strategies. Deception includes two basic forms: mimicry, in which the morph seems very much like something else (e.g., a male morph looks like a female or vice versa), and crypsis, in which the morph is very difficult to detect (e.g., due to protective coloration or small size or hiding in the weeds). Data for three of the mating systems (for isopods, sunfish and damselflies) are comparable to the lizard data presented in Chapters 2 and 3, and we use Dynamo (Sandholm et al. 2012) to compute haploid dynamics (even though two of the systems have detailed genetic data on diploid inheritance: isopods and damselflies). For the other systems the payoff data here are far less detailed.

The isopod species *Paracerceis sculpta* inhabits spongocoels—round cavities in marine sponges—and males have three morphs (Shuster and Wade 1989, 1992). The large α males pursue a despotic strategy but are vulnerable to the female mimic strategy β, which is defeated by the cryptic γ strategy. However, when γ's prevail, α's invade and guard harems of females who seem positively attracted to α males (or possibly the reverse, with αs particularly good at finding spongocoels with lots of females).

The fitness bar graph in Figure 7.8 indicates that the payoff matrix is a generalized RPS with two apostatic vertices (α and γ are type ii) and one type i vertex (β). The α vertex is apostatic because the despotic α-males wrestle with each other using their large posterior-facing horns, the winner ejecting the loser until only one α-male remains in a spongocoel containing many females. These conflicts reduce α fitness and leave lots of room for β- and γ-males. The γ vertex is apostatic due to a general property of cryptic strategies—they lose their effectiveness when they become too common, a classic form of negative frequency dependence.

Shuster has also shown that there is a fairly simple genetic basis for the three male morphs. The traits are controlled the Alternative Male Strategy (AMS) locus with three alternative alleles (α, β, γ). One β allele generates β phenotypes regardless of the other copy, while one γ allele generates γ phenotypes even if the other copy is α. The alleles are also found in females but have no visible effect on their phenotype. Analysis of published payoffs (Sinervo and Calsbeek 2006) with replicator dynamics using the Dynamo package (Sandholm et al. 2012) yields the stable interior equilibrium (0.024, 0.726, 0.250) as shown in Panel B of Figure 7.8.

Similar generalized RPS mating systems are often seen in fish. A parental (i.e., care-giving) but despotic strategy is vulnerable to invasion by sneaker males, but sneakers suffer from negative frequency dependence as parental males become

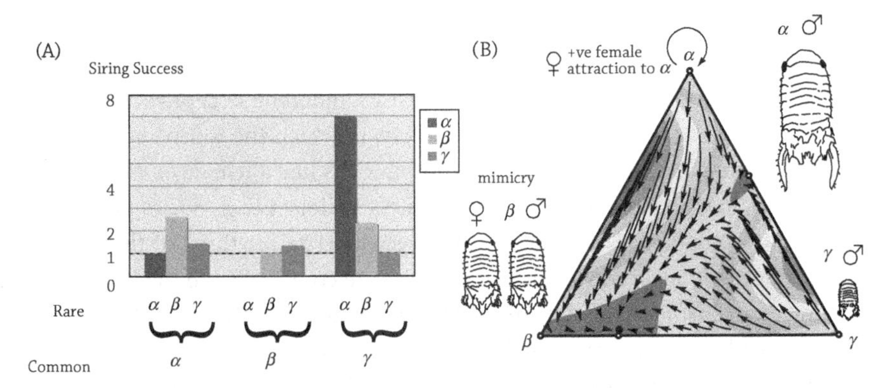

Figure 7.8 (A) Payoffs for male isopod strategies, and (B) evolutionary dynamics.

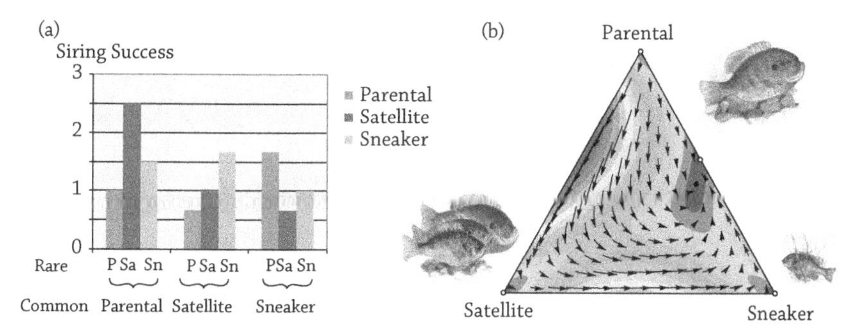

Figure 7.9 (a) Payoffs from siring success as estimated for parent, satellite, and sneaker males of the blue gilled sunfish, and (b) evolutionary dynamics. Illustrations courtesy of Krista Anandakuttan.

more aware, and females may show preference for the male morphs that care for their young.

Figure 7.9 illustrates the game for Bluegill sunfish. The males come in three different size morphs (Neff et al. 2003). A large territorial parental male actively courts females in his territory, and then defends a nest into which the female deposits eggs. While these males are aggressive, they are also cooperative with neighboring parental males in jointly defending against egg predators (other fish species). A medium-sized satellite male mimics females and gains fertilization success by interrupting a territorial male while he courts a female. Often this interruption results in the satellite male squirting his sperm onto the eggs, and mixing it with the sperm of the territorial male. Finally, sneaker males are very small and agile. When the female begins to lay the eggs, the sneaker quickly squirts an ejaculate of sperm in a strafing run, trying to make it between the mating parental male and female. As with the isopods, we have two cryptic and one despotic male strategy, although the latter here has a cooperative aspect. Our analysis of paternity data reported in Neff et al. (2003), using methods outlined in Chapter 3, is summarized in the bar graphs shown in Figure 7.9. We obtain the other mixed case of generalized RPS, with parental as the only apostatic vertex.

There are likely to be many examples of such RPS systems in fish where high payoffs to a despotic male type leave space for two different cryptic strategies. For example the swordtail, *Xiphophorus nigrensis*, exhibits a female mimic and a diminutive male sperm bomb (Figure 7.10, Panel A). The genetics of the game are known and arise from three alternative alleles at a diploid autosomal chromosome and, as with isopods, genetic dominance dictates the phenotypes of heterozygous gene combinations.

Recent theoretical analyses suggest that three types are not the only stable configuration. While the theory suggests stable games can be found in configurations of any prime number (Allesina and Levine 2011), three students

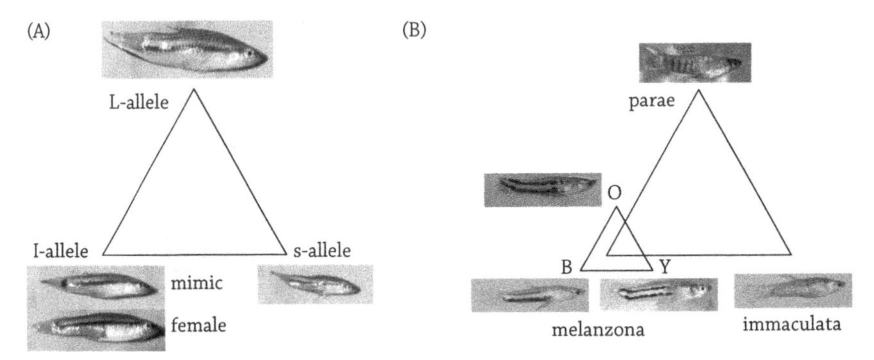

Figure 7.10 Two other fish species, *Xiphophorus nigrensis* with three morphs that might exhibit RPS dynamics (A) and *Poecilia parae* (B) with five morphs that might also exhibit higher dimension RPS dynamics colloquially referred to as rock-paper-scissors-lizard-spock.

(Carla Sette, James Lamb and Dustin Polgar) investigated this proposition in a 2015 class project. Using simulations, they found that stable (e.g., with interior stable equilibrium) three-strategy games were quite common, stable four-strategy games were slightly more rare, and stable five-strategy games were extremely rare, contrary to published theory. Of the 10,000 three-strategy games with randomly generated payoffs (with payoff entries uniformly drawn in the interval (−1,1), 707 (7.07%) had internal equilibria. Of the 10,000 four-strategy games, 110 (1.10%) had internal equilibria. Of those stable games, 114 of a possible 440 (25.9%) three-strategy edge games had internal equilibria. Only 11 (0.11%) of the five-strategy games had internal equilibria. Of those stable games, 9 of the 55 (16.4%) four-strategy edge games and 34 of the possible 110 (30.9%) three-strategy edge games had stable internal equilibria. In Figure 7.10, Panel B depicts the only five-morph male mating game of which we are aware. In *Poecilia parae*, a fish related to common guppies, one finds variation on the size of the males arising from a locus on the Y-chromosome. However, in one of the size types, referred to as melanzona, three color variants in orange, blue, and yellow can be found on the Y-chromosome (Lindholm et al. 2004).

There is a 5 strategy intransitive game called RoPaScLiSp (lizard and spock are the names of the strategies beyond rock, paper, and scissors) in which each strategy beats two others but loses to the remaining two strategies. The 5×5 payoff matrix is shown in Figure 7.11. It is difficult to visualize the 4-dimensional simplex for this game, but it seems intuitive that it has cycles. Some readers may be able to visualize the five 3-d faces corresponding to zeroing out one of the five strategies, but everyone can visualize the 2-d faces obtained by zeroing out two of the strategies. There are $\binom{5}{2} = 10$ such 2-d faces, which are all shown in Figure 7.11. In five cases, one strategy dominates the other two, while in the other five cases, we have a true RPS face game with closed loops.

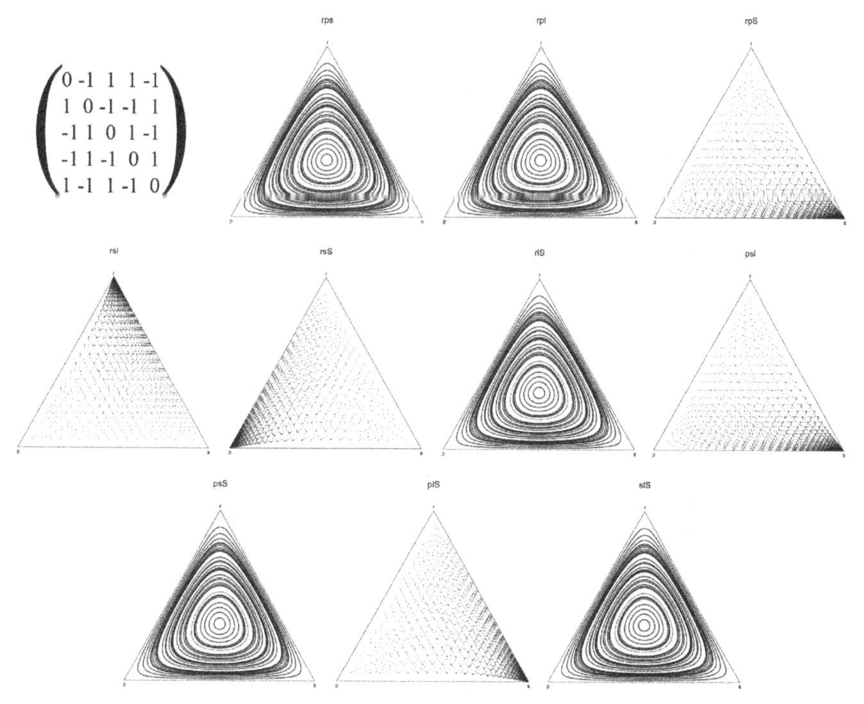

Figure 7.11 Replicator dynamics for the 10 two-dimensional (3×3) face games for the rock-paper-scissors-lizard-spock game with payoffs given in the upper left.

The student project found that none of the 4-d face games has a stable interior equilibrium, and conjectured about the implications for long run evolution. If RoPaScLiSp evolves from RPS, which seems plausible, there should be some intermediary 4×4 game with a stable interior equilibrium. The idea is that RPS might evolve into a game like RPSD (recall from Chapter 3 that this four-strategy game does have an interior equilibrium and that it reduces to RPS on one face), and then later RPSD might evolve into RoPaScLiSp or some other similar 5×5 game with a stable interior rest point. Cycles in the RoPaScLiSp game exhibit apparent chaos; a formal test can be found in (Sprott 2003). Chapter 14 will have more to say about long run evolution.

The next two examples are birds that seem to have RPS mating systems, but one of the morphs in each example is so rare that the game is played out very close to an edge of the simplex. The ruff, *Philomacus pugnax*, is a lekking shorebird. Some males defend territories each of about two square meters. That is barely enough space to do a little dance to attract females, but one morph with dark plumage, called Resident, defends its tiny piece of turf with vigor. A light colored morph, called Satellite, does not defend territory but lurks close to Residents and offers "appeasement" displays, perhaps to avoid aggressive attacks (Lank et al. 1995). Satellites sometimes fly off to another lek, while Residents are said to remain on

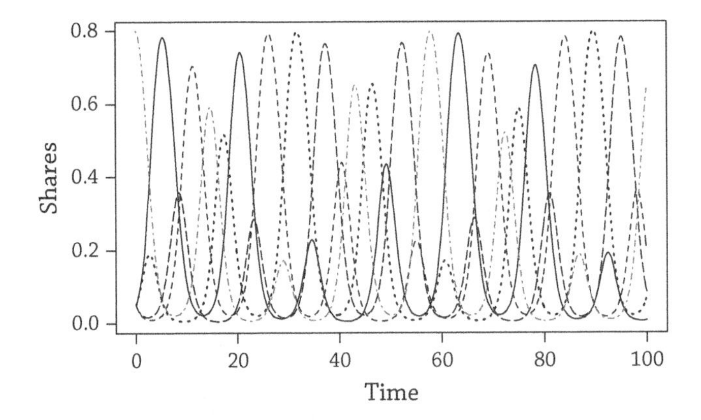

Figure 7.12 Cycles in the RoPaScLiSp game.

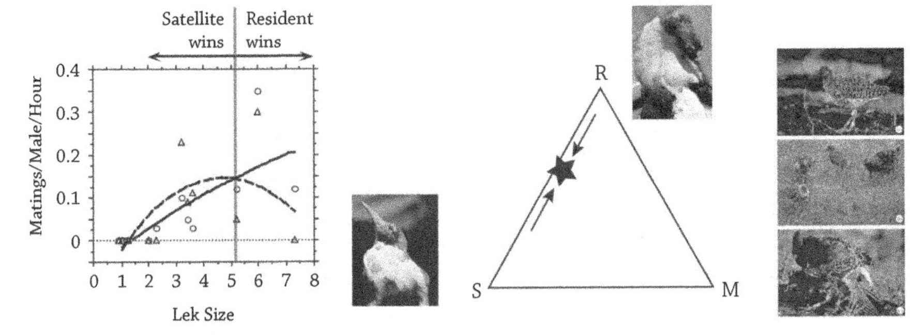

Figure 7.13 Pairwise payoffs for Resident (R) and Satellite (S) male morphs as a function of lek size in the Ruff, *Philomacus pugnax*. Because the female mimic (M) occurs at very low frequency, it is difficult to estimate its FD payoffs. Photos courtesy of D. Lank and from Jukema and Piersma 2006.

their chosen territory for the whole season. Both morphs use the puffy feathers on their neck, in the ruff-display, to attract females and intimidate (or appease) rivals. The third morph, called a Nape-neck, lacks the ruff (van Rhijn 1991) and has plumage similar to a female although is slightly larger (Figure 7.13). Thus Nape-necks are female mimics. The overall frequency of Nape-necked males is so low that it is difficult to estimate their fitness. Females observe all the males in the lek prior to choosing a copulation partner, and new evidence indicates that even Nape-necks are chosen at least occasionally (Jukema and Piersma 2006).

Since females prefer larger leks (Höglund et al. 1993), additional Satellites may draw more females to a lek, analogous to putting ads on a larger billboard. As shown in Figure 7.13, the data suggest that Resident fitness increases steadily in lek size, while Satellite fitness has decreasing marginal returns that eventually turn negative. Thus a Satellite male in a lek that has more than five males should leave and find a lek with fewer males, which seems consistent with observations.

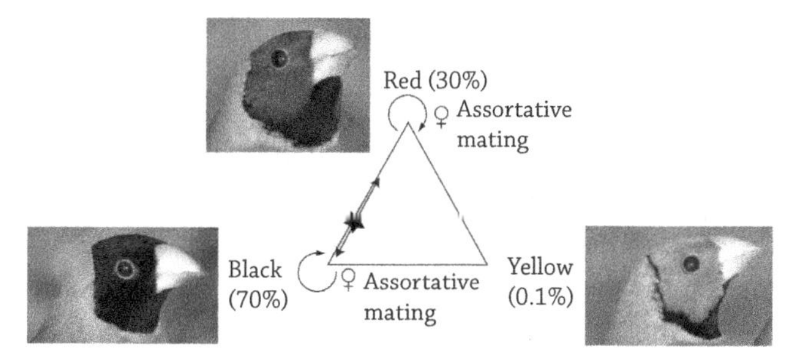

Figure 7.14 The Gouldian finch of Australia may reflect a more ancient RPS system that is now degenerate and in the process of speciating into two sets of monomorphic types (red and black) and ejecting the third type (yellow) from the game.

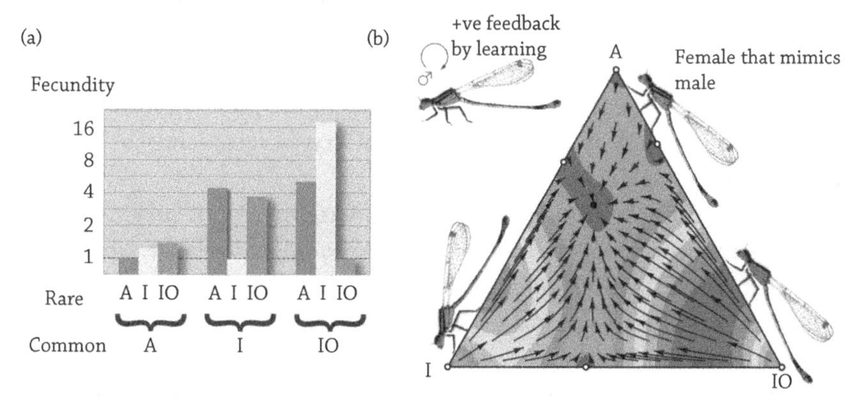

Figure 7.15 Payoffs for female damselflies (fecundity) when morphs are common or rare in the male's memory of search image for female types.

Several other bird species exhibit two male morph systems, often with a rare third morph. Often, one of the male morphs mimics juveniles, a strategy called juvenile plumage maturation; Greene et al. (2000) describe payoffs for a two-morph system in the Lazuli bunting, a passerine bird, where the aggressive males benefit from nearby juvenile plumage males. The Gouldian finch is an example of near-degenerate RPS, where red and black male morphs have substantial shares while yellow morphs are rare. As noted in Figure 7.14, this system appears to be headed towards speciation—red females prefer red males and black females prefer black males, and there is even evidence of infertility in cross-morph mating (Pryke and Griffith 2009). This sort of speciation will be discussed in Chapter 14.

Generalized RPS games also appear in female strategies, notably for the damselflies of Europe and North America. An example of a purely apostatic RPS game, documented by Svensson et al. (2005), is played by females of the damselfly species *Ischnura elegans*. In Figure 7.15 we have re-expressed the data as a payoff matrix, where the apostasis is thought to arise via male learning. Males eventually learn to spot the more common female morph, presumably via a process similar to that discussed in the next section for predators. Rare cryptic female morphs, which are harassed the least because few males have learned to recognize them as female, gain the highest fecundity, while females of the most common type get harassed by males to such an extent that their energy reserves are depleted, reducing their fecundity.

7.4 PREDATORS LEARN

As suggested by some of the previous examples, three-strategy RPS interactions can be part of a larger game. The next section will focus on predator-prey systems where the prey have three alternative strategies and the predators can do better if they learn more about the prey. As preparation, the present section reviews evidence on how predators learn (a) to spot cryptic prey, and (b) to avoid conspicuous but toxic prey.

How do predators learn to exploit common forms of cryptic prey? And what are the consequences for prey evolution? These questions motivate experiments reported by Bond and Kamil (1998, 2002). They put the western blue jay, *Cyanocitata cristatta,* in a virtual-reality environment, where the birds can learn to recognize alternative forms of cryptic prey. As shown on the left side of Figure 7.16, a variety of abstract forms ("digital moths") are presented on computer screens against a camouflaging background. The bird receives a positive reward (a meal worm) if it pecks on the screen image of cryptic prey but it receives a negative reward (a time out before seeing the next image and potential mealworm) if it pecks an area of the screen where there is no cryptic prey. The algorithm for the screen display continually generates new variants of virtual prey, and decreases the prevalence of virtual prey types that the bird pecks. Panel B of the figure confirms that crypsis is apostatic: the birds learned to target common types of cryptic virtual prey, greatly lowering fitness for those types. New cryptic variations will continue to appear, but none becomes common for very long in the presence of predators that learn to spot them.

The evolution of aposematic signals—conspicuous signals for toxic (or other defended) prey—is more difficult to understand than the evolution of crypsis. Conspicuous signals would seem to *attract* a predator, and by themselves offer no deterrence to predation. Predators will investigate conspicuous signals, sample the prey, and in so doing kill it or injure it. The sampled individual pays a high cost while the benefits (of teaching the predator) are realized mainly by other members of the species. Thus aposematism involves altruism.

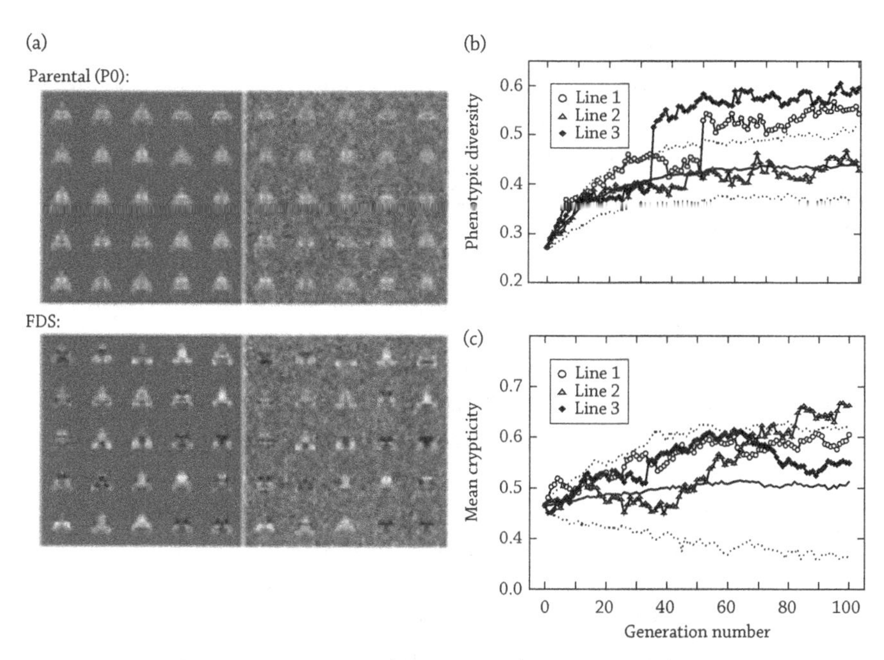

Figure 7.16 A. Samples of virtual prey (digital moths) for blue jays shown on uniform gray (left) and cryptically textured (right) backgrounds. P0 is the initial screen and FDS is screen after 100 generations of frequency-dependent selection. Random mutations in prey coloration that are conspicuous are pecked at and thus immediately de-selected, while the more cryptic mutations are left to "breed" and become more common. Changes in (b) phenotypic variance and (c) mean crypticity across successive generations in three experimental lines (plotted in heavy black lines and symbols, Lines 1–3), contrasted with distribution of values from two sets of control lines (plotted in black). Source: Kamil (1998).

R.A. Fisher (1930) first realized that evolution would favor this form of altruism if kin were clustered, for example, by females that lay clumps of eggs, as is the case for Monarch butterflies and their conspicuous larvae that feed on toxic milkweed. Clusters of conspicuous progeny generate a form of positive frequency dependent selection: nearby kin benefit from the death of an individual when it generates a deterrent image of "toxic prey" in the predator's memory (Endler and Mappes 2004). The cost of such altruism is greatly reduced by rapid "one-trial" learning that toxic food evokes in many animals (including humans: after a bout of food poisoning at a restaurant, you are unlikely to try that restaurant or that food item again any time soon!). The chapter-end exercises include construction of a simple model of one-trial learning.

Even once established, an aposematic system is susceptible to invasion by "cheaters" who display the conspicuous signal but spare themselves the cost of actually becoming toxic. For example, Viceroy butterflies look very much like Monarchs, but Viceroys eat whatever they want, not just toxic milkweed. Such cheaters are referred to as Batesian mimics after the Victorian-era English naturalist Henry Bates, who first described the phenomena.

Batesian mimicry (B) is the prey's best response to morph A (aposematic, i.e., conspicuously toxic) in the presence of predators that learn quickly, while the best response to B when it becomes common and targeted by predators evidently is crypsis (C). As noted earlier, crypsis becomes ineffective when common, and the best response is A, especially when clumped. Thus we should have a form of RPS among prey morphs.

Does this conjecture describe actual predator/prey relations? Early experiments (Mappes and Alatalo 1997) confirmed Fisher's idea that gregarious aposematic prey gain an advantage through single-trial learning of predators, in which clustered and obvious but noxious forms have a survival advantage over dispersed noxious forms. An obstacle to more direct tests is a widespread innate predator aversion to certain colors, which in avian systems are usually yellow or red (Brodie III and Janzen 1995). If these colors are used to signal toxicity, this innate aversion is confounded with the emergence of true aposematism.

To avoid the confound, Lindström et al. (2001, 2004) create novel worlds where the signal is carried by abstract geometrical shapes rather than colors, as in Figure 7.17. A great tit (*Parus major*) inspects the floor of a novel world aviary during learning trials. Rare or common geometrical objects are attached to rewards (undefended prey) or to a combination of rewards + noxious taste (defended prey).

These experimental data verify that birds generate a generalized rock-paper-scissors game among aposematic-Batesian-cryptic prey types (7.18), where the aposematic vertex is true RPS while the Batesian and cryptic vertices are both apostatic.

7.5 A CO-EVOLUTIONARY MODEL OF PREDATORS AND PREY

Although various aspects of signaling and mimicry can be found in the literature (e.g., Kokko et al. 2003, Endler and Mappes 2004, Mappes et al. 2005), we have not seen an explicitly co-evolutionary model of aposematism. Here we put one together, crossing ideas from Section 7.1 with ideas and data just presented.

Three alternative strategies are available to prey:

A = aposematic—develop a toxin or other adaptation that makes predation undesirable, along with a conspicuous signal of undesirability, as with

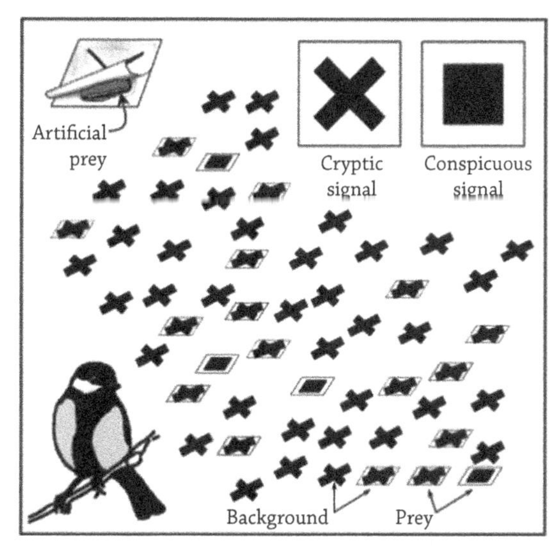

Figure 7.17 Novel world experiment, where a bird is challenged with paper prey filled with rewards (batesian, cryptic) or a bitter taste (aposematic), presented on a background that makes the prey stand out (aposematic, batesian), or be cryptic. Reprinted with permission of M. Joron.

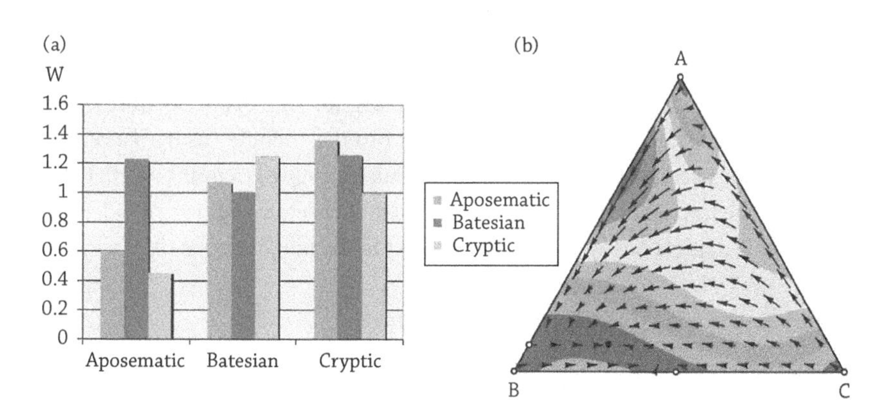

Figure 7.18 Payoffs to prey morphs. Beginning with the data in Lindström et al. (2001), we computed payoffs when rare and common. We also subtracted an imputed cost $c = 0.4$ to the production of the toxin.

Monarch butterflies that concentrate milkweed toxin and display Halloween colors; and

B = Batesian mimic—save the cost of carrying the toxin but still display the conspicuous signal of undesirability, as with Viceroy butterflies.

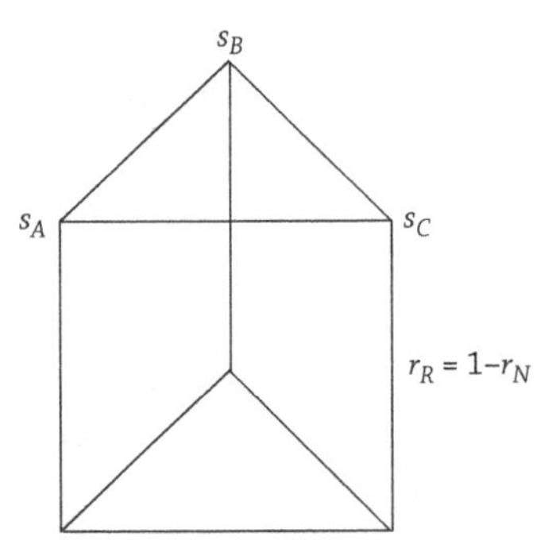

Figure 7.19 State space for co-evolutionary model with trimorphic prey and bimorphic predators.

C = cryptic—use camouflage or other adaptations to hide from predators; examples of butterflies with cryptic wing patterns abound.

The predator population has two alternative strategies:

N = naive—ignore conspicuous signal, and cope with toxic prey after the fact, such as by spitting out Monarchs after tasting; and
R = responsive to signal—avoid prey with the conspicuous signal.

The state space is then the prism shown in Figure 7.19. As in Section 7.1, the prey population state $s = (s_A, s_B, s_C)$ is a point in the triangle (using barycentric coordinates as usual), while the predator population state is a point $r \in [0,1]$ that represents the fraction of predators responsive to the signal (i.e., $r = r_R = 1 - r_N$). The prism is depicted as standing on end, so r is the height above the naive base triangle.

Consider the following prey fitness functions.

$$w_A(r,s) = b - k_A - a_A(1-r), \tag{7.5}$$

$$w_B(r,s) = b - k_B - a_A(1-r) - a_B r, \tag{7.6}$$

$$w_C(r,s) = b - a_C r - \alpha_C s_C, \tag{7.7}$$

where b is base fitness, and k_i is the fitness cost of strategy i relative to crypsis, so $k_A > 0 = k_C$ while $k_B < k_A$ and may have either sign. That is, the A strategy is the most costly (e.g., Monarchs restrict diet to milkweed), the mimic strategy is cheaper, and the cryptic strategy may be more or less costly than the mimic strategy. The term $a_A(1-r)$ captures predation, which for A morphs occurs at some rate $a_A > 0$ per N-predator and rate 0, of course, for R-predators. To the extent that the mimicry is imperfect, the B morphs face some R-predation; hence the negative own-population effect $-a_B r$, where $a_B \geq 0$. Since R-predators target mainly C-prey, this specialization may improve their effectiveness; hence the term $-a_C r$. Finally, as we have seen, crypsis tends to be less effective when common, hence the final term $-\alpha_C s_C$. (Implicitly we assume that the conspicuous strategists, A and B, are as easy to target when rare as when common, an assumption we relax below in a more complicated model.)

The payoff difference functions will be handy for later analysis. They are:

$$\Delta w_{A-B} = a_B r - k_o, \tag{7.8}$$

$$\Delta w_{B-C} = ar + \alpha_C s_C - k_1, \tag{7.9}$$

$$\Delta w_{A-C} = (a + a_B) r + \alpha_C s_C - k_2 \tag{7.10}$$

where $k_o = k_A - k_B > 0$, $k_1 = k_B + a_A \geq 0, k_2 = k_A + a_A > 0$, and $a = a_A - a_B + a_C > 0$. The locus $\Delta w_{A-B} = 0$ is a plane parallel to the triangular base that cuts the prism at height $r^* = \frac{k_o}{a_B}$; for lower r (which includes the entire prism if $r^* > 1$), the prey strategy B is fitter than A. Likewise, C beats B for states (r, s) below the locus $\Delta w_{B-C} = 0$, which is a plane that cuts the $s_C = 1$ edge of the prism at height $r = \frac{k_1 - \alpha_C}{a}$ and cuts the $s_C = 0$ face of the prism at height $r = \frac{k_1}{a}$. Finally, A beats C for states (r, s) above the locus $\Delta w_{A-C} = 0$, which is a plane that cuts the $s_C = 1$ edge of the prism at height $r = \frac{k_2 - \alpha_C}{a + a_B}$ and cuts the $s_C = 0$ face of the prism at height $r = \frac{k_2}{a + a_B}$.

The expression for the equilibrium fraction of responsive predators is revealing, $r^* = \frac{k_o}{a_B} = \frac{k_A - k_B}{a_B}$. It shows the difficulty of Batesian as opposed to Mullerian mimicry in this model. The denominator is the incremental advantage of A over B in reduced predation and the numerator is incremental cost.

Now consider the predator fitness functions

$$u_R(r, s) = \tilde{b} + (d_C + \delta) s_C \tag{7.11}$$

$$u_N(r, s) = \tilde{b} + d_C s_C + d_B s_B - d_A s_A \tag{7.12}$$

with payoff difference

$$\Delta u_{R-N} = d_A s_A - d_B s_B + \delta s_C. \tag{7.13}$$

Here $\tilde{b}$ is base fitness for predators and the term $-d_A s_A$ reflects the toxicity (or analogous defenses) of the A type. The parameters $d_B, d_C > 0$ reflect the net nutritive value of finding and consuming B and C prey, and $\delta \geq 0$ is the predator fitness counterpart of a_C, reflecting any advantage of specialization in detecting cryptic prey. The locus $\Delta u_{R-N} = 0$ is a vertical plane cutting the prism, and R beats N for prey distributions with smaller shares of the B morph.

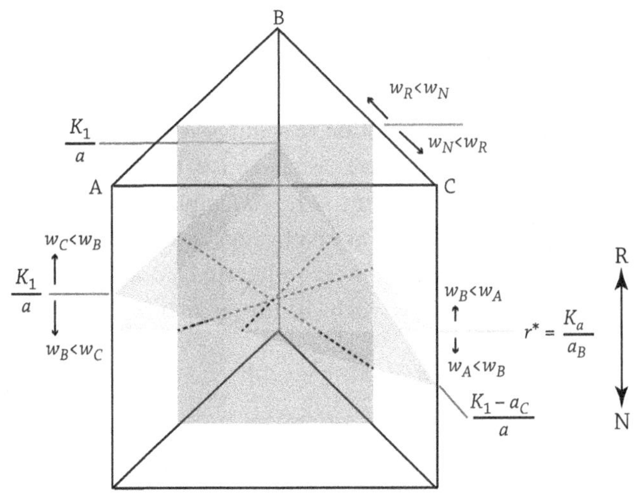

Figure 7.20 Sectoring the state space for coevolutionary model with trimorphic prey (A, B, C) and bimorphic predators (R, N). The plane for Δw_{A-B} is in light grey, Δw_{B-C} is in medium grey, and Δw_{R-N} is in darker grey and vertical, while Δw_{A-C} is not shown as it is redundant in the calculation of the interior NE.

Equilibrium is defined by 4 linear equations in 4 unknowns (r, s_A, s_B, s_C):

$$0 = \Delta u_{R-N} = d_A s_A - d_B s_B + \delta s_C \tag{7.14}$$

$$0 = \Delta w_{A-B} \Longrightarrow r = r^* \equiv \frac{k_o}{a_B} \tag{7.15}$$

$$0 = \Delta w_{B-C} = ar + \alpha_C s_C - k_1 \Longrightarrow s_C = \frac{k_1 - ar}{\alpha_C} \tag{7.16}$$

$$s_A + s_B + s_C = 1. \tag{7.17}$$

From the second and third of these four equations we get the equilibrium value

$$s_C^* = \frac{k_1 - ar^*}{\alpha_C}. \tag{7.18}$$

Plugging this and the simplex identity into the first of these linear equations, we have at equilibrium

$$0 = d_A s_A - d_B[1 - s_A - s_C^*] + \delta s_C^*$$

whose solution is

$$s_A^* = \frac{d_B}{d_A + d_B} - \frac{\delta + d_B}{d_A + d_B} \frac{k_1 - ar^*}{\alpha_C}. \tag{7.19}$$

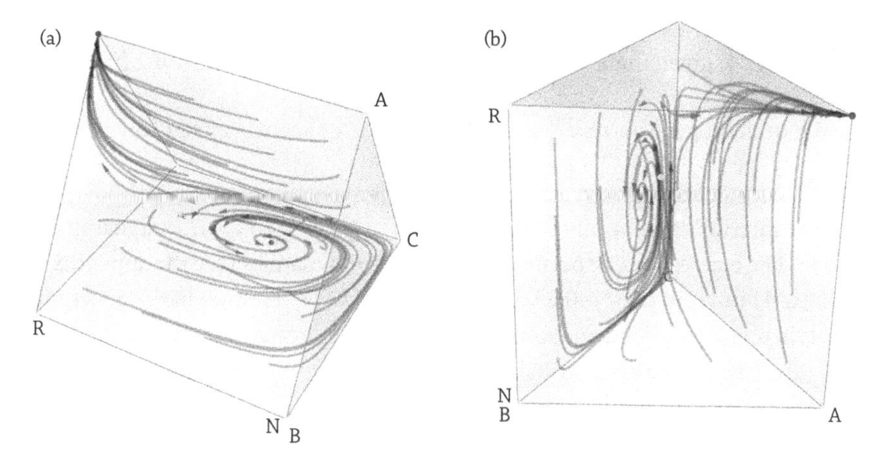

Figure 7.21 Dynamo output for the predator-prey model. Parameters are as in the text, with $\beta = 1$. Panel b shows the prism on end.

Hence

$$s_B^* = \frac{d_A}{d_A + d_B} + \frac{\delta - d_A}{d_A + d_B} \frac{k_1 - ar^*}{\alpha_C}. \tag{7.20}$$

For example, parameters $(k_A, k_B, a_A, a_B, a_C, \alpha_C, d_A, d_B, d_C, \delta) = (.5, 0, 1.5, 0.75, 1, 1, 3, 3, 3, 1)$ yield composite parameters $(k_0, k_1, k_2, a) = (0.5, 1.5, 2, 1.75)$ and interior equilibrium $r^* = 2/3$ and $(s_A^*, s_B^*, s_C^*) = (5, 7, 6)/18$.

We obtain replicator dynamics as follows. Let a new parameter $\beta > 1$ (or < 1) represent the extent to which predators learn faster (or slower) than prey adapt. Using general expressions developed in Chapter 3 and the Delta functions just derived, we have

$$\dot{r} = \beta r(1-r)\Delta u_{R-N} = \beta r(1-r)[d_A s_A - d_B s_B + \delta s_C] \tag{7.21}$$

$$\dot{s_A} = s_A[s_B \Delta w_{A-B} + s_C \Delta w_{A-C}] = s_A s_B(a_B r - k_0) + s_A s_C(ar + \alpha_C s_C - k_1)$$

$$\dot{s_B} = s_B[s_A \Delta w_{B-A} + s_C \Delta w_{B-C}] = -s_A s_B(a_B r - k_0) + s_B s_C(ar + \alpha_C S_C - k_1)$$

$$\dot{s_C} = s_C[s_A \Delta w_{C-A} + s_B \Delta w_{C-B}] = -s_A s_C(ar + \alpha s_C - k_1) - s_B s_C(ar + \alpha_C S_C - k_1).$$

Figure 7.21 shows a simulation of these equations using Dynamo (Sandholm et al. 2012). The example parameters just given (and $\beta = 1$) imply the own-population effect matrices $\begin{pmatrix} 0 & 0 \\ 0 & 0 \end{pmatrix}$ for predators, and $\begin{pmatrix} 0 & 0 & 0 \\ 0 & 0 & 0 \\ 0 & 0 & -1 \end{pmatrix}$ for prey, since (so far) we only have the negative impact of cryptic on cryptic $-\alpha_C = -1$. From equations (7.5)–(7.7), we see that the impact of predator state $(r, 1-r)$ on the prey state (s_A, s_B, s_C) is captured in the matrix $\begin{pmatrix} b - k_A & b - k_A - a_A \\ b - k_B - a_B & b - k_B - a_A \\ b - a_c & b \end{pmatrix}$

$=\begin{pmatrix} 5.5 & 4 \\ 5.25 & 4.5 \\ 5 & 6 \end{pmatrix}$, given the usual parameter values and arbitrarily setting $b=6$ and $\tilde{b}=5$. For the same parameter values, the reverse impact of prey state on predator state is $\begin{pmatrix} \tilde{b} & \tilde{b} & \tilde{b}+d_C+\delta \\ \tilde{b}-d_A & \tilde{b}+d_B & \tilde{b}+d_C \end{pmatrix} = \begin{pmatrix} 5 & 5 & 9 \\ 2 & 8 & 8 \end{pmatrix}$.

Is the interior equilibrium $r^*=2/3$ and $(s_A^*, s_B^*, s_C^*) = (5,7,6)/18$ dynamically stable? The eigenvalues associated with replicator dynamics (see the appendix to Chapter 4) turn out to be $-0.25 \pm 0.26\sqrt{-1}$ and $5/18$. Thus we have converging spirals in one plane, but deviations from that plane increase over time, as depicted in Figure 7.22. There are two stable equilibria, one on the prism face $s_A=0$ not far from the interior equilibrium, and another (with a larger basin of attraction) at the corner $s_A=1, r=1$ where the aposematic morph has completely taken over and 100% of predators respond to the signal.

Such complete signaling success is boring and unrealistic, so we will improve the model via own-population effects. Suppose, as seems reasonable, that Batesian mimics benefit from abundant aposematic models, captured by appending the term $\alpha_B s_A$ to the fitness expression for morph B. Conversely, when Batesian mimics become common, the aposematic morph benefits from subtle distinctions between their signals. Predators will have greater incentive to learn to tell them apart from the B's, especially when the A's are clustered (Fisher's effect). With clustering, the effect would be present even when the A's themselves are rare, so we could append the term $\alpha_A s_B$ to their fitness.

Rewriting the fitness equations to incorporate these effects, we obtain

$$w_A(r,s) = b - k_A - a_A(1-r) + \alpha_A s_B, \tag{7.22}$$

$$w_B(r,s) = b - k_B - a_A(1-r) - a_B r + \alpha_B s_A, \tag{7.23}$$

$$w_C(r,s) = b - a_C r - \alpha_C s_C. \tag{7.24}$$

Applying the same methods and same parameters described above, (augmented by $\alpha_A = \alpha_B = 1$) the only change is to the own-population effects matrix, which now is $\begin{pmatrix} 0 & 1 & 0 \\ 1 & 0 & 0 \\ 0 & 0 & -1 \end{pmatrix}$. There is still a unique interior equilibrium, which now is $\mathbf{s}^* = (2,3,3)/8$ for the prey and $r^* = 1/2$ for the predator. The relevant eigenvalues for replicator dynamics are approximately -0.39 and $-0.017 \pm 0.21\sqrt{-1}$. Thus we see that the interior equilibrium now is indeed dynamically stable, as shown in Figure 7.22.

As a sidelight reprising some of this chapter's themes, note that the prey own-population effect matrix just given, regarded as a separate game, would imply that the cryptic morph is dominated. Including the fitness impact of predators at

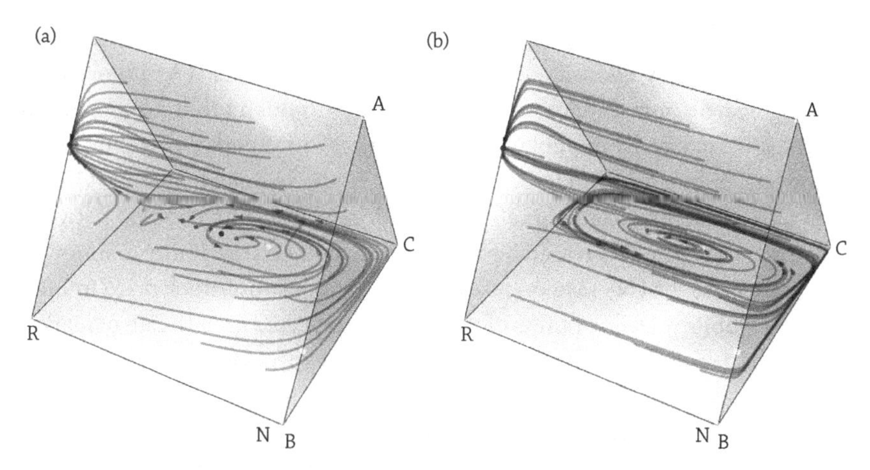

Figure 7.22 Dynamo output for the modified predator-prey model, with $\alpha_A = \alpha_B = 1$, reveals an internal rest point for $\beta = 1$ (a) and for $\beta = 5$ (b).

$r^* = 1/2$ yields the prey 3×3 payoff matrix $w = \begin{pmatrix} 4.75 & 5.75 & 4.75 \\ 5.875 & 4.875 & 4.875 \\ 5.5 & 5.5 & 4.5 \end{pmatrix}$. This is almost an apostatic RPS game in that each morph is its own worst reply, and B is the best reply to A while C is a better reply to B than is B itself, and A is a better reply to C than is C itself, which seems consistent with the experimental observations on birds (Figure 7.18). Unfortunately, however, C is still never a best reply. Of course, we should not hold constant $r = r^* = 1/2$, but rather should let it vary according to the given Δu_{R-N}. The interested reader may be able to find plausible parameter values that give a nice RPS-like projection of dynamics from the prism to the prey simplex, so that each morph is a best reply at some point (and worst reply at some other point) of the cycle.

It is also useful to trace the impact of differing learning speeds β, not just the neutral ($\beta = 1$) case covered in the simulations or the extreme cases ($\beta \to 0$ and $\beta \to \infty$) implicitly considered in the previous paragraph. Figure 7.22 shows the impact of increasing β from 1 to 5; trajectories are compressed towards the plane Δu_{R-N}. To capture $\beta = 5$ in *Dynamo* (Sandholm et al. 2012) for this model, one simply multiplies the matrix defining the impact of prey on predators by β obtain $\begin{pmatrix} 25 & 25 & 45 \\ 10 & 40 & 40 \end{pmatrix}$. If there were non-zero predator own-population effects, they also would have to be multiplied by β.

7.6 DISCUSSION

The model just described for aposematism has other potential applications. For example, consider financial markets with two populations of participants. On the industry side, a population of firms sells shares of stock and attempts to increase earnings by choosing among three broad strategies: (A) innovate by investing heavily in research and development, (B) pretend to innovate, but in fact invest more heavily in public relations and executive perks, or (C) don't try to innovate, but instead aim to be the low cost leader with high volume. On the other side of the financial market, a population of investors chooses between two broad strategies: (R) research firms to estimate their true ability to innovate (or to cut costs), or (N) save on research costs and just use stock price to infer future profitability. Define fitness of a firm as the expected growth rate of market capitalization, and define fitness of an investor as expected return on shares of stock.

Equations (7.5)–(7.11), although developed for a much different application, are not an unreasonable description of how such a financial market might behave over time. Readers familiar with financial markets are invited to find appropriate values of the parameters $(k_A, k_B, a_A, a_B, a_C, \alpha_C, d_A, d_B, d_C, \delta, \beta)$, and to consider whether other own-population effects have a significant role.

Throughout this chapter we have hinted that once an RPS cycle evolves, it will bring in other populations of players which might benefit from the predictability of cycles. In mating system games among males, females can benefit by evolving rules of active mate selection or of sperm sorting. The predator-prey game will select among predators for those that can better discriminate the Batesian mimic from aposematic model. We will develop some of these ideas more fully in Chapter 14, and readers may see applications in the social and virtual worlds. Chapter 8 will explore the contrast between slow genetic evolution and rapid learning processes that might be better described by best response dynamics than by replicator dynamics.

Another underlying theme in this chapter may have applications beyond biology. Cooperation, aggression, and deception are three intimately related social behaviors that have a role in many models that readers may wish to construct. We will see them again in Chapter 13, among other places.

APPENDIX

Here we find algebraic expressions for the interior equilibrium of the modified predator-prey game of Section 7.5. Recall that the payoff difference functions including the three own-population effects are:

$$\Delta w_{A-B} = 0 = a_B r - k_o + \alpha_A s_B - \alpha_B s_A, \tag{7.25}$$

$$\Delta w_{B-C} = 0 = ar + \alpha_C s_C - k_1 + \alpha_B s_A, \tag{7.26}$$

$$\Delta w_{A-C} = 0 = (a + a_B)r + \alpha_C s_C - k_2 + \alpha_A s_B. \tag{7.27}$$

The difference function for predators remains unchanged:

$$\Delta u_{R-N} = 0 = d_A s_A - d_B s_B + \delta s_C. \tag{7.28}$$

which yields (using the substitution $s_B = 1 - s_A - s_C$):

$$s_C = \frac{d_B - (d_A + d_B)s_A}{\delta + d_B}. \tag{7.29}$$

Multiplying equation (7.25) by α_C, multiplying equation (7.26) by α_A, and subtracting, we obtain an equation of the form $0 = r + intercept + slope s_A$:

$$0 = (a\alpha_A + \alpha_B \alpha_c)r + \alpha_A \alpha_C - \alpha_A k_1 - \alpha_C k_o + (\alpha_A \alpha_B - \alpha_A \alpha_C - \alpha_B \alpha_C)s_A. \tag{7.30}$$

Subtracting equation (7.26) by equation (7.27) we obtain:

$$0 = ar - k_1 + k_2 - \alpha_A + (\alpha_A + \alpha_B)s_A + \alpha_A s_C. \tag{7.31}$$

We can simplify this by using equation (7.29) to substitute for s_C:

$$0 = ar - k_1 + k_2 \alpha_A + \frac{\alpha_A d_B}{\delta d_B} + \left(\alpha_A + \alpha_B - \frac{d_A + d_B}{\delta d_B} \right) s_A. \tag{7.32}$$

thus obtaining a second equation of the form $0 = r + intercept + slope s_A$.

We can solve the two equations with the form $0 = r + intercept + slope s_A$ by multiplying equation (7.30) by a, multiplying equation (7.32) by $(a\alpha_A + \alpha_B \alpha_C)$, and subtracting to obtain:

$$s_A^* = \frac{-a(\alpha_A \alpha_C - \alpha_A k_1 - \alpha_C k_o) - (-k_1 + k_2 \alpha_A + \frac{\alpha_A d_B}{\delta + d_B})}{a(\alpha_A \alpha_B - \alpha_A \alpha_C - \alpha_B \alpha_C) + (a\alpha_A + \alpha_B \alpha_C)(\alpha_A + \alpha_B - \frac{d_A + d_B}{\delta + d_B})} \tag{7.33}$$

which upon rearrangement is:

$$s_A^* = \frac{(\delta + d_B)(a\alpha_A k_1 + a\alpha_C k_o - a\alpha_A \alpha_C + k_1 - k_2 \alpha_A) - \alpha_A d_B}{(\delta + d_B)(2a\alpha_A \alpha_B - a\alpha_A \alpha_C - a\alpha_B \alpha_C + a\alpha_A^2 + \alpha_A \alpha_B \alpha_C + a\alpha_B^2)) - d_A - d_B}, \tag{7.34}$$

Using s_A^*, we can obtain the other shares at equilibrium from Equations (7.25), (7.29), and (7.30), and we obtain:

$$r^* = \frac{\alpha_A k_1 - \alpha_C k_o - \alpha_A \alpha_C + (\alpha_A \alpha_C + \alpha_B \alpha_C - \alpha_A \alpha_B)s_A^*}{a\alpha_A + \alpha_B \alpha_C}, \tag{7.35}$$

$$s_C^* = \frac{d_B - (d_A + d_B)s_A^*}{\delta + d_B}, \tag{7.36}$$

$$s_B^* = \frac{\alpha_B s_A^* + k_o - \alpha_B r^*}{\alpha_A}. \tag{7.37}$$

EXERCISES

1. Recall the 3×3 matrices U_a, U_b, and S presented in Section 7.2. Using standard techniques, verify that each of them has only one NE, that it is interior, and that in each case it is not at the centroid, but in fact is $(1,1,2)/4$. Using either eigenvalue techniques or the definition of ESS, verify that this NE is dynamically stable (and hence an ESS) for matrix S but that it is unstable for U_a and U_b.
2. Verify the statement in Section 7.2 that trajectories are predicted to run counterclockwise for S and U_a, but run clockwise for U_b.
   Hint: Of course, this assumes the usual convention on labeling the triangle vertices that RPS (or 123) are encountered in counterclockwise sequence. Observe that the two U matrices are the same except that the labels are switched for the second and third strategies. Argue that this label switch corresponds to flipping the simplex over in a particular way, and that the flip maps counterclockwise orbits to clockwise orbits.
3. The theoretical section of this chapter showed how rotating an equal fitness line past a vertex of type (i) changed cycling: if the rotation is against the RPS flow, then the vertex becomes apostatic (type ii), while if the rotation is with the flow, then the vertex becomes locally dominant, breaking RPS. What happens if you rotate an equal fitness line past a vertex of type (ii)?
   Hint: it depends whether the rotation is with or against the flow. In one direction, you get back a type (i) RPS vertex, and in the other direction you get a locally dominant vertex.
4. Recall from Chapter 3 that continuous best response dynamics (CBRD) are given by the system of ordinary differential equations $\dot{s} = b(s) - s$ wherever the best response function $b : S \rightarrow \{e^1, e^2, e^3\}$ is uniquely defined. Show that there is, in fact, a unique best response almost everywhere in the simplex, and the exceptions lie on segments of the $\Delta u = 0$ lines. Then show that trajectories for CBRD are piecewise linear, with kinks at those exceptional segments. Finally, show that, starting at any point other than the NE, the CBRD trajectories for the given U matrices converge to a particular triangle that lies strictly inside the simplex. That triangle is called the Shapley polygon, and the time average over it (a particular point in its interior) is called the TASP.
   Hint: You may use either analytic methods or a simulation package to show the trajectories.
5. Develop replicator dynamics for the five-morph male mating system for the fish *Poecilia parae*, using the schema suggested in Figure 7.10.
   Hint: This is actually an original research project. One possible way to proceed is to write out a full rock-paper-scissors-lizard-spock game for the

three melanzona morphs, immaculata, and parae. Then apply standard replicator dynamics. As done in the case of analyzing the simplex edge games for the RPS, analyze the RPS "edge" games for the 4-plex and 5-plex systems.

6. Develop the following model of abrupt learning by predators. The predator has latent variable $z \geq 0$ representing her current aversion to prey emitting the conspicuous signal. That variable decays exponentially at rate $\rho > 0$ and jumps by a unit when tasting a toxic prey emitting that signal. That is, $z' = \mathbf{1}_A + z\exp(-\rho\tau)$, where z' is the updated value after tasting prey, $\mathbf{1}_A$ is the indicator function for the current prey being type A (both toxic and emitting the conspicuous signal), z is the previous value, and τ is the elapsed time since the previous update. The behavioral assumption is that the predator avoids tasting conspicuous prey when z exceeds some threshold $T > 0$, and otherwise tastes any prey she can find. Thus a single bad experience abruptly switches behavior when $z \in [T-1, T]$. Show that the switch lasts for up to $\tau_{max} = \rho^{-1}\ln(\frac{T+1}{T})$. (Hint: consider the case where z is just a tad below T when the bad experience occurs.) Then show that if the parameters T and ρ are both small, then the max time in avoidance mode R is quite long, and so is the average time. (Hint: Under vanilla assumptions the average will be at least half the max). Finally, show that the model implies that, for given underlying parameters, there is some composite parameter $\epsilon > 0$ such that the fraction r of predators responsive to the signal is confined to the interval $[0, 1-\epsilon]$.
7. Prove or disprove the following conjecture regarding the co-evolutionary model on the prism. When morph C (cryptic) vanishes, the state space reduces to a rectangle and the dynamics are equivalent to those of the Buyer-Seller game described in Chapter 3.
8. Show that the interior equilibrium for the general predator-prey model (including own-population effect coefficients $\alpha_A, \alpha_B > 0$) is given by:

$$r^* = \frac{\alpha_A k_1 - \alpha_C k_o - \alpha_A\alpha_C + (\alpha_A\alpha_C + \alpha_B\alpha_C - \alpha_A\alpha_B)s_A^*}{a\alpha_A + \alpha_B\alpha_C},$$

$$s_A^* = \frac{(\delta + d_B)(a\alpha_A k_1 + a\alpha_C k_o - a\alpha_A\alpha_C + k_1 - k_2\alpha_A) - \alpha_A d_B}{(\delta + d_B)(2a\alpha_A\alpha_B - a\alpha_A\alpha_C - a\alpha_B\alpha_C + a\alpha_A^2 + \alpha_A\alpha_B\alpha_C + a\alpha_B^2)) - d_A - d_B},$$

$$s_B^* = \frac{\alpha_B s_A^* + k_o - \alpha_B r^*}{\alpha_A},$$

$$s_C^* = \frac{d_B - (d_A + d_B)s_A^*}{\delta + d_B}.$$

For what ranges of parameter values do these expressions define a point in the interior of the prism? For which subranges is the interior equilibrium dynamically stable?

NOTES

Early biological work on RPS

Maynard Smith was not the first to write about the theory for the RPS game. Sewall Wright wrote out a set of equations with three haploid frequency dependent strategies in an RPS intransitivity. As noted in Chapter 2 exercises, Franjo Weissing in his PhD research also speculated about the existence of 3 alternative types of yeast species but did not present direct evidence of frequency dependent selection because the data were limited to pairwise not three-way interactions. After the lizard example was discovered, the bacterial example outlined in Chapter 6 was next described (Kerr et al. 2002; Kirkup and Riley 2004). Such games may reflect one of the most common games in nature given that this game is played out by bacteria inside their hosts, providing a powerful mechanism for preserving diversity.

The first clear biological RPS, with empirical evidence presented for its existence based on frequency dependent selection, was the North American common lizard. Reminiscing on the moment of its discovery, Barry Sinervo recalls sitting in his office in Bloomington, Indiana, where he and Curt Lively were writing equations to describe the unusual system of three males. Based on observed behaviors associated with throat color, they could see aggressive orange-throated males, blue-throated mate-guarding males, and yellow-throated female mimics. As they wrote out a set of equations describing the payoffs for the three strategies, simultaneously both of them recognized the game as one they had read about in Maynard Smith's 1982 book, *Evolution and the Theory of Games*. They extracted the frequency dependent payoffs of males associating with females, and by inference paternity success (Sinervo and Lively 1996, refs Chapter 3), which were later confirmed with DNA paternity studies (Zamudio and Sinervo 2000, refs Chapter 3).

Models of aposematism

The Endler and Mappes (2004) article features a diploid model in discrete time, and focuses on the relative visibility v of two morphs, essentially A and C above, but with no B. Moreover, the analysis treats the fraction of responsive predators d (or r in our notation) as an exogenous constant. However, it does consider a variety of positive and negative frequency dependent terms. Translated into the terminology of Part I of our book, the main results of the article consist of finding regions of parameter space (v, b) where each of our three types of 2x2 interactions

(HD, CO, DS) prevails. Kokko et al. (2003) consider all three kinds of prey, and look for regions in parameter space (degree of toxicity of B relative to A, and relative abundance) where predators will have higher fitness with R than with N. Our model may be the first to consider co-evolution of the two populations in the context of aposematism.

Economic Cycles

Masking and Tirole (1988) have one of the few modern micro economic models of price cycles. See Romer (2010) for a recent survey of macro economic cycles.

BIBLIOGRAPHY

Akerlof, George A., and Robert J. Shiller. *Animal spirits: How human psychology drives the economy, and why it matters for global capitalism.* Princeton University Press, 2010.

Allesina, Stefano, and Jonathan M. Levine. "A competitive network theory of species diversity." *Proceedings of the National Academy of Sciences* 108, no. 14 (2011): 5638–5642.

Bond A. B., and A. C. Kamil. "Apostatic selection by blue jays produces balanced polymorphism in virtual prey." *Nature* 395 (1998): 594–596.

Bond A. B., and A. C. Kamil. "Visual predators select for crypticity and polymorphism in virtual prey." *Nature* 415 (2002): 609–613.

Bomze, Immanuel M. "Lotka-Volterra equation and replicator dynamics: A two-dimensional classification." *Biological Cybernetics* 48, no. 3 (1983): 201–211.

Bomze, Immanuel M. "Lotka-Volterra equation and replicator dynamics: New issues in classification." *Biological Cybernetics* 72, no. 5 (1995): 447–453.

Brodie III, E. D., and F. J. Janzen. "Experimental studies of coral snake mimicry: Generalized avoidance of ringed snake patterns by free-ranging avian predators." *Functional Ecology* 9, no. 2 (1995): 186–190.

Cason, Timothy N., Daniel Friedman, and Ed Hopkins. "Cycles and instability in a rock-paper-scissors population game: A continuous time experiment*." *The Review of Economic Studies* 81, no. 1 (2014): 112–136.

Cason, Timothy N., Daniel Friedman, and Florian Wagener. "The dynamics of price dispersion, or Edgeworth variations." *Journal of Economic Dynamics and Control* 29, no. 4 (2005): 801–822.

Edgeworth, Francis Ysidro. "The pure theory of monopoly." In *Papers relating to political economy* (1925).

Endler, John A., and Johanna Mappes. "Predator mixes and the conspicuousness of aposematic signals." *The American Naturalist* 163, no. 4 (2004): 532–547.

Fisher, Ronald Aylmer. *The genetical theory of natural selection.* Clarendon Press, 1930.

Friedman, Daniel, and Ralph Abraham. "Bubbles and crashes: Gradient dynamics in financial markets." *Journal of Economic Dynamics and Control* 33, no. 4 (2009): 922–937.

Greene, Erick, Bruce E. Lyon, Vincent R. Muehter, Laurene Ratcliffe, Steven J. Oliver, and Peter T. Boag. "Disruptive sexual selection for plumage coloration in a passerine bird." *Nature* 407, no. 6807 (2000): 1000–1003.

Gross, Mart R. "Sunfish, salmon, and the evolution of alternative reproductive strategies and tactics in fishes." In R. Wooton, and G. Potts, eds., *Fish Reproduction: Strategies and Tactics*, pp. 55–75. London: Academic Press, 1984.

Gross, Mart R. "Evolution of alternative reproductive strategies: frequency-dependent sexual selection in male bluegill sunfish." *Philosophical Transactions of the Royal Society B: Biological Sciences* 332, no. 1262 (1991): 59–66.

Harlow, Arthur A. "The hog cycle and the cobweb theorem." *Journal of Farm Economics* 42, no. 4 (1960): 842–853.

Hoffman, Moshe, Sigrid Suetens, Uri Gneezy, and Martin A. Nowak. "An experimental investigation of evolutionary dynamics in the Rock-Paper-Scissors game." *Scientific Reports* 5 (2015): 8817–8827.

Höglund, Jacob, Robert Montgomerie, and Fredrik Widemo. "Costs and consequences of variation in the size of ruff leks." *Behavioral Ecology and Sociobiology* 32, no. 1 (1993): 31–39.

Jukema, Joop, and Theunis Piersma. "Permanent female mimics in a lekking shorebird." *Biology Letters* 2, no. 2 (2006): 161–164.

Kerr, Benjamin, Margaret A. Riley, Marcus W. Feldman, and Brendan JM Bohannan. "Local dispersal promotes biodiversity in a real-life game of rock-paper-scissors." *Nature* 418, no. 6894 (2002): 171–174.

Kirkup, Benjamin C., and Margaret A. Riley. "Antibiotic-mediated antagonism leads to a bacterial game of rock-paper-scissors in vivo." *Nature* 428, no. 6981 (2004): 412–414.

Kokko, Hanna, Johanna Mappes, and Leena Lindström. "Alternative prey can change model-mimic dynamics between parasitism and mutualism." *Ecology Letters* 6, no. 12 (2003): 1068–1076.

Lach, Saul. "Existence and persistence of price dispersion: An empirical analysis." *Review of Economics and Statistics* 84, no. 3 (2002): 433–444.

Lindholm, A. K., R. Brooks, and F. Breden. "Extreme polymorphism in a Y-linked sexually selected trait." *Heredity* 92, no. 3 (2004): 156–162.

Lindström, Leena, Rauno V. Alatalo, Anne Lyytinen, and Johanna Mappes. "Strong antiapostatic selection against novel rare aposematic prey." *Proceedings of the National Academy of Sciences* 98, no. 16 (2001): 9181–9184.

Mappes, Johanna, Nicola Marples, and John A. Endler. "The complex business of survival by aposematism." *Trends in Ecology and Evolution* 20, no. 11 (2005): 598–603.

Maskin, Eric, and Jean Tirole. "A theory of dynamic oligopoly, II: Price competition, kinked demand curves, and Edgeworth cycles." *Econometrica 79, no. 2* (1988): 571–599.

Minsky, Hyman P. *Can it happen again?: Essays on instability and finance.* ME Sharpe, 1982.

Neff, Bryan D., Peng Fu, and Mart R. Gross. "Sperm investment and alternative mating tactics in bluegill sunfish (*Lepomis macrochirus*)." *Behavioral Ecology* 14, no. 5 (2003): 634–641.

Plott, Charles R. "Principles of market adjustment and stability." *Handbook of Experimental Economics Results* 1 (2008): 214–227.

Pryke, Sarah R., and Simon C. Griffith. "Postzygotic genetic incompatibility between sympatric color morphs." *Evolution* 63, no. 3 (2009): 793–798.

Romer, Christina D. "Business cycles." *The concise Encyclopedia of economics* 26 (2008): 2010.

Shuster, S. M., M. J. Wade. "Equal mating success among male reproductive strategies in a marine isopod." *Nature* 350 (1991): 606–661.
Shuster, S. M. "The reproductive behavior of α-, β-, and γ-male morphs in *Paracerceis sculpta*." *Evolution* 43, no. 8 (1989): 1683–1698.
Sinervo, Barry, and C. M. Lively. "The rock-paper-scissors game and the evolution of alternative male strategies." *Nature* 380 (1996): 240–243.
Sinervo, Barry, and Ryan Calsbeek. "The developmental, physiological, neural, and genetical causes and consequences of frequency-dependent selection in the wild." *Annual Review of Ecology, Evolution, and Systematics* (2006): 581–610.
Sprott, Julien Clinton. *Chaos and time-series analysis.* Vol. 69. Oxford University Press (2003).
van Rhijn, J. G. *The Ruff.* London: T. & A. D. Poyser, 1991.
Zamudio, Kelly R., and Barry Sinervo. "Polygyny, mate-guarding, and posthumous fertilization as alternative male mating strategies." *Proceedings of the National Academy of Sciences* 97, no. 26 (2000): 14427–14432.

8

Learning in Games

How does human strategic behavior evolve?

This chapter begins with some general perspectives on that big question. Then, in Section 8.2, it zooms in on a tiny but revealing example: a population of six people in a laboratory earning payoffs based on a particular 2×2 matrix. Here the big question reduces a manageable empirical task: fitting a parametric model to the observed population dynamics.

The empirical task here is dual to that in Chapter 3. There we knew the dynamical process (a noisy version of replicator dynamics) but had to estimate the payoff matrix. Here we know the monetary payoff matrix (indeed, we can control it in the laboratory) but have to estimate the dynamical process. We seek an empirical model that, given any payoff matrix, can explain and predict how the distribution of strategy choices changes over time in a human population.

Sections 8.3 through 8.6 assemble a standard empirical model of learning, called weighted fictitious play. They show how to fit that model to choices made by human subjects in a laboratory experiment, and explain the model's empirical successes and shortcomings.

The rest of the chapter zooms back out to broader questions. The fitted model pertains to simultaneous choice in discrete time, but Section 8.7 considers asynchronous choice in continuous time. It revisits the laboratory RPS game discussed in Chapter 7, and compares the observed behavior to the predictions of learning models and replicator dynamics. Remaining sections introduce other empirical models of interactive human learning, and scan the current research frontiers.

An appendix sketches some new ideas on learning in continuous time. As usual, the concluding notes fill in some details and points to the wider literature. The exercises include one routine question and some starters for ambitious readers.

8.1 PERSPECTIVES ON LEARNING AND EVOLUTION

What are the basic forces that cause humans to change their strategic behavior over time? That is, why might shares increase or decrease of the strategies that humans choose?

We see at least five different evolutionary processes at work.

- **Survival and fecundity** across human generations. This classic sort of gene or meme evolution surely is important in the long run. Often it can be captured nicely in replicator dynamics.
- **Entry and exit.** Even within a single human generation, shares of alternative strategies may change because some individuals leave the relevant population and other individuals join it with a different strategy distribution. This process can be important in analyzing alternative strategies in the workplace, or financial market strategies, or product cycles.
- **Resource redistribution.** Even with the same individual participants, the shares can change as some individual firms or people gain control over more resources, and other individuals see their holdings dwindle. For example, the more successful firms in an industry (or fund managers in financial markets) attract more capital and increase their market shares at the expense of less successful firms or managers.
- **Imitation or coercion.** Individual participants may switch strategies in order to imitate successful peers, or in response to peer pressure. That pressure might be subtle or, at the opposite extreme, it might take the form of conquest and annexation. The latter is important in the political sphere, and sometimes in the religious sphere as well.
- **Belief learning.** A given individual may accumulate experience that changes her beliefs about effectiveness of alternative strategies. This may increase the shares of the strategies that she (and others) now believe to yield higher payoff, and decrease shares of other strategies.

The last process is the focus of this chapter. Belief learning doesn't require transmission as such, but in some cases beliefs change from observing others, and here learning shades into imitation with horizontal transmission of memes. In the last section we will say a bit more about such gray areas.

Figure 8.1 presents a dynamic framework for modeling the belief learning process. A payoff function, denoted W on the right side of the triangle, specifies the outcome of strategic interaction given current strategies chosen by the players. Depending on the institution in which the game is played (sometimes called the mechanism or format), each player observes some portion of the current

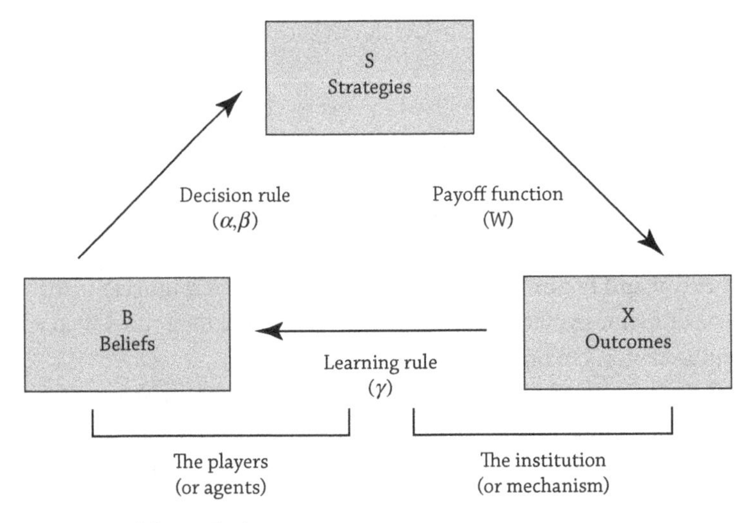

Figure 8.1 Schema for learning in games.

outcome, perhaps only her own payoff or perhaps the entire strategy profile and payoff vector. To the extent that the observed outcome is not fully anticipated, the player revises her beliefs via some learning rule, as depicted in the bottom side of the triangle. The left side of the triangle is a decision rule that specifies how the player chooses her strategy given her current beliefs.

For most of this chapter, we assume that the same game is played repeatedly in discrete time. Beliefs typically will change as experience accumulates. Changes in beliefs typically will induce changes in actions and outcomes, hence further changes in beliefs. Thus we have a dynamic process that may or may not eventually settle down to an equilibrium.

The research goal is to find specifications of the learning rule and the decision rule that capture as much actual human behavior as possible. Laboratory data are especially helpful because they come with control over the institution (hence the information available to human players) and the payoff function. In the lab we can fit learning models to observed behavior and use the fitted models to predict subsequent behavior. We seek models that describe behavior well even for a new game with new institutions.

8.2 AN EMPIRICAL EXAMPLE

A simple laboratory example will help bring the task into focus. Suppose that six human subjects interact strategically over 10 periods, each period choosing one of two alternative actions, A or B. Their payoffs are governed by the 2x2 matrix

$$\mathbf{w} = \begin{pmatrix} w_{AA} & w_{AB} \\ w_{BA} & w_{BB} \end{pmatrix} = \begin{pmatrix} 5 & -1 \\ 3 & 3 \end{pmatrix}. \tag{8.1}$$

The institution here involves playing the field, also known as mean matching: the payoff is the average over all possible matches of a given player with everyone (including himself). Thus, if fraction $s = s_A$ of the population plays A, and $1 - s = s_B$ plays B, then any player choosing A gets $w_A = \mathbf{e_1 w} \cdot \mathbf{s} = (1,0)\mathbf{w} \cdot (s, 1 - s) = w_{AA}s + w_{AB}(1 - s) = 5s + (-1)(1 - s) = 6s - 1$, and any player choosing B gets $w_B = \mathbf{e_2 w} \cdot \mathbf{s} = (0,1)\mathbf{w} \cdot (s, 1 - s) = w_{BA}s + w_{BB}(1 - s) = 3s + 3(1 - s) = 3$. Hence $\Delta w = w_A - w_B = 6s - 4$, with root $s^* = 2/3$. We know from Chapter 4 that $s^* = 2/3$ defines a mixed NE that equates the payoffs of actions A and B. Since the slope of Δw is positive, we also know that s^* is an upcrossing and thus separates the basins of attraction for the two pure NE, one with $s = 1$ (all A) and the other with $s = 0$ (all B).

But what will humans actually do? They might try for the all-A equilibrium since it gives them a higher payoff, or they might go for the all-B equilibrium because it is safer (e.g., no danger of negative payoffs) and has the wider basin of attraction. Or perhaps they might hover in a neighborhood of the mixed NE, with no clear trend.

Table 8.1 displays actual choices in a run of 10 periods by six subjects recruited to play this game. They received about thirty cents per point, and played several other runs of 10 periods each. Most of them took home \$20–30 for about 2 hours participation.

Our task now is to use behavior observed in period $t = 1$ to predict what will happen in period $t = 2$ and to compare actual behavior to this prediction; then to use observed behavior in these two periods to predict what will happen in period $t = 3$; ... ; and finally to use the data observed in the first nine periods to predict what will happen in period $t = 10$. We want to build a model whose predictions are as accurate as possible.

8.3 LEARNING RULES

Figure 8.1 breaks the prediction task into two parts. Leaving decision rules to the next section, we consider in this section various learning rules, that is, various models of how people change their beliefs as experience accumulates.

Perhaps the simplest learning rule, which goes back at least to Cournot (1838), is that each player always updates beliefs to conform to the most recent observation. In the Table 8.1 example, the Cournot learning rule specifies that

Table 8.1. CHOICES IN A COORDINATION GAME. THE DATA COME FROM LABORATORY SESSION EVGAME 21, GROUP 1, PART 2, RUN 3, AT THE UNIVERSITY OF CALIFORNIA, SANTA CRUZ, AS PART OF A STUDY REPORTED IN FRIEDMAN (1996) AND CHEUNG AND FRIEDMAN (1997).

	Choice of Player:	**Payoff of Player:**
***t*=**	**1 2 3 4 5 6**	**1 2 3 4 5 6**
1	B A B A B A	3 2 3 2 3 2
2	B B B B B A	3 3 3 3 3 0
3	B A B B B A	3 1 3 3 3 1
4	B A B B A A	3 2 3 3 2 2
...	...	...
9	B B B B A A	3 3 3 3 1 1
10	B A B B A A	3 2 3 3 2 2

each player believes she will face $s(t) = 0.5$ in period $t = 2$ because the most recently observed state was $s(t-1) = 3/6 = 0.5$ in period 1. Likewise each player believes she will face $s(t) = 1/6$ in period $t = 3$ because that was the actual state in period 2.

Using the notation $Es(t) = \hat{s}(t)$ to denote beliefs regarding the state a player faces in period t, we have the following general formula for the **Cournot Learning Rule**:

$$\hat{s}(t+1) = s(t) \text{ for periods } t = 1, 2, 3, ..., T-1. \quad \textbf{(8.2)}$$

There are situations where the Cournot rule is hard to beat. That tomorrow's weather will be the same as today's is a pretty good guess, and hard to beat consistently unless you have access to the National Weather Service or its data. That tomorrow's exchange rate (for the Euro, say, against the U. S. dollar) will be the same as today's is a forecast that turns out to be surprisingly hard to beat, even with access to the best data and modern forecasting techniques (see, for example, Meese and Rogoff (1983) or Engel and West (2005).

But there are other situations where the Cournot learning rule is too naive. If you noticed that $s(t)$ is usually near 0 except for occasional one-period jumps to $s = 1/2$, then you probably wouldn't subscribe to the Cournot formula $\hat{s}(9) = 1/2$ in period 6 just because $s(8) = 1/2$. In such situations it might be more reasonable to believe that the state will revert to its long-run average. George Brown (1951) and Julia Robinson (1951) developed this idea, which they called fictitious play.

The general formula for the **Fictitious Play Learning Rule** is

$$\hat{s}(t+1) = \frac{1}{t}\sum_{k=0}^{t-1} s(t-k) \text{ for periods } t = 1,2,...,T-1. \tag{8.3}$$

That is, the forecast of time $t+1$ play is the simple average of all prior observations. In the example, $\hat{s}(3)$ is the average of the choices in periods 1 and 2, namely $\frac{1}{2}(\frac{1}{2}+\frac{1}{6}) = \frac{1}{3}$.

There are also situations in which it might be reasonable to specify something in between Cournot and Fictitious Play. A player might believe that the most recent observation is the most revealing, but that previous observations still are worth taking into account. One way to do this is to say that the impact on beliefs fades by some fraction each period, according to a recency parameter $0 \leq \gamma \leq 1$. For example, if $\gamma = 0.6$, then the next most recent observation has only 60% of the impact of the current observation, and the observation two periods earlier has only $(0.6)^2 = 36\%$ of the impact.

This learning rule is called **Weighted Fictious Play**, and the general formula is

$$\hat{s}(t+1) = \frac{1}{\sum_{k=0}^{t-1}\gamma^k}\sum_{k=0}^{t-1}\gamma^k s(t-k) \text{ for periods } t = 1,2,...,T-1, \tag{8.4}$$

where $\frac{\gamma^n}{\sum_{k=0}^{t-1}\gamma^k}$ is the fractional weight at time t placed on an observation made at time $t-n$.

Clearly as $\gamma \to 1$, the fractional weight converges to $1/t$, as in fictitious play. Likewise as $\gamma \to 0$, the weight on observation t (where $n = 0$) converges to one and the weight on earlier observations converges to zero, as in the Cournot learning rule. Thus weighted fictious play includes as special cases both the Cournot model and the fictitious play learning model. The intermediate cases are parametrized by the free parameter $0 < \gamma < 1$, to be estimated from the data.

The last section of this chapter includes a brief discussion of lab techniques for directly eliciting beliefs, and notes the limitations. The data considered so far, however, capture what people do, not what they believe. To estimate the recency parameter γ here, we must specify how beliefs translate into actions. That is, we must specify a decision rule.

8.4 DECISION RULES

Suppose that, for whatever reason, you believe that you face state $\hat{s}(t+1) = 0.5$ in period $t+1 = 5$ in the game of Table 8.1. Which action should you pick, A or B?

Given your beliefs, you expect payoff $6\hat{s}-1=3-1=2$ if you pick A, and payoff 3 if you pick B. Since $3>2$, it seems that B is the better choice.

This illustrates the simplest decision rule—simply pick a strategy i that maximizes expected payoff $\hat{w}_i = \mathbf{e_i w \hat{s}}$. More formally, the **Best Response Decision Rule** is

$$Pr[a(t+1)=i]=0 \text{ if } \hat{w}_i(t+1) < \max_j \hat{w}_j(t+1). \tag{8.5}$$

Thus the player always picks a strategy expected to give the highest payoff and never picks one that is expected to give a lower payoff. If several strategies are tied for highest expected payoff, the Best Response rule doesn't say which of these will be chosen, but we will assume below (unless otherwise noted) that they are equally likely, that is, that ties for best are broken randomly.

The Best Response rule is logical, but it is too sharp for some purposes. For instance, in the example game it says that the probability of playing A is 0.0 if $\hat{s}=.666$ but the probability jumps to 1.0 when $\hat{s}=.667$. Humans are seldom that sensitive to tiny differences. To smooth out such discontinuities, researchers often use decision rules that assign positive probabilities to each possible strategy, with higher probabilities assigned to strategies with higher expected payoffs.

Perhaps the simplest smooth decision rule is called **Luce's Choice**. It can be used when all the expected payoffs $\hat{w}_i$ are positive, and says

$$Pr[a(t+1)=i]=\frac{\hat{w}_i(t+1)}{\sum_j \hat{w}_j(t+1)}. \tag{8.6}$$

In the example, when $\hat{s}=0.5$, we have $\hat{w}_A=2$ and $\hat{w}_B=3$, so Luce's Choice rule says the probability of the player choosing A is $\frac{2}{2+3}=0.4$, versus $\frac{3}{2+3}=0.6$ for B.

There are situations where actual choices are more like Best Response than in Luce's Choice, and other situations where the choices are more random. To accommodate these situations, analysts sometimes raise the expected payoffs in (8.6) to some positive power d. This gives us the **Power Decision Rule**

$$Pr[a(t+1)=i]=\frac{\hat{w}_i(t+1)^d}{\sum_j \hat{w}_j(t+1)^d}. \tag{8.7}$$

As d gets large, the largest expected payoff dominates the expressions in (8.7). As a result, the expressions converge to those in the Best Response rule (8.5) as $d \to \infty$, and to converge to equal probabilities as $d \to 0$. Of course, we are back to Luce's Choice when $d=1$.

The Power rule doesn't work when some of the expected payoffs are negative. To avoid the problem, researchers often use a slight variant on the same idea. It uses the logit function with decisiveness parameter $\beta \in [0,\infty)$, analogous to the power d. The **Logit Decision Rule** (known to computer scientists as the Soft-Max rule) is

$$Pr[a(t+1)=i]=\frac{e^{\beta\hat{w}_i(t+1)}}{\sum_j e^{\beta\hat{w}_j(t+1)}}. \tag{8.8}$$

For example, if $\beta=2$ and if $\hat{s}=0.5$ so again $\hat{w}_A=2$ and $\hat{w}_B=3$, then the logit rule predicts that A will be chosen with probability $\frac{e^{2\cdot 2}}{e^{2\cdot 2}+e^{2\cdot 3}}\approx\frac{54.6}{54.6+403.4}\approx 0.12$, while B will be chosen approximately 88% of the time.

When $\beta=0$, the logit decision rule says that all strategies are equally likely. No matter what the sign of the expected fitnesses, the logit rule converges to the Best Response rule as the decisiveness parameter $\beta\to\infty$.

What if a player is biased towards one or another choice? For instance, in the 10-period run partly shown in Table 8.1, players 1 and 3 always chose B and player 6 always chose A, despite the fact that they all observed the same history of play when choosing actions. To account for this possibility, we introduce a new parameter α_i indicating the degree of bias towards action i. Then the logit choice rule becomes

$$Pr[a(t+1)=i]=e^{\alpha_i+\beta\hat{w}_i(t+1)}/\sum_j e^{\alpha_j+\beta\hat{w}_j(t+1)}. \tag{8.9}$$

With no loss of generality, we can set $\alpha_n=0$ in a game with n pure strategies. To see this, just multiply the numerator and denominator of equation (8.9) by $e^{-\alpha_n}$ and note that all α's are simply shifted downward by the amount α_n without changing the quotient.

In the present example there are only two choices, A and B. Set $\alpha_B=0$ and simplify notation by dropping the A subscript on the remaining α. To illustrate, suppose that $\alpha=\alpha_A=5$ and other parameter values are as before. The model then predicts action A with probability $\frac{e^{5+2\cdot 2}}{e^{5+2\cdot 2}+e^{0+2\cdot 3}}\approx\frac{8301}{8301+403.4}\approx 0.95$. Another player with an equally strong bias against A, represented by $\alpha=-5$, would be predicted to choose A with probability only $\frac{e^{-5+2\cdot 2}}{e^{-5+2\cdot 2}+e^{0+2\cdot 3}}\approx\frac{0.37}{0.37+403.4}\approx 0.001$.

8.5 ESTIMATING A MODEL

The last two sections converted a vague question into a well-posed problem. The prediction question raised at the end of Section 8.2 now comes down to a statistical exercise—how to estimate the free parameters α,β, and γ from data like that in Table 8.1.

Three implementation issues must be dealt with before we proceed to estimate the parameters. First, what are the payoffs, as human subjects perceive them? We will assume here that earning money dominates other possible considerations when subjects decide whether to play A or B. Thus we will take the monetary payoffs in the matrix w as the perceived payoffs, and use the parameter α to allow for possible bias. We return to this point in the concluding section.

Second, what constitutes a best fit of the parameters to the data? We will use the maximum likelihood (ML) procedure below, but another option is least squares, a popular fitting procedure which choses parameters that minimize the sum of squared errors. For our sort of data, the two procedures usually give similar results.

Third, do we fit the parameters separately for each individual player, or assume that one size fits all? For simplicity, we begin with the latter, sometimes called the representative agent assumption, and then go on to discuss individual fits.

We are now ready to proceed with ML estimation. The estimation procedures are essentially the same as in Chapter 4: write out LL, the log likelihood of the observed data as a function of the parameter vector, and find fitted parameter values that maximize LL.

The first step here is to streamline the expression of LL as a function of the parameter vector. Write $\alpha = \alpha_A - \alpha_B$ as before, and set

$$u = \alpha + \beta \Delta w(\hat{s}(\gamma)). \tag{8.10}$$

Note that the expression for u is linear in parameters α and β. For matrix games, the expression Δw is always linear in its argument $\hat{s}$, and by equation (8.4) the argument $\hat{s}$ is itself linear in the fractions $s(t-k)$ of A and B choices observed in each previous period, although their coefficients (weights) are nonlinear in the parameter γ.

Equation (8.10) is very helpful because (as you are invited to confirm in Exercise 8.1) the likelihood ratio is e^u when the observed action is A and is e^{-u} when it is B. The joint likelihood of all observed data just multiplies these ratios across observations. Of course, LL simply sums their logs, $\pm u$, over all observations.

With these facts in mind, it is straightforward to estimate the parameters. Pick a particular value of γ between 0 and 1, and use it and the observed data to calculate $\hat{s}$ and Δw in period 1, then in period 2, then in period 3, ... and finally in period 9. Add up the resulting expressions for $u_t = \alpha + \beta \Delta w(\hat{s}_t(\gamma))$ for periods $t = 2, 3, \ldots, 10$ to get the expression for $\mathrm{LL}(\alpha, \beta, \gamma)$. Use standard optimization techniques find the LL-maximizing parameters α and β and the corresponding maximized value $\mathrm{LL}(\gamma)$. Then search across a grid of γ values between 0 and 1 to obtain the grid point $\hat{\gamma}$ that yields the maximum $\mathrm{LL}(\gamma)$. Finally, report $\hat{\gamma}$ and the previously determined values of α and β that maximize this $\mathrm{LL}(\hat{\gamma})$.

8.6 RESULTS

Standard statistical packages, such as Stata, SAS, S and R, automate most or all of this procedure. We used Stata on the data from Table 8.1, along with the other observations that session—a total of 12 subjects, each observed in 6 runs of the coordination game with money payoffs as in equation (8.1). Each run consisted of 10 periods, but (as noted earlier) we didn't try to predict behavior in the first period. Thus we used $9 \times 6 \times 12 = 648$ observations. Stata gave us the following parameter estimates (indicated with a $\hat{\cdot}$) and standard errors (indicated after the $\pm$ symbol):

$$\hat{\alpha} = 0.864 \pm 0.228, \quad \hat{\beta} = 0.891 \pm 0.123, \quad \text{and} \quad \hat{\gamma} = 0.43. \qquad \textbf{(8.11)}$$

The estimated α is positive, and is economically and statistically significant. It indicates that the typical human player in our experiment had a moderate bias towards playing strategy A, which yields the higher payoff in equilibrium. The estimated β also is positive and significant, indicating that the typical player was fairly responsive to the perceived payoff advantage.

The estimated γ is almost halfway between 0 and 1, suggesting that the typical player's perceptions were a bit closer to Cournot than to fictitious play, so recency was important but not overwhelming. Stata can't readily give us a standard error for the estimate of γ because (due to the nonlinear way γ enters the expressions for u) LL is formed separately for each grid value.

Earlier we noted that the choice bias α might differ across individual subjects. It is also possible that some individuals respond more strongly than others to the same perceived payoff differential Δw. To check this possibility, one can also estimate β separately for each individual subject. We could even consider separate individual estimates of γ to see whether some individuals give more weight than others to the more recent observations. Of course, with six subjects, we then would have to estimate $3 \cdot 6 = 18$ parameters from the data instead of just 3. It takes far more data than shown in Table 8.1 to get plausible estimates of that many free parameters.

The data discussed so far are from just 1 of 30 laboratory sessions conducted to examine alternative learning models. Techniques similar to those just discussed led to the following general conclusions, distilled from Friedman (1996) and Cheung and Friedman (1997, 1998).

First, six players seems to be about the smallest human population size to which evolutionary game theory might apply. In populations with two to four players, there is evidence that individual players often chose actions in order to influence others players' future behavior, but that they very seldom engaged in such behavior (dubbed "Kantian" in Friedman 1996) when the population size was eight players or more.

Second, individuals differ in how they learn from experience. Fitting the three-parameter model to individual subjects enabled classifying most of them as short memory (Cournot), intermediate, or long memory (fictitious play). About a quarter of the players were not classified because the γ estimate was too diffuse, due to invariant play (e.g., always choosing A) or inconsistent play or a small sample size. About a tenth of the players had estimated γ's less than 0 or greater than 1. The remaining 65% of players included roughly equal numbers of short and long memory classifications and about half as many intermediates.

The study considered three other types of 2x2 games besides the CO game given in Table 8.1. These included a HD-type game with one population as well as two asymmetric two-population games. It turned out that the distribution of γ estimates was essentially the same across all four games, an important vindication of the learning model.

The same study also estimated a representative agent version of the learning model by imposing the restriction that all three parameters were the same across all individual subjects. Of course, the restriction saves on free parameters, but it was inconsistent with the data. The standard Chi-squared test for maximum likelihood estimates rejected the restriction at $p=0.05$ in every session, and at minuscule p values in most sessions.

Equally important, the representative agent restriction creates a specification error. The representative agent estimates for γ varied significantly across games, mainly because Cournot behavior was not detected in the aggregate. The finer-grained individual estimates show that we actually have γ distributions that seem almost invariant across games.

Even though it is contrary to the evidence, the representative agent restriction allows a fair comparison of the learning model to replicator dynamics. It is natural to include three free parameters in the replicator model to represent choice bias and two adjustment speeds (depending on the information feedback). It turns out that the restricted learning model yields slightly more accurate predictions than the replicator model for the single-population games, and much more accurate predictions in the two-population games.

Returning to the unrestricted learning model, what can we say about the stability of the decision parameters α and β? Theory suggests that the bias parameter α will change as the game matrix changes, and the data confirm the suggestion. In particular, for the coordination game in Table 8.1, almost all players' estimated α's were either near zero or positive, indicating a preference for the "Kantian" strategy A. In the other three games, the α distributions were more bell-shaped and included a fair number of negative biases. As predicted, the decisiveness parameter β tended to be larger in coordination games than in games with only a single NE in the interior, and in all games the β estimates were consistently positive.

The institutions used in the 30 sessions also varied the information feedback. The example game involved mean matching (MM), but some sessions used an alternative treatment, random pairwise (RP), in which each player was matched each period with just one of the possible opponents. History of play so far in the run was displayed on players screens (Hist) in the example session, but some other sessions suppressed that information (No Hist). Theory suggests that players will respond more strongly to differences in expected payoffs when the expectation is based on better information, and indeed we consistently got higher β estimates in MM than in RP and in Hist than in No Hist. Again as suggested by theory, these information treatments had no systematic impact on the estimated bias parameter α.

Theory suggests a subtle effect of the information treatments on the learning parameter γ. When contemporaneous evidence is weak, rational players will rely more heavily on older evidence. The data confirmed this suggestion as well: γ estimates either differ insignificantly or are higher in RP than in MM and in No Hist than in Hist.

8.7 LEARNING IN CONTINUOUS TIME

The preceding analysis assumes that the data are observed at evenly spaced discrete points of time $t=0,1,2,\ldots$. But in human society, many important games are played continuously. Contributions of time and money to neighborhood or charitable organizations (like PTA or the Red Cross) are more or less continuous, as are contributions of effort in team sports such as soccer and basketball or or joint ventures such as coauthored books and articles. Of course, pollution abatement or using restraint in fishing in common waters generate flow costs in real time, and benefits accrue continuously from public goods such as the Internet, clean air, and roads.

What difference (if any) does it make whether a game is played in discrete versus continuous time? To get started, we must formalize the idea of flow payoffs. For example, consider the game defined by the matrix $\mathbf{w}$ in equation (8.1). If the current state is $(s(t),1-s(t))$, then a player currently choosing the action A earns $w_A=(1,0)\mathbf{w}\cdot(s,1-s)=6s-1$ points *per unit time* if the game is played in continuous time, and over a time interval $[a,b)$ earns a total of $\int_a^b[6s(t)-1]dt$ points.

In general, a player's overall payoff is the integral of her flow payoff over the entire time interval over which the game is played. As noted in the Appendix, the integrand should be discounted by factor e^{-rt} if the time interval is long relative to the interest rate $r > 0$, but discounting is negligible in the games we will consider here.

The RPS laboratory experiment of Cason et al. (2014) mentioned in Chapter 7 was designed to compare discrete time to continuous time. Recall that in some treatments, the human subjects' choices defined strategy shares that followed very nice cycles in the appropriate direction. Now we will dig a little deeper to describe and analyze models of how individual humans adjust their strategies in continuous versus discrete time.

To recap, each subject in that experiment faced the same 3×3 payoff matrix, either the theoretically stable matrix $S = \begin{pmatrix} 36 & 24 & 66 \\ 96 & 36 & 30 \\ 24 & 96 & 36 \end{pmatrix}$, or the unstable matrix $U_a = \begin{pmatrix} 60 & 0 & 66 \\ 72 & 60 & 30 \\ 0 & 72 & 60 \end{pmatrix}$ (or the matrix U_b, a variant of U_a with the second and third rows and columns switched.) In M treatments, human participants could pick any mixed strategy, while in P treatments they could pick only a pure strategy at a simplex vertex. Player j's flow payoff $w(j,t) = \mathbf{a}^j\mathbf{w} \cdot \mathbf{s}(t)$ is determined by that player's chosen strategy $\mathbf{a}^j$, the payoff matrix $\mathbf{w} = \mathbf{S}, \mathbf{U_a}$, or $\mathbf{U_b}$ and the average $\mathbf{s}(t)$ of all players' current strategies. The discrete time treatment (D) allowed players to change their strategies only at a discrete set of times $t = \Delta t, 2\Delta t, ..., K\Delta t$, but there also were two novel continuous time treatments, one with instantaneous adjustment (CI) and the other with steady slow movement towards the player's chosen target (CS).

What does theory predict for the continuous time treatments? One possible prediction comes from the continuous replicator dynamic

$$\dot{s}_i = c(w_i(\mathbf{s}(t)) - \bar{w}(\mathbf{s}(t))s_i, \ i = 1,2,3, \tag{8.12}$$

where the positive rate constant c should be smaller in the CS treatment than in the CI treatment. For any positive rate constant, replicator dynamics yield cycles that spiral out to the simplex boundary (mathematicians call this a heteroclinic limit cycle) with the U matrix treatments, but that converge to the interior NE point $s^* = (1,1,2)/4$ under S matrix treatments. Predictions of limiting behavior are similar for the discrete replicator dynamic, at least for time interval Δt that is small relative to the rate constant c.

What do the learning models predict? To extend them to continuous time, denote the logit function by $\psi(\cdot)$, in other words, write the weight on strategy i in the logit response rule (8.8) as $\psi_i(\mathbf{s}|\beta) = \frac{e^{\beta \mathbf{e}^i \mathbf{w} \cdot \mathbf{s}}}{\sum_j e^{\beta \mathbf{e}^j \mathbf{w} \cdot \mathbf{s}}}$. (Here we ignore the complications of personal biases introduced in (8.9) by setting aggregate $\alpha = 0$,

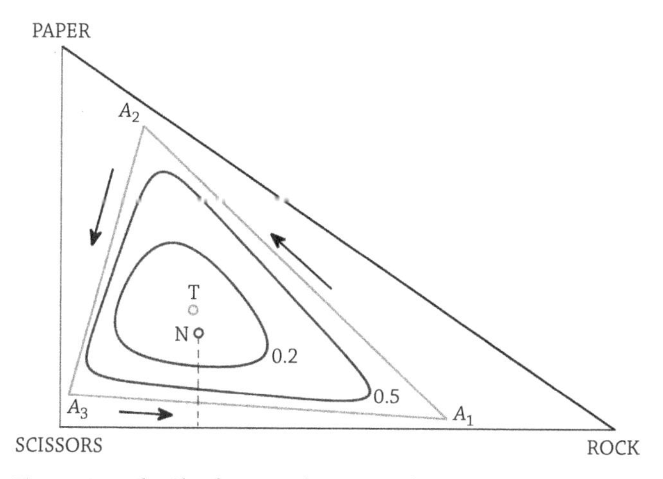

Figure 8.2 The Shapley triangle $A_1A_2A_3$ for game U_a with NE (N). Also illustrated are orbits for the logit dynamics for precision parameter values $\beta = 0.2$ and 0.5.

as seems reasonable in an essentially symmetric game like RPS.) Assuming that all players see the current state $\mathbf{s}(t)$, belief updating is very straightforward: we simply have $\hat{\mathbf{s}}(t) = \mathbf{s}(t)$. The model then says that on average each player adjusts her current weight a_i on action i steadily towards $\psi_i(\mathbf{s}|\beta)$ at some rate $c > 0$ that depends on the player's reaction speed and the treatment (CI vs. CS). Noting that the average of the a_i's over all players is s_i, we obtain

$$\dot{s}_i = c(\psi_i(\mathbf{s}|\beta) - s_i), \quad i = 1,2,3 \tag{8.13}$$

as the extended logit learning model with precision parameter β.

The predictions of the logit learning model differ noticeably from those of the replicator model for the U matrix treatments. Cason et al. cite known results showing that in the BR limit ($\beta \to \infty$) there is a limit cycle for the matrix U_a given by the Shapley triangle shown in Figure 8.2. That is, starting from any interior point outside this triangle, the trajectory spirals in slowly (or out slowly) and ultimately converges to the triangle itself. Likewise, trajectories that begin inside the triangle (except at the steady state) spiral outward and ultimately converge to the Shapley triangle. For $\beta < \infty$ there is a smooth limit cycle with rounded corners that lies inside the Shapley triangle, as shown in the two smaller closed loops in Figure 8.2. Again, the model predicts that trajectories inside the limit cycle spiral out to it, and trajectories starting outside the limit cycle spiral inwards towards it. Predicted behavior is exactly the same in the U_b matrix treatment except that the spirals are clockwise instead of counterclockwise.

For the S matrix, the logit learning model predicts pretty much the same thing as the replicator model in the CI and CS treatments: the trajectories will all spiral inward and ultimately converge to an interior steady state s^*. (A qualification is in order. As will be explained in Section 8.8, the interior steady state depends on the precision parameter β and is exactly the NE only in the limit $\beta \to \infty$.)

In the discrete time treatment D, the prediction depends on the way beliefs are formed. For Δt small relative to the rate constant c, the Cournot assumption that $\hat{\mathbf{s}}(t + \Delta t) = \mathbf{s}(t)$ leads to predictions that are quite similar to those for the continuous time treatments. However, if discrete time choices are synchronized and, except when Δt is on the order of human reaction time, the Cournot assumption has no compelling logic. Everyone could completely change their strategies from one time period to the next, and in a game like RPS where it pays not to be predictable, it is very hard to guess the next period's state.

So which predictions turned out to best describe the experimental results? As we saw in Chapter 7, the human subjects produced very nice cycles in the CS time treatment with U matrices. The trajectories did not tend to converge to the heteroclinic cycle, however; instead they seemed to move towards the limit cycles shown in Figure 8.2 for β in the range 0.2–0.5. As in earlier experiments, there was no clear evidence for cycles in the discrete time D treatments; the state seemed to wander aimlessly in a large neighborhood of the interior NE.

The main surprise in the experiment was the persistence of cycles for the Stable matrix in the continuous time treatments CI and CS. These cycles had smaller amplitude than those for corresponding Unstable matrices, but they never disappeared. An appendix to Cason et al. argues that all the main features of the observed data can be accounted for using a discrete version of the logit learning dynamics, where the step size is treatment-specific in a natural way.

8.8 OTHER MODELS OF LEARNING

The literature on learning in games has become quite large in recent decades. Dozens of different models have been tested in hundreds or perhaps thousands of laboratory sessions. We close this chapter by mentioning a few of the highlights.

Alternative procedures

Many of the specific procedures described in this chapter can be altered with little impact on the main results. For example, in single-population mean matching, it is natural to match a player only against the other players, and not match her against herself. The theoretical complication is that each player in general faces a different state vector $s(t)$. We are not aware of any evidence that including or excluding self-matches makes a practical difference in populations with more than 6 players.

We already noted that least squares procedures can be used instead of maximum likelihood. Other estimation procedures worth checking include fixed effects or random effects at the session level or (in representative agent specifications) at the individual level. Recently many studies cluster standard errors at the session level to control for within-group dependence, and clustering at the individual subject level makes sense for learning models when feasible. One can have more confidence in results that are robust in the sense that they are confirmed in all reasonable econometric specifications.

In the logit decision rule formula (8.8), one can replace the expressions e^u in the numerator and denominator by a more general expression $F(u)$, where F is any convenient increasing positive function. This more general version is called a quantal response model. A popular quantal response model, called probit, sets $F =$ the cumulative normal distribution function. Standard statistical packages usually can run probit regressions, at least when there are only two alternative actions. Another advantage of probit is that its coefficient estimates can be directly interpreted as marginal probabilities, while the logit coefficients are more awkward to interpret. The papers mentioned in Section 8.6 use probit, although the results were also checked using Logit. It is difficult to find any empirical examples where findings with logit differ appreciably from findings with probit.

The models discussed so far treat players' beliefs as latent variables, not observable by the researcher (and perhaps not explicit in the player's own mind). An alternative procedure is to use the Becker-DeGroot-Marschak (or similar) mechanism to elicit subjects' beliefs, and to use the elicited beliefs as explanatory variables. There is some controversy as to possible biases implicit in the mechanisms and (we believe more importantly) in eliciting beliefs at all. There seems to be a Heisenberg uncertainty principle of the mind—as in quantum mechanics (but for a very different reason!), the very act of eliciting beliefs often changes behavior. See Rutström and Wilcox (2009) for further discussion.

Theories of initial behavior

Learning models as such do not predict behavior in the first period. But it can be important. For example, the models described in previous sections imply that behavior will settle down the the neighborhood of the pure NE with $s = 1$ if the initial period is well inside its basin of attraction, say $s(1) > 0.8$, but will settle down near $s = 0$ if the initial period is well inside its basin of attraction, say $s(1) < 0.4$.

Therefore, it would be quite helpful to have a supplementary model that predicts initial behavior. Stahl and Wilson (1994) and Nagel (1995) first proposed such models, called level-k. The basic idea is that level-0 players are random, equally likely to chose any strategy. Each level-1 player believes that everyone else is level-0, so $\widehat{s^1}$ is the mean of the uniform distribution. Each level-2 player believes that everyone else is level-1, so $\widehat{s^2}$ is the best response to that mean state. Level-3 players in turn best respond to that belief, and so on. There are many variants, in which level-k may or may not know the true distribution of lower-level players, k is truncated at 2 or 3, additional types of players are added (e.g., some who play their part of a NE strategy), and so on.

Level-k models are usually fit to the data using ML methods to estimate the proportions of each type, along with decision parameters similar to our β. They have had mixed success in predicting behavior when human subjects first confront a new game in the lab. See Crawford et al. (2013) for a recent discussion of the evidence.

Reinforcement learning and EWA

The learning models presented so far assume that players form beliefs and then decide what actions to take. When beliefs are not observed, it is not obvious from the data whether this description is accurate, or whether players simply react directly to experience. One possible approach, not much explored as far as we know, is to see whether behavior changes when players receive more or less or different feedback information from the institution in Figure 8.1. With reinforcement learning, behavior should change only when players receive different information about their own payoffs, but with belief learning, behavior can change when players receive different information about the behavior or payoffs of other players.

The possibility that players simply react directly to experience is called *reinforcement learning*, and it goes back to 1950s-era models in psychology. It combines the bottom and left sides of the triangle in Figure 8.1 into a direct action rule, bypassing beliefs. A basic reinforcement model works as follows. A player in a discrete time game has at time t a specified attraction $A_i(t) \geq 0$ for each alternative strategy $i = 1,...,n$. He chooses strategy i according to Luce's choice rule (8.6) with the expected payoff w replaced by attraction A. If action i is chosen, then its attraction is updated via $A_i(t+1) = A_i(t) + w_i(t)$; other attractions $j \neq i$ are not updated, so $A_j(t+1) = A_j(t)$.

There are many variants—for example, one can use the power rule (8.7) or logit instead of Luce's choice, so it might seem that there is no practical difference between belief-based learning models and reinforcement learning models. Yet one important distinction remains. In standard belief learning models, updating of beliefs and hence choice probabilities is the same for strategies not chosen that period as for the strategy actually chosen. By contrast, as we have just seen, reinforcement learning treats them differently, and does not update the unchosen attractions.

In a series of papers in the late 1990s, Camerer and Ho proposed and tested a hybrid model called EWA, for experience-weighted attraction. The key parameter is the updating weight $\delta \in [0,1]$ put on all unchosen strategies j relative to the chosen strategy i that period. Their model reduces to a variant of reinforcement learning when $\delta = 0$ and to a variant of belief learning when $\delta = 1$. Due to its flexibility, the EWA model has become the most popular empirical model for learning in games, and has commanded a large share of the literature since 2001.

Steady state behavior

What happens when learning has run its course? One possibility is convergence to NE, and that was the focus of many early studies of learning in games. However, such convergence is not possible under some of the leading decision rules. Take the logit rule (8.8), for example. For a chosen $\beta > 0$ and $\alpha = 0$, one looks for a fixed point—that is, a mixed strategy s that reproduces itself under (8.8) when anticipated. The equations are

$$s_i = \frac{e^{\beta \hat{w}_i(s)}}{\sum_j e^{\beta \hat{w}_j(s)}}, \quad i = 1, \ldots, n. \tag{8.14}$$

A solution $s^*(\beta)$ to (8.14) is called a logit equilibrium, and more generally, when the exponential function is replaced an arbitrary non-negative increasing function F, it is called a *quantal response equilibrium* (QRE). As $\beta \to \infty$, it is easy to see that $s_i^*(\beta) \to 0$ unless strategy i is a best response to $s^*(\beta)$. With more care, one can verify the intuitive result that $s^*(\beta)$ traces out a curve from the centroid $s_c = (\frac{1}{n}, \ldots, \frac{1}{n})$ of the simplex to a NE as β increases from 0 to ∞.

For present purposes, the main point is that, for fixed $\beta \in (0, \infty)$, if a learning model using (8.8) ever settles down, it will not be to a NE, but rather to the appropriate QRE. See McKelvey and Palfrey (1995) and Chen et al. (1997) for the original articles. See also Selten and Chmura (2008) for alternative specifications of steady state behavior, including payoff sampling equilibrium and action sampling equilibrium as well as impulse balance equilibrium. The last concept uses a handy trick for fitting recalcitrant data—consistent with the behavioral economics idea of loss aversion, let the weight on negative payoffs (or losses relative to some target payoff) be about twice the weight on gains.

8.9 OPEN FRONTIERS

Learning in games has attracted considerable attention in the last two decades, from empirical researchers as well as theorists, but many open questions remain. Here we mention three of the largest.

Generalizable learning

Standard learning models offer out-of-sample predictions in the narrow sense that they use behavior already observed to predict the next action. But there is an important broader sense that remains under researched. How does what a person learned in one context carry over to a new context? In particular, what can we predict about behavior with a different payoff function or different institution with different information conditions? The discussion late in Section 8.6 touched on such matters, but did not take it up systematically.

Available evidence suggests that such carry over (or transfer) is more difficult than one might suspect; see for example Rick and Weber (2010). A particularly striking example is found in Page (1998). He found that subjects readily learned to accept an offer to switch doors in a 100-door version of the notorious Monty Hall puzzle (Friedman 1998), but were typically unable to transfer their insight to the original 3-door setting. Of course, teachers have long recognized the difficulty in imparting generalizable knowledge as opposed to simple facts. For that reason among many others, one of the most important open research questions in this area is how and when people are able to leverage learning from one strategic context to another.

Ecologies of learning styles

Virtually all studies that look for it find that humans have very heterogeneous learning processes. Again, Section 8.6 raised this point but didn't take it very far, nor do other papers we know about.

It may be worth building higher-level evolutionary models. Instead of shares representing actions, consider shares of learning styles. If there are lots of Cournot learners, for example, when will it be advantageous or disadvantageous to be a fictitious play learner? If everyone else is eager to try new strategies, it may be more advantageous to learn from others' mistakes and triumphs. On the other hand, if everyone else follows a beaten path, it may be especially advantageous to sample new ideas.

Optimality and machine learning

As we will see in Chapter 10, computer scientists have made major strides in recent decades in formulating learning models that home in on optimal behavior fairly rapidly even when the number of pure strategies is very large. These "machine learning" models assume that fitness is a constant function of strategy choice, or that the function occasionally shifts exogenously; so far they do not account for frequency dependence (i.e., strategic interaction). Of course, these models are not intended to describe actual human behavior, so they do not directly address the concerns of the present chapter. There seems to be a wonderful opportunity to draw on the machine learning literature to construct models of learning in games, and to find conditions under which the hybrid models approximate the behavior of human experts.

8.10 APPENDIX: TOWARDS MODELS OF LEARNING IN CONTINUOUS TIME

How should a player adapt in a game played in continuous time? An idealized robot playing against slower opponents might always choose a best response, and instantaneously adjust its action whenever the state $s(t)$ changes. Human players are not that fast, and the implementation of the game might impose explicit lags or switching costs even on the fastest robots. It is not obvious how best to even formalize what happens when instantaneous robots play against other robots with the same capabilities.

In this appendix, we focus on a simpler, but neglected, question. How should human players form beliefs $\hat{s}(t)$ about the average state they are likely to face in the near future, and how should they act on these beliefs?

Recall that in discrete time, weighted fictitious play beliefs took the form $\sum_{k=0}^{t-1} \gamma^k s(t-k)$ divided by the normalization factor $\sum_{k=0}^{t-1} \gamma^k = \frac{1-\gamma^k}{1-\gamma}$. The continuous time analogue of the weight γ^k is e^{-rk}, where $r = -\ln \gamma$ is called the interest rate. The corresponding normalization factor is $\int_0^t e^{-ru} du = \frac{1-e^{-rt}}{r}$. In continuous time, weighted fictitious play thus takes the form

$$\hat{s}(t) = \frac{r}{1-e^{-rt}} \int_0^t s(t-u) e^{-ru} du. \tag{8.15}$$

In the limiting case as the interest rate $r \to 0$, we see (using L'Hospital's rule) that the normalization factor converges to t and we get the (unweighted) fictitious play expression $\hat{s}(t) = \frac{1}{t} \int_0^t s(t-u) du$. As the interest rate gets larger, the more recent observations get greater weight. This can be quantified via the half-life $h(r)$ as follows. Solving the equation $\frac{1}{2} = e^{-rh}$, we see $h(r) = \frac{\ln 2}{r} \approx \frac{0.7}{r}$ is the lag time over which a given observation loses half of its influence on beliefs. For example, if the interest rate is $r = 0.07$ per second, then the half life is about $0.7/0.07 = 10$ seconds, so current beliefs respond about half as strongly to the state observed 10 seconds ago as they do to the current state.

Modeling beliefs as in (8.15) allows us to predict behavior. In discrete time, one predicts the players' choices next period given their current beliefs. In continuous time, however, there is no analogue of "next period" so we instead try to predict *when* a player will change strategy. More specifically, with only two pure strategies A and B, we try to predict, given current beliefs $\hat{s}$, the probability per unit time $h_{AB}(\hat{s})$ that a player currently choosing A will switch to B, and also the reverse transition rate (or intensity) $h_{BA}(\hat{s})$.

To put it in more general terms, the empirical task is to estimate the Markov transition rate matrix

$$\begin{bmatrix} h_{AA}(\hat{s}) & h_{AB}(\hat{s}) \\ h_{BA}(\hat{s}) & h_{BB}(\hat{s}), \end{bmatrix}, \tag{8.16}$$

where the probability density of remaining in action A is $h_{AA}(\hat{s}) = 1 - h_{AB}(\hat{s})$, and $h_{BB}(\hat{s}) = 1 - h_{BA}(\hat{s})$ is the expression for remaining in B. The data are the times $0 \leq t_{A1} \leq t_{A2} \leq \ldots$ at which a player switches from action A (to B) and the times $0 \leq t_{B1} \leq t_{B2} \leq \ldots$ at which a player switches from action B (to A), along with the states $s(t_{Ai})$ and $s(t_{Bi})$ observed at those times (and immediately after).

Very little work has yet been published reporting estimates from such data. In the broader literature (e.g., medical studies and process engineering), the standard model is the relative risk or Cox model, which here takes the form

$$h_{BA}(t) = \overline{h}_{BA}(\tau)\exp[\alpha + \beta\Delta w(\hat{s}(t))], \tag{8.17}$$

where τ is the time elapsed since the previous event t_{Ai} or t_{Bi}. That is, the Cox model decomposes the transition rate from B to A into two components: a nonparametric component $\overline{h}_{BA}$ that depends only on the time elapsed since the last event, and a parametric component analogous to the logit model in equation (8.9). (Presumably $\overline{h}_{BA}(\tau) = 0$ at $\tau = 0$ and increases monotonically to some asymptote.) The other transition rate $h_{AB}(t)$ is the product of its own baseline rate $\overline{h}_{AB}(\tau)$ and the reciprocal $\exp[-\alpha - \beta\Delta w(\hat{s}(t))]$ of the parametric component in (8.17).

The estimation strategy is to maximize the partial likelihood function formed using the parametric components, and let the non-parametric component follow the data. We hope to see this approach used in future research.

EXERCISES

1. Verify the claim of Section 8.5 that the likelihood ratio is e^u when the observed action is A and is e^{-u} when it is B, where $u = \alpha + \beta\Delta w(\hat{s}(\gamma))$ and $\alpha = \alpha_A - \alpha_B$. Hint: Begin with equation (8.9), which here reads

$$Pr[a = A] = e^{\alpha_A + \beta\hat{w}_A} / [e^{\alpha_A + \beta\hat{w}_A} + e^{\alpha_B + \beta\hat{w}_B}] \tag{8.18}$$

$$Pr[a = B] = e^{\alpha_B + \beta\hat{w}_B} / [e^{\alpha_A + \beta\hat{w}_A} + e^{\alpha_B + \beta\hat{w}_B}]. \tag{8.19}$$

Multiply these equations by $1 = e^{-[\alpha_B + \beta\hat{w}_B]} / e^{-[\alpha_B + \beta\hat{w}_B]}$ to obtain numerator e^u in the first equation and numerator 1 in the second equation. Conclude that the likelihood ratio is $e^u/1 = e^u$ when A is observed and is $1/e^u = e^{-u}$ when B is observed.

2. Consider the 3×3 matrix $w = \begin{pmatrix} 1 & -10 & -0.1 \\ 1.999 & 2 & 2 \\ 0 & 0 & 0 \end{pmatrix}$. Show that under continuous replicator dynamics and also BR learning dynamics (the continuous logit dynamics with precision $\beta = \infty$) the only EE (stable steady states) are $s^* = (1,0,0)$ and $s^{**} = (0,1,0)$. Show, however, that the basin of attraction for s^* is tiny while the basin for s^{**} is almost the entire simplex under BR dynamics, while the reverse is true under replicator dynamics. This example (adapted from Golman and Page 2010) is another way to make the point that dynamics matter.
3. Build an ecologies of learning styles simulation model as follows. Code a set of (α, β, γ) weighted fictitious play parameters as binary strings. Look up the literature on the genetic algorithm (GA), or similar simulation procedures for evolving strategies. Pick a set of simple games (e.g., the set of four games mentioned in Section 8.6). Run the GA algorithm on those games, and let the strings evolve. Verify or disconfirm our conjecture that two or more distinct clusters of learning styles will evolve, perhaps one with low αs and high βs and γs, and another cluster with the reverse.
4. Find a data set recording behavior of a population of players who can switch freely in continuous time between two alternative strategies, A and B. Develop ML estimates of the parameters in (8.16)–(8.17).

NOTES

We thank Luba Petersen for providing the Stata estimates reported in the first part of Section 8.6, and thank Ciril Bosch and Jacopo Magnani for working out approach to continuous time estimation sketched in the Appendix.

For background on Luce's Choice Rule, see for example Luce (1977, 2005).

The American mathematician Julia Robinson (1919–1985), one of two independent inventors of fictitious play learning in games, should not be confused with the prominent British economist Joan Robinson (1903–1983). Sadly, some widely read literature perpetuates this confusion.

Section 8.6 neglects many crucial details that can be found in the original articles. Here we mention that the criterion for short memory was that the player's estimated γ parameter was significantly ($p = .05$) less than 1.0 but insignificantly different from 0; for long memory it was the opposite; and for intermediate memory it was both significantly less than 1.0 and also significantly greater than 0. Friedman (1996) interprets $\gamma < 0$ as "fickle" because the sign of the weight on a given earlier observation flips from one period to the next, and iterprets $\gamma > 1$ as "imprinting" because first impressions matter most. Fudenberg and Levine (2014) is a good recent reference on recency.

There is a vast literature on learning in games beyond the articles mentioned in the text. A good introduction to it is the book *Behavioral Game Theory* by Colin Camerer. A recurrent tension in behavioral economics arises in how best to explain persistent apparent deviations from NE (or from optimality). Proponents of prospect theory and other generalizations of expected utility theory regard the deviations as a basic fact that should be accommodated, while proponents of models of adaptation and learning regard such deviations as arising from slow or obstructed learning processes. See the last part of Friedman (1998), and also Erev and Roth (2014).

BIBLIOGRAPHY

Brown, George W. "Iterative solution of games by fictitious play." *Activity analysis of production and allocation* 13, no. 1 (1951): 374–376.

Camerer, Colin. *Behavioral game theory: Experiments in strategic interaction*. Princeton University Press, 2003.

Cason, Timothy N., Daniel Friedman, and Ed Hopkins. "Cycles and instability in a rock-paper-scissors population game: A continuous time experiment*." *The Review of Economic Studies* (2014): 112–136.

Chen, Hsiao-Chi, James W. Friedman, and Jacques-Francois Thisse. "Boundedly rational Nash equilibrium: A probabilistic choice approach." *Games and Economic Behavior* 18, no. 1 (1997): 32–54.

Cheung, Yin-Wong, and Daniel Friedman. "Individual learning in normal form games: Some laboratory results." *Games and Economic Behavior* 19, no. 1 (1997): 46–76.

Cheung, Yin-Wong, and Daniel Friedman. "A comparison of learning and replicator dynamics using experimental data." *Journal of Economic Behavior and Organization* 35, no. 3 (1998): 263–280.

Cournot, Antoine-Augustin. *Recherches sur les principes mathématiques de la théorie des richesses par Augustin Cournot*. chez L. Hachette, 1838.

Crawford, Vincent P., Miguel A. Costa-Gomes, and Nagore Iriberri. "Structural models of nonequilibrium strategic thinking: Theory, evidence, and applications." *Journal of Economic Literature* 51, no. 1 (2013): 5–62.

Engel, Charles, and Kenneth D. West. "Exchange rates and fundamentals." *Journal of Political Economy* 113, no. 3 (2005): 485–517.

Erev, Ido, and Alvin E. Roth. "Maximization, learning, and economic behavior." *Proceedings of the National Academy of Sciences* 111, Supplement 3 (2014): 10818–10825.

Friedman, Daniel. "Equilibrium in evolutionary games: Some experimental results." *The Economic Journal* 106, no. 434 (1996): 1–25.

Friedman, Daniel. "Monty Hall's three doors: Construction and deconstruction of a choice anomaly." *American Economic Review* 88, no. 4 (1998): 933–946.

Fudenberg, Drew, and David K. Levine. "Recency, consistent learning, and Nash equilibrium." *Proceedings of the National Academy of Sciences* 111, Supplement 3 (2014): 10826–10829.

Golman, Russell, and Scott E. Page. "Basins of attraction and equilibrium selection under different learning rules." *Journal of Evolutionary Economics* 20, no. 1 (2010): 49–72.

Luce, R. Duncan. *Individual choice behavior: A theoretical analysis.* Courier Corporation, 2005.

Luce, R. Duncan. "The choice axiom after twenty years." *Journal of Mathematical Psychology* 15, no. 3 (1977): 215–233.

Meese, Richard A., and Kenneth Rogoff. "Empirical exchange rate models of the seventies: Do they fit out of sample?." *Journal of International Economics* 14, no. 1 (1983): 3–24.

McKelvey, Richard D., and Thomas R. Palfrey. "Quantal response equilibria for normal form games." *Games and Economic Behavior* 10, no. 1 (1995): 6–38.

Nagel, Rosemarie. "Unraveling in guessing games: An experimental study." *The American Economic Review* 85, no. 5 (1995): 1313–1326.

Page, Scott E. "Let's make a deal." *Economics Letters* 61, no. 2 (1998): 175–180.

Rick, Scott, and Roberto A. Weber. "Meaningful learning and transfer of learning in games played repeatedly without feedback." *Games and Economic Behavior* 68, no. 2 (2010): 716–730.

Robinson, Julia. "An iterative method of solving a game." *Annals of Mathematics* 54, no. 2 (1951): 296–301.

Rutström, E. Elisabet, and Nathaniel T. Wilcox. "Stated beliefs versus inferred beliefs: A methodological inquiry and experimental test." *Games and Economic Behavior* 67, no. 2 (2009): 616–632.

Selten, Reinhard, and Thorsten Chmura. "Stationary concepts for experimental 2 × 2-games." *The American Economic Review* 98, no. 3 (2008): 938–966.

Stahl, Dale O., and Paul W. Wilson. "Experimental evidence on players' models of other players." *Journal of Economic Behavior and Organization* 25, no. 3 (1994): 309–327.

9

Contingent Life-Cycle Strategies

Behavior need not be rigidly pre-programmed. It can be stochastic, for example, a particular genome may map to a mixture of behaviors, and that can be adaptive when the environment or population distribution is unknown. The biological literature refers to this as *bet hedging*, and points to examples such as the fire-bellied toad in which females produce eggs with tremendous variation in size. If the optimum egg size fluctuates unpredictably from year to year, then this mixed strategy will ensure that at least some progeny are close to the optimum (Kaplan and Cooper 1984).

Behavior and strategies can also be *plastic*, or responsive to cues about current circumstances. The upper-left panel of Figure 9.1 illustrates an extreme example. The largest female blue-headed wrasse will change her sex and become male when large parental terminal phase males are removed, naturally or experimentally (Warner and Swearer 1991). In addition, females can also change into the smaller intermediate phase males. The lower-left panel illustrates spadefoot toads in tadpole stage. The presence of energy-rich shrimp is used as a local cue to trigger transformation of an omnivorous type tadpole into the carnivore type. The upper-right panel shows how dung beetles turn into horned and aggressive males called majors or submissive minor males with a short horn, depending on how much dung was provided to them by their parents. By contrast, the lower-right panel illustrates the rigid (not plastic) strategies of male isopods: as explained in Chapter 7, they develop into α, β and γ males according to the genes acquired at conception that determine their fate.

There is a recent biological literature about non-strategic plasticity, for example, Leimar (2005), Leimar et al. (2006), and Svardal et al. (2011), which analyzes the relative advantages of evolving rigid versus plastic strategies in terms of the frequency and amplitude of environment fluctuations. There also is a separate literature on contingent strategic behavior, for example, Maynard Smith (1982, Chapter 8) and Broom et al. (2004), but the contingencies considered have been somewhat limited. In the case of female mate choice, however, there is

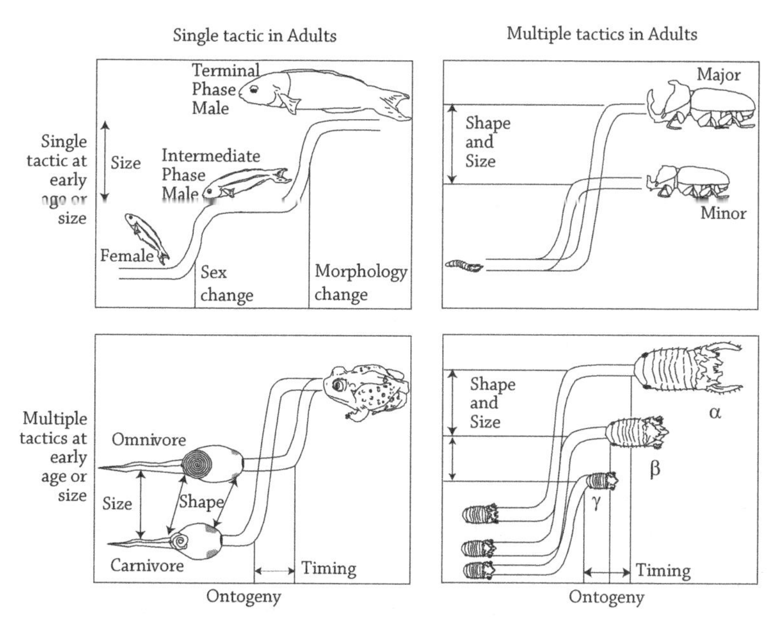

Figure 9.1 Three plastic strategies (blue-headed wrasse, dung beetles, spadefoot toad larva), contrasted with a fixed strategy (isopod males).

a rich developing field of context-dependent choice in which females adopt a flexible strategy in choice of mate that will enhance progeny fitness in subsequent generations. The case of side-blotched lizard females who choose rare sires, discussed in Chapter 7, reflects context dependent choice. It can be analyzed in terms of best response dynamics (Alonzo and Sinervo 2001, 2007). Qvarnström (2001) reviews other cases of context-dependent mate choice, but the theory is not well developed.

In this chapter we take a more unified view of plastic strategies. Putting aside bet-hedging (which might be considered plastic from the gene's perspective but not from that of an individual organism), we analyze the strategic implications of responding to various sorts of contingencies. We begin with the standard Hawk-Dove game, as in Figure 9.2, for simplicity and concreteness. The contingencies or cues that we consider fall into four different categories: the individual's role (e.g., resource holder vs. intruder), the environmental state (e.g., high or low resource value), the population state (e.g., the current phenotype shares), and the current payoffs to different behaviors.

Sometimes behavior at one stage of life (e.g., educational attainment as a young student) affects the opportunities and payoffs later. There is a classic game-theoretic literature on such games, but it mainly focuses on equilibrium outcomes defined stage by stage, and has little to say about adaptation and

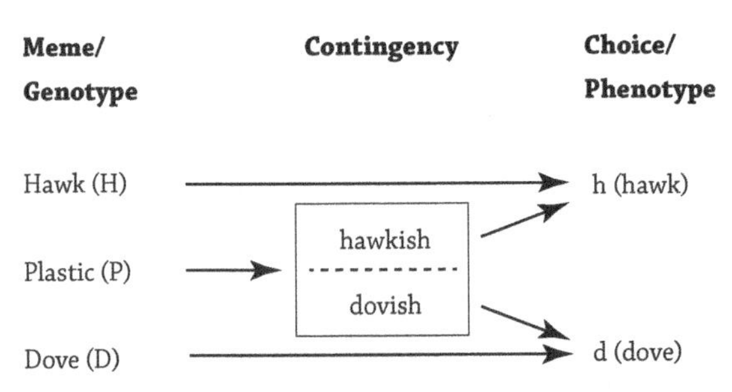

Figure 9.2 Hawk-Dove game with plasticity. Strategy P results in behavior h (hawk) in some contingencies and behavior d (dove) in other contingencies.

dynamic stability. There is also a classic biological literature on fitness over the life cycle, but it largely ignores strategic interaction.

The rest of the chapter sketches an evolutionary approach to strategic interaction over the life cycle, including the longer-term benefits and costs of plastic strategies. A long section applies the approach to elephant seals. It includes new data and modeling techniques. The concluding discussion offers some thoughts on further applications, followed by the usual chapter-end notes and exercises. Technical details are collected in an appendix.

9.1 HAWKS, DOVES, AND PLASTICITY

It will be helpful to begin with a familiar static strategic situation, involving aggressive (hawk) versus cooperative (dove) behavior. Recall from Chapter 2 that with resource value $v > 0$ and conflict cost $c > v$, the basic HD payoff matrix is $\begin{pmatrix} \frac{v-c}{2} & v \\ 0 & \frac{v}{2} \end{pmatrix}$ with payoff difference $\Delta_{h-d}W = \frac{1}{2}(v - s_h c)$, and that the solution of equation $\Delta_{h-d}W(s_h) = 0$ yields the equilibrium fraction of hawk play $s_h^* = \frac{v}{c}$.

Maynard Smith (1982) and later authors such as Broom et al. (2004) discuss what we call *role contingent* plasticity. Maynard Smith considers two mutually exclusive roles, resource owner or intruder, and introduces the plastic strategy B (for bourgeois) which calls for h if owner and d if intruder. Of course, there are also the rigid strategies H (play h in either role) and D (always play d). Suppose for concreteness that $c = 6$ and $v = 4$ and that the two roles are equally likely. Then, for example, the fitness of B against H is $W_{BH} = 0.5W_{hh} + 0.5W_{dh} = 0.5(-1) + 0.5(0) = -0.5$. For the symmetric game with pure strategies $\{H, D, B\}$, similar calculations yield the payoff matrix $\begin{pmatrix} -1 & 4 & 1.5 \\ 0 & 2 & 1 \\ -0.5 & 3 & 2 \end{pmatrix}$. Using the methods

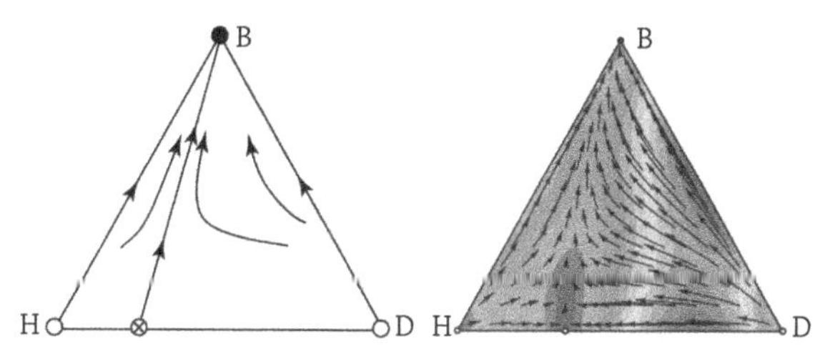

Figure 9.3 Phase portrait for HDB game (Panel A) and replicator dynamics (Panel B) as drawn by Dynamo (Sandholm et al. 2013).

of Chapter 4, it is straightforward to see that the Nash equilibria for this game include the replicator saddle point $(2,1,0)/3$ and the sink (and ESS) $(0,0,1)$. Figure 9.3 illustrates. Absent other considerations, once the bourgeois strategy invades, it should eventually reach fixation.

Broom et al. 's Marauder strategy M specifies h in the role of intruder and d in the role of owner. They develop an expanded symmetric h-d game with four pure strategies: $\{H,D,B,M\}$. Their development includes additional payoff-relevant parameters, and they analyze replicator dynamics for the resulting 4x4 payoff matrix.

Such role-contingent plasticity models of course require mutually exclusive roles that the players can immediately detect. Behavior adapts on the usual time scale, as shares of the role-contingent strategies evolve.

A second sort of plasticity, *state contingent,* is in the spirit of Leimar (2005) and Lively (1986). In state-contingent models, the environment fluctuates and strategies are contingent on the environmental state. For example, suppose that with equal probability 0.5 the resource value can either be high at $v=4$, or low at $v=2$, while the cost of conflict remains constant at $c=6$. The organism reliably detects the true state and responds by playing, for example, *hd*—that is, *h* in the high state and *d* in the low state. Against the rigid strategy $H=hh$ the payoff to *hd* is

$$W(hd,hh)=p^{hi}W^{hi}(h,h)+p^{lo}W^{lo}(d,h)=0.5\frac{4-6}{2}+0.5\cdot 0=-0.5.$$

The other plastic strategy is, of course, to play *d* in the high state and *h* in the low state, denoted *hd*. The remaining strategy is $D=dd$ (i.e., rigid dove). Straightforward calculations for those strategies yield the following payoff matrix for the expanded game with pure strategy set $\{hh,hd,dh,dd\}$: $\begin{pmatrix} -1.5 & 0.5 & 1 & 3 \\ -0.5 & 0 & 2 & 2.5 \\ -1 & 1 & 0 & 2 \\ 0 & 0.5 & 1 & 1.5 \end{pmatrix}$.

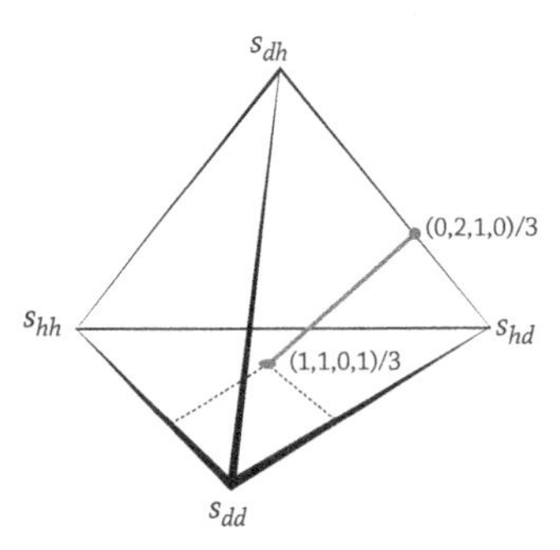

Figure 9.4 Equilibrium set in state-contingent Hawk-Dove game.

Equilibrium in this example is characterized by $s_{hh} + s_{hd} = s_h^{hi*} = \frac{2}{3}$ and $s_{hh} + s_{dh} = s_h^{lo*} = \frac{1}{3}$. Both conditions are met by a range of mixed strategies, or shares. The share s_{dh} can range from 0 to $\frac{1}{3}$, as long as the other shares adjust appropriately. To spell it out formally, the set of NE is the continuum $\{(s_{hh}, s_{hd}, s_{dh}, s_{dd}) = (\frac{1}{3} - z, \frac{1}{3} + z, z, \frac{1}{3} - z) : z \in [0, \frac{1}{3}]\}$, and each point in this set is neutrally stable. Figure 9.4 illustrates.

By definition, plastic strategies of this sort require that the organism be able to detect the environmental state and respond to it. For example, the cue to the environmental state might be the amount of the resource or its color or taste when v is high, and the response might be via a boost to endocrine levels. The model assumes that observed behavior reacts very quickly to the revealed state, but the shares of state-contingent plastic strategists (and other strategists) evolve on the usual time scale.

A third sort of plasticity, first suggested in Sinervo (2001), is *frequency contingent* (e.g., play h when the share s_H of H strategists is below some threshold $\hat{s}_H$). Such strategies can speed convergence to equilibrium following a shock. To illustrate, suppose that in the low state of the previous example, the share s_F of frequency-dependent strategists is at least $\frac{1}{3}$ and the threshold is $\hat{s}_H = \frac{1}{3}$, but some sort of shock to the system depletes hawkish play. Then frequency-contingent hawks can restore the equilibrium $s_h^{lo*} = \frac{1}{3}$ immediately, in a single period. Sinervo (2001) discusses a more complicated RPS game, where the frequency-contingent strategy plays paper if $s_R \geq \hat{s}_R$ and otherwise plays scissors; such strategies can gain higher fitness than either P or S in a variable environment.

The requirements for such plastic strategies are similar to their state-contingent analogues. Here the organism might sample behavior (or morphs) in its vicinity

Contingency	**Example**	**Source**
Role	Resource Endowment ("Bourgeois")	Maynard Smith (1980) Broom et al. (2004)
State	Resource Value	Leimar (2005), Lively (1986), Pfennig (1992)
Frequency	$s_H \lessgtr \overline{s}_H$	Sinervo (2001)
Payoff	$W_i \lessgtr W_j$	herein, Alonzo & Sinervo (2001, 2007) Sinervo et al. (2007)

Figure 9.5 Four possible types of plasticity.

and adjust its endocrine levels. Again, observed behavior of the plastic type reacts quickly, but their shares evolve more slowly via standard replicator dynamics.

Figure 9.5 summarizes all three contingencies considered so far and introduces a fourth sort of plasticity called *payoff contingent,* for example, play h when it gives a higher payoff than d and otherwise play d. This sort of plasticity is seldom discussed explicitly in the literature (see the chapter notes), but the required cues (e.g., outcomes of territorial disputes) seem no less observable than for the other types of contingencies. Observed behavior again adapts rapidly when there is a reasonably large share of payoff-contingent strategists.

To illustrate, suppose again that there are two equally likely states for the resource value, $v = 4$ (hi) and $v = 2$ (lo). Besides the rigid hawk ($\mathrm{H} = hh$) and rigid dove ($\mathrm{D} = dd$) strategies, let there be a single plastic strategy (P) that picks the action with higher payoff. Denote the shares by $s_H + s_D + s_P = 1$. We do not need an augmented matrix to describe behavior, but do need to keep track of the fraction x of the plastic strategists who end up playing hawk. Thus the fraction of hawk play is $s_h = s_H + xs_P$, while the fraction of dove play is $s_d = 1 - s_h = s_D + (1-x)s_P$.

The idea is that x adjusts rapidly in each state. For example, suppose that initially $s_P = \frac{1}{4}$, while $s_H = \frac{1}{2}$ and so $s_D = \frac{1}{4}$. Then in the low state we have $\Delta_{h-d}W = \frac{1}{2}(v - s_h c) = 1 - 3s_h = 1 - 3(s_H + xs_P) \leq 1 - 3s_H = -\frac{1}{2} < 0$, that is, d yields the higher payoff for any $x \in [0,1]$. Hence x quickly adjusts to 0 in the low state, as all P strategists choose d, and they and the D strategists receive payoffs $\frac{1}{2}$ higher than the rigid H strategists. In the high state, $\Delta_{h-d}W = \frac{1}{2}(v - s_h c) = 2 - 3s_h = \frac{1}{2} - \frac{3x}{4} \gtrless 0$ when $x \lessgtr \frac{2}{3}$. Thus the fraction x of P strategists quickly adjusts to equalize h and d payoffs, and settles at $x = \frac{2}{3}$. Thus, when $(s_H, s_D, s_P) = (\frac{1}{2}, \frac{1}{4}, \frac{1}{4})$, the short-run equilibrium is $x^{hi} = \frac{2}{3}, x^{lo} = 0$.

We just saw that in this short-run equilibrium, P and D both earn $\frac{1}{2}$ more payoff in the low state than does H, while all earn the same payoff in the high state. Since the low state occurs with probability $\frac{1}{2}$, we conclude that rigid strategy H suffers a payoff disadvantage $\frac{1}{4}$ relative to D and P. Therefore, on the time scale

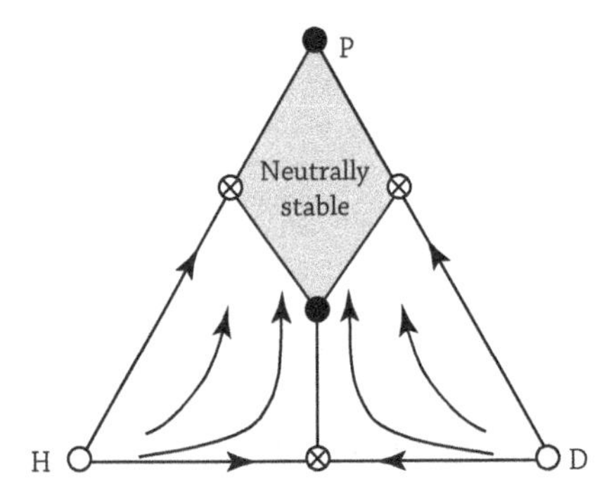

Figure 9.6 HDP Phase Portrait, $a = 0$

that replicator dynamics operate, the H share will fall until its payoff disadvantage disappears at $s_H = \frac{1}{3}$. By similar reasoning, when the initial fraction of D is above $\frac{1}{3}$, these dedicated doves will fare poorly in the high state, and they will receive lower lifetime fitness than the H and P strategists. We conclude that in long-run equilibrium we must have $s_D \leq \frac{1}{3}$ and $s_H \leq \frac{1}{3}$, so $s_P \geq \frac{1}{3}$. As before, there is a whole range of shares that satisfy these inequalities, and all of them are neutrally stable, as shown in Figure 9.6.

9.2 COSTLY PLASTICITY

The preceding examples suggest that plasticity should be widespread, perhaps even universal. But those examples assume that plastic strategies are just as easy to implement as rigid strategies, contrary to intuition and evidence. Generally speaking, plasticity bears some cost (Lively 1986). Rapid adjustments via the endocrine system are tough on the metabolism, for example, and cue detection is seldom perfect.

To see the consequences of such costs, suppose that strategy P incurs a modest payoff reduction $a \in (0, \frac{1}{4}]$ relative to either rigid strategy. Then the equilibrium shares are unique, and given by $s_H = s_D = \frac{1+2a}{3}$ and $s_P = \frac{1-4a}{3}$. Deriving these expressions is straightforward but a bit messy; however, it is easy to check that they yield equal lifetime fitnesses, W_i^L, $i = H, D, P$. For example,

$$W_H^L = \frac{1}{2}[W_h^{hi}(s_H + s_P) + W_h^{lo}(s_H)] = \frac{2 + 10a}{3};$$

the first equality uses the facts that the two states are equally likely and that (in equilibrium) the plastic strategists will all choose h in the high state and will

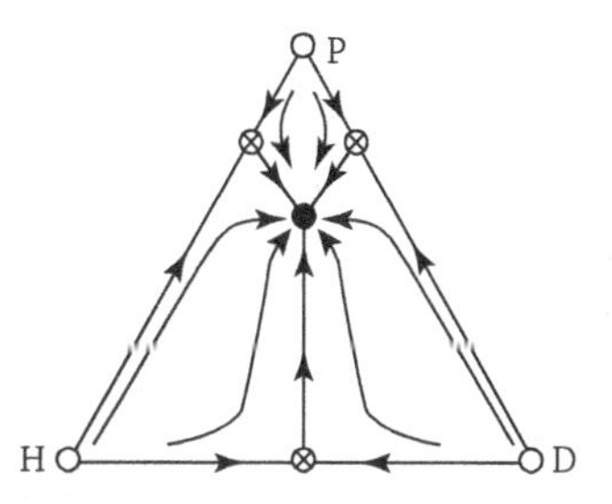

Figure 9.7 HDP Phase Portrait, $a \in (0, \frac{1}{4})$

all choose d in the low state. The second equality simply evaluates the relevant matrix entries at the given states and simplifies. Similarly one confirms that $W_D^L = \frac{1}{2}[W_d^{hi}(s_H + s_P) + W_d^{lo}(s_H)] = \frac{2+10a}{3}$ and that $W_P^L = \frac{1}{2}[W_h^{hi}(s_H + s_P) + W_d^{lo}(s_H)] - a = \frac{2+10a}{3}$.

With a little more work, one can check that when any of the three strategies has a larger share than in equilibrium, it will receive a lower lifetime fitness than an alternative strategy. Intuitively, stability of the unique equilibrium seems clear; an exercise invites you to establish it analytically. The phase portrait is shown in Figure 9.7.

If the cost disadvantage is too large, so $a \geq \frac{1}{4}$, then the plastic strategy can't invade and it disappears; the equilibrium is simply $s_H = s_D = \frac{1}{2}$. On the other hand, if the plastic strategy has a cost advantage, so $a < 0$, then it will go to fixation $(s_P = 1)$ in this example.

The example readily generalizes to more complicated games. In short-run equilibrium, a payoff-contingent plastic strategy tends to equalize payoffs across the pure strategies it can emulate. In the longer run, under replicator dynamics (or other monotone dynamics) its share increases when it earns higher net payoffs on average than non-extinct alternative strategies, and its share shrinks (possibly to zero) when alternative strategies earn higher average net payoffs.

Potential applications abound. Consider, for example, two or more different lines of work or different locales. Sometimes one line or locale offers better pay, sometimes another, so workers who are committed to a particular line or locale will see their fortunes fluctuate. Some workers, however, may be able and willing to switch rapidly to the best paying locale. If the plasticity cost (of retraining or relocation) is not too high, then they will shift labor supply from low to high demand locales, thereby smoothing out the fluctuations to some degree. In the limit where plasticity cost goes to zero, the share of plastic (mobile) workers will increase sufficiently in long-run equilibrium to completely eliminate the pay differences across locales.

9.3 CLASSIC LIFE-CYCLE ANALYSIS

Sometimes observed strategies appear to be dominated when looked at in isolation, but make perfect sense when considered in the context of the player's past and future. Menopause is a striking example. In most mammals, female fertility declines gradually with age, but never disappears. In humans, however, women's fertility ends abruptly with menopause, typically many decades before death. Why would Nature squander the opportunity? Or, to put the question less metaphorically, why wouldn't women whose menopause comes later (or not at all) have more descendants on average and displace those whose fertility ends in early middle age?

An appealing answer is that menopause encourages women to help their daughters raise the grandchildren. Looked at over the life cycle, women who have their own kids when relatively young and who later invest in grandkids may have more surviving descendants than those who bear more of their own children late in life (e.g., Williams 1957; Hawkes et al. 1997).

Fitness calculations are tricky for lifetime strategies. Economists (and engineers) realized long ago that special calculations are needed to compare lifetime costs and benefits received at different times of life, especially when lifetimes differ. (For example, present values must be annuitized to compare, say, a five-year computer replacement policy to a two-year replacement policy.) In biology there is a classic literature on life-cycle fitness in nonstrategic settings, summarized in textbooks such as Charlesworth (1980). The focus is demography, the dynamics of an age-structured population.* A typical question: Under which conditions does a species that reaches maturity quickly tend to displace another species competing for the same resources that has greater fitness (higher survival rates or greater fecundity) but takes longer to mature (De Roos et al. 1992)?

The standard mathematical tool for dealing with such questions is the Leslie matrix based on age structure or, with slightly greater generality, the Leftkovitch matrix based on stage structure. For example, consider a population with life stages 1 = newborn, 2 = juvenile, 3 = young adult, 4 = prime adult, and 5 = old. The demographic equation is

$$\begin{bmatrix} N_{1,t+1} \\ N_{2,t+1} \\ N_{3,t+1} \\ N_{4,t+1} \\ N_{5,t+1} \end{bmatrix} = \begin{bmatrix} P_{1,1} & 0 & F_3 & F_4 & F_5 \\ P_{1,2} & P_{2,2} & 0 & 0 & 0 \\ 0 & P_{2,3} & P_{3,3} & 0 & 0 \\ 0 & 0 & P_{3,4} & P_{4,4} & 0 \\ 0 & 0 & 0 & P_{4,5} & P_{5,5} \end{bmatrix} \begin{bmatrix} N_{1,t} \\ N_{2,t} \\ N_{3,t} \\ N_{4,t} \\ N_{5,t} \end{bmatrix}. \tag{9.1}$$

The left-hand side of the equation is the stage-structured population vector at time $t+1$, and the right-hand side is the Leftkovitch matrix L multiplied by the time t population vector. The top row of L contains fecundities F_i, the offspring per capita at each stage, and so determines the new newborn population $N_{1,t+1}$. (For diploid organisms, the fecundities represent offspring per capita for females surviving to that stage.) The possibility that some newborns in time t retain that status in time $t+1$ is captured in the matrix entry $P_{1,1} \geq 0$. The other entries indicate transition rates $P_{i,j} \geq 0$ from stage i to stage j, and are subject to the constraints $P_{i,i}+P_{i,i+1} \leq 1$ and $P_{i,j}=0$ for the other j's. That is, individuals at stage i either remain in that stage, or transition to the next stage, or (with probability $1-P_{i,i}-P_{i,i+1} \geq 0$) disappear from the population (i.e., do not survive). The Leslie matrix imposes the further constraints that $P_{i,i}=0$, each stage lasts at most one time period, and so can be considered an age. Thus diagonal entries are 0 in a Leslie matrix, but can be positive in a Leftkovitch matrix.

Standard matrix theory tells us that long-run dynamics are determined as follows. The asymptotic growth factor λ is the leading eigenvalue of L, that is, the root of its characteristic equation with largest real part. (Biologists refer to that characteristic equation as the Euler-Lotka equation.) The corresponding eigenvector gives the long-run demographic profile (i.e., the shares of each stage). In particular, the answer to the question posed earlier in this section is that species with largest λ will displace rival species with smaller λ, where λ is the specific function of transition rates and fecundities defined by the characteristic equation.

9.4 STRATEGIC LIFE-CYCLE ANALYSIS: TWO PERIODS

Classic life-cycle analysis neglects strategic interaction, and a major goal of this chapter is to show how that gap might be filled.

To get started, in this section we will consider only two life stages, juvenile (index=1) and adult (index=2), one immediately following the next in discrete time. For the moment we also assume that only adults generate payoff (or fitness). In this case, the Leslie (= Lefkowitz) matrix is very simple: $L=\begin{pmatrix} 0 & f \\ p & 0 \end{pmatrix}$, where $f>0$ is the fecundity of adults and $p \in (0,1)$ is the probability that juveniles survive to become adults.

Absent strategic considerations, we can treat f and p as constants, and solve the characteristic function $0=\lambda^2-pf$ to obtain

$$\lambda=\sqrt{pf}. \tag{9.2}$$

It follows that if the population of adults is $N_{2,t}$ in period t, then two periods later it will be $N_{2,t+2}=pfN_{2,t}$. The asymptotic growth rate is the growth factor less 1.0, for example, $\frac{N_{2,t+1}-N_{2,t}}{N_{2,t}}=\frac{\lambda N_{2,t}-N_{2,t}}{N_{2,t}}=\lambda-1=\sqrt{pf}-1$. We conclude that,

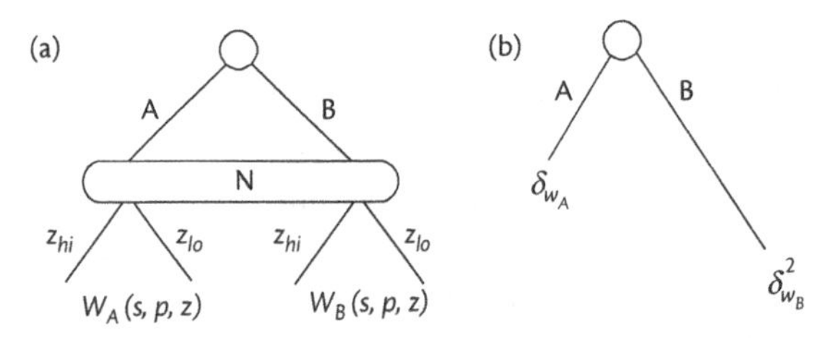

Figure 9.8 (A) Extensive form when juvenile player picks strategy A or B and Nature independently picks z_{hi} or z_{lo}. (B) Extensive form when player picks early maturation strategy A or late maturation strategy B.

after a few periods, the population grows exponentially at a constant rate as long as $f > 1/p$ and that it shrinks exponentially if $f < 1/p$.

What is different when players interact strategically? The population strategy shares could affect each strategy's fecundity f, or survival probability p, or both. Suppose, for example, that juveniles can choose either strategy A or strategy B, resulting in survival probabilities $p = (p_A(s), p_B(s))$, where $s = (s_A, s_B)$ denotes the cohort shares of the two choices. The impact on fecundity $f_i(s,z)$ could depend on some random event z. For example, in the male elephant seal application developed in a later section, z could be the harem size, large or small. In the education example mentioned in the introduction to this chapter, z could be the demand for educated workers, again either large or small. Figure 9.8A summarizes the situation, following the game theory convention that random moves are written as if chosen by player N (for "Nature"). The contingent payoffs are $W_A(s,p,z) = p_A(s)f_A(s,z)$. Their expected values are E_zW_A and E_zW_B. For example, if z_{hi} and z_{lo} are the only possibilities and are equally likely, then $E_zW_A = 0.5p_A(s)[f_A(s,z_{hi}) + f_A(s,z_{lo})]$.

The key insight here is that these expected values are the relevant payoffs in a large population, and they depend only on the strategy share vector s, just as in earlier chapters. The life-cycle variables p and f enter payoffs only via their product. Thus, just as in earlier chapters, the equilibria and dynamics will be determined by the payoff difference $\Delta_{A-B}(s) = E_zW_A - E_zW_B$. We conclude, for example, that in NE all surviving strategies will have identical expected payoffs. Strategy A, for instance, might have a higher survival rate than strategy B, but this will be exactly compensated at an interior equilibrium by a lower fecundity.

Thus the analysis of this simple life cycle case boils down to something familiar from earlier chapters. The same is true for more elaborate cases with the same logical structure. With three or more alternative strategies, one still computes the overall Δ_{i-j}'s and analyzes them as in previous chapters. We have assumed so far that the random event z is independent of the player's choice, but readers are invited to show that the analysis above also covers the case where z is correlated with players' choices.

Life-cycle considerations begin to bite when different strategies receive payoffs at different life stages. For example, suppose that students choosing less education can start earning a paycheck sooner. Or consider salmon breeding strategies. Dominant hooknose breeding males (B) return late in life after a costly migration, and they work hard to scour out a nesting site. Females are attracted to these nests and the attending hooknose male and begin to lay their eggs. An alternative strategy (A) used by jack male salmon is to return earlier at smaller size and to forgo any nesting behavior. Jack males squirt their sperm into the mix as the hooknose male fertilizes the eggs being laid by the females.

Figure 9.8B illustrates the simplest possible case, where strategy A receives its payoff $W_A(s)$ in the first stage of the life cycle while strategy B receives its payoff only in the second and final stage. The corresponding L matrix for B is as above (with subscript B attached to p and f), while for A it is $L_A = \begin{pmatrix} f_A(s) & 0 \\ p_A(s) & 0 \end{pmatrix}$.

Given shares $s = (s_A, s_B)$ of the two strategies, the growth factors λ_A, λ_B are easily computed. From equation (9.2) we have $\lambda_B = \sqrt{p_B(s) f_B(s)}$. The characteristic equation for L_A is $0 = (f_A(s) - \lambda_A)(-\lambda_A)$, so we have simply $\lambda_A = f_A(s)$. Strategy A is dominant if $\lambda_A > \lambda_B$ for all s, and strategy B is dominant in the reverse case. There will be a stable interior equilibrium if the fecundity of either strategy (or the survival probability of B) is sufficiently negatively frequency dependent that the expression $f_A^2(s_A) - p_B(s_A) f_B(s_A)$ has a root at some downcrossing $s_A^* \in (0,1)$.

Note that the survival probability of strategy A is irrelevant in this example since the strategy has zero fecundity in the second stage. Also, we can easily have $f_A < f_B$ at an equilibrium in which A has positive share (or even is dominant), for example, where $p_B = 1$, $f_A = 1.1$ and $f_B = 1.21$. The intuition is that a 10% per period growth rate for A can balance 21% growth that takes *two* periods for B.

The general principle is that, in a life-cycle context with fitness for different strategies received disproportionately at different stages of life, one must apply the delta function analysis not to fitness received any particular stage, but rather to the growth factors, since they summarize fitness across all stages. For example, life-cycle NE s^* can be characterized by the statement that all surviving strategies i have maximal growth factors (i.e., $\lambda_i(s^*) = \max_j \lambda_j(s^*)$ if $s_i^* > 0$).

To tie up a remaining loose end, note that in Figure 9.8B the discount factor δ applies to payoffs received the first stage, and δ^2 applies to those in the second period. Classical game theorists usually regard the discount factor as a parameter determined outside the game, and they sometimes describe it in terms of an interest rate (or a subjective impatience rate) r via $1 + r = 1/\gamma$. Here $r = \lambda - 1$ is naturally interpreted as the growth rate, so we can set $\lambda = 1/\gamma$, and define the discount factor via $\gamma = 1/\lambda$. As in classical game theory, the intuition for discounting is that payoffs earned earlier are more valuable to the extent that they can be reinvested and compounded.

Readers trained in classical game theory may wonder whether trees like those in Figures 9.8A and 9.8B should be thought of as extensive form game trees or just decision trees. In a single-population evolutionary game, the distinction is moot: individuals take the population shares as given when they select a strategy, so (as in a decision tree) only one payoff function need be written out at the terminal nodes. The trees will look just like those for typical extensive form games when two or more distinct populations co-evolve, but the payoff vector at each terminal node will in general be a function of the population state variable (capturing "playing the field" effects) as well as the path through the tree (capturing specific matching effects). The next section and the appendix say a bit more about such matters.

9.5 STRATEGIC LIFE-CYCLE ANALYSIS: MORE GENERAL CASES

Things get even more interesting when the environment, and thus the specific game to be played at a later stage, is not known at an earlier stage. A *fixed* lifetime strategy can't respond to the unfolding contingencies, while a *plastic* lifetime strategy is capable of switching between two or more stage game actions once the specific stage game is revealed. Of course, such flexibility usually comes at some price.

Consider, for example, the economic analysis of when to launch a new product. By waiting, a firm maintains greater flexibility in responding to actual and anticipated competition and to changes in the overall size of the market. But waiting longer shortens the revenue stream and reduces the opportunity to build consumer loyalty. Indeed, the cost of early plasticity can be prohibitive. Consider, for example, an ideologically flexible U.S. politician who receives a high payoff early in his career by taking a centrist position in a statewide race, but who later seeks to win national primary elections where the key voting block punishes politicians who ever deviated from extreme right (or left) positions.

To deal with such possibilities, we sketch a fairly general model of life cycle strategies. Here the sketch will be mainly verbal; see the appendix for some mathematical formalities. We contemplate a population of individuals who all go through a finite number of life stages. Guided by genes and/or memes, at each stage each individual chooses an action, possibly contingent on something observable at that time. Her action has three consequences: (a) it affects the payoff in the current stage, (b) it affects the actions she can choose (and perhaps the contingencies she can respond to) in subsequent stages, and (c) it affects her probability of surviving to the next stage, and perhaps the discount factor that applies.

A strategy specifies a feasible action (possibly contingent) at each stage of life. Good strategies should balance current and future payoffs, in other words, at each stage they should maximize the sum of the current payoff and the expected discounted payoff over remaining life stages.

Two conceptual issues require immediate discussion. First, as explained in the appendix, the discount factor can incorporate the probability of surviving to the next stage as well as time preference (or opportunities to reinvest payoffs).

Second, payoffs in the present as well as future stages depend on the actions of the rest of the population, so it is not obvious how to specify the expectation. Classical game theory looks for equilibrium in the sense that the expectations are taken with respect to the actual strategies that will be taken, using Bayes rule to update where necessary. A more adaptive perspective is that expectations are based on historical averages, possibly giving more weight to more recent observations.

Under either interpretation, if things settle down, it should be to what we will call a *perfect lifetime equilibrium* (PLE)—a strategy share distribution and set of expectations such that at every stage each player (i) maximizes the sum of current and expected discounted future payoff, and (ii) expectations are correct on average given the strategy distribution.

In PLE, all strategies with positive share produce the same maximal lifetime fitness as summarized in the λ's. It follows that any surviving plastic strategies will also equalize the remaining lifetime payoffs across all actions that they actually employ. Presumably each plastic strategy has a fitness handicap relative to surviving rigid strategies; otherwise it should tend to displace the rigid strategies it can mimic. If so, to equalize post-handicap fitnesses, the rigid strategies present in equilibrium must bring lower lifetime fitness than the pre-handicap fitness of a plastic strategy.

We know of no formal proof, but it seems reasonable to conjecture that if strategies that yield greater lifetime fitness (i.e., larger λ) become more prevalent over time, then stable steady states are PLE. For example, we suspect that this is generally true for replicator dynamics.

On the other hand, imitation dynamics might have non-PLE stable steady states because payoffs at some life stages may be more prominent than at other life stages. Consider, for example, a high school dropout with an opportunity to join a criminal gang. The high short run expected payoff may be quite prominent and encourage imitation even when the lifetime payoff is lower than that of now taking a minimum wage job.

9.6 APPLICATION: MALE ELEPHANT SEALS

Having a particular application in mind will help us navigate new abstract theory. To that end, consider the life cycle of male elephant seals, *Mirounga angustirostris*. Each winter, females gather together on certain beaches and most give birth to a single pup, a "weaner." After a few months of very intensive nursing, the weaners are ready to swim out to sea. In the last few weeks of January and the first week or two of February, the moms (and the newly matured females) become sexually receptive before also swimming away.

About half the weaners are males. They all follow the same strategy for the first few years, at juvenile stage: stay in the water, try to avoid sharks and other hazards, and try to gain body mass and skills. Usually in year 4, 5 or 6 (though some as late as year 13) they come ashore during a breeding season for the first time. Each year, each grown male adopts one of four strategies. Alphas—typically 9 to 13 years old and weighing up to 2.5 tons—find a concentration of females on the beach and threaten or attack any other male that gets too close. Betas—often a bit smaller than the alphas—linger at the edge of a harem, not challenging the alpha but chasing off outsiders. Gammas also spend the breeding season on the beach but they do not defend any territory and just seek uncontested copulations when higher-ranking males are inattentive. Finally, "surf monsters" spend most of the breeding season just offshore, hoping to mate with a receptive female when she returns to sea.

Due to a population bottleneck in the first part of the 20th century, males are very closely related, so establishing paternity via DNA samples is prohibitively expensive. Fortunately, data from another elephant seal species in the southern hemisphere, *A. leonina*, suggests that paternity is proportional to copulations, which are relatively easy to observe since most action takes place in early daylight hours and in early afternoon. We therefore will assume that the fitness of each strategy (α, β, γ, Sm) is, up to sampling error, proportional to the observed copulation rates of the males following that strategy.

Our data include a fairly extensive sample of copulations recorded during the 2010 and 2011 breeding seasons. Each record includes the date and time, the identity and strategy of the male, the shares of male strategies in the neighborhood of the harem (at that time—shares change by the hour as γs leave or arrive from other harems), and the size of the harem to which the female belongs.

The first step in the data analysis is to count the number of copulations for each male observed each day, normalized to a daily rate c. For example, if seal X1, a β male, is observed in two successful copulations in observation i, a 2.5-hour period beginning 8am, he is credited with a daily rate of 3.44 because the observation period covers 0.58 of the daily expected copulations, and $2/0.58 = 3.44$. The actual unit of analysis is the per capita payoff to each male strategy in a given

Table 9.1. MLE OF COPULATION RATES

Strategy	κ	95%*CI*	θ	95%*CI*	**mean** $(\kappa\theta)$
α	1.765	[1.435–2.171]	0.268	[0.211–0.340]	0.47
β	1.631	[1.328–2.004]	0.177	[0.139–0.225]	0.29
γ	1.709	[1.390–2.101]	0.022	[0.017–0.028]	0.04
Sm	5.041	[4.055–6.268]	0.025	[0.020–0.031]	0.13

harem, so if there are two β males and the second, seal X2, gets zero copulations during that period, then the observation is recorded as $c_{\beta,i} = 3.44/2 = 1.72$.

The next step is to get a sense of the distribution of payoffs for each strategy or, to use jargon properly, for each action chosen in a given season. We use the overall sets of observed copulations to fit the shape and scale parameters (κ,θ) of the gamma distribution. The parameter estimates maximize the likelihood function

$$L_j = \prod_{i=1}^{n_j} \frac{1}{\Gamma(\kappa)\theta^{\kappa}} c_{j,i}^{\kappa-1} e^{\frac{-c_{j,i}}{\theta}}, \tag{9.3}$$

where i indexes the daily normalized copulation rates observed for males following strategy $j = \alpha, \beta, \gamma, Sm$. Results using the Matlab routine *gamfit* are collected in Table 9.6 including centered 95% confidence intervals as well as point estimates. The last column shows the mean copulation rate implied by the point estimates. Figure 9.9 plots the data (i.e., empirical cumulative distributions) against the fitted gamma cumulative distributions.

From this first cut we see that, except for surf monsters, the strategies all have about the same shape parameter but differ in scale. Not surprisingly, alphas' mean overall copulation rate of 0.47 is considerably higher than betas' (0.29) and an order of magnitude higher than gammas' (0.04). The empirical distributions are reasonably close to the fitted gamma distributions, taking into account that large fractions of the sampled data are (and should be) $c_{j,i} = 0$.

So far, so good. The next step is more analytical, and seeks to characterize strategic interdependence, that is, how observed payoffs for each strategy depend on the strategy shares $\mathbf{s} = (s_\alpha, s_\beta, s_\gamma)$ and on harem size h. We hypothesize the structural model

$$\begin{aligned} \tilde{\kappa}_j &= \tau_{j0} + \tau_{j1}s_j + \tau_{j2}s_i + \tau_{j3}s_i s_j, \\ \tilde{\theta}_j &= a_j + b_j h. \end{aligned} \tag{9.4}$$

That is, we hypothesize that the scale parameter $\tilde{\theta}$ of the gamma distribution describing copulations for strategy j is a linear function of harem size, while the shape parameter $\tilde{\kappa}$ is a linear function (plus an interaction term) of own strategy

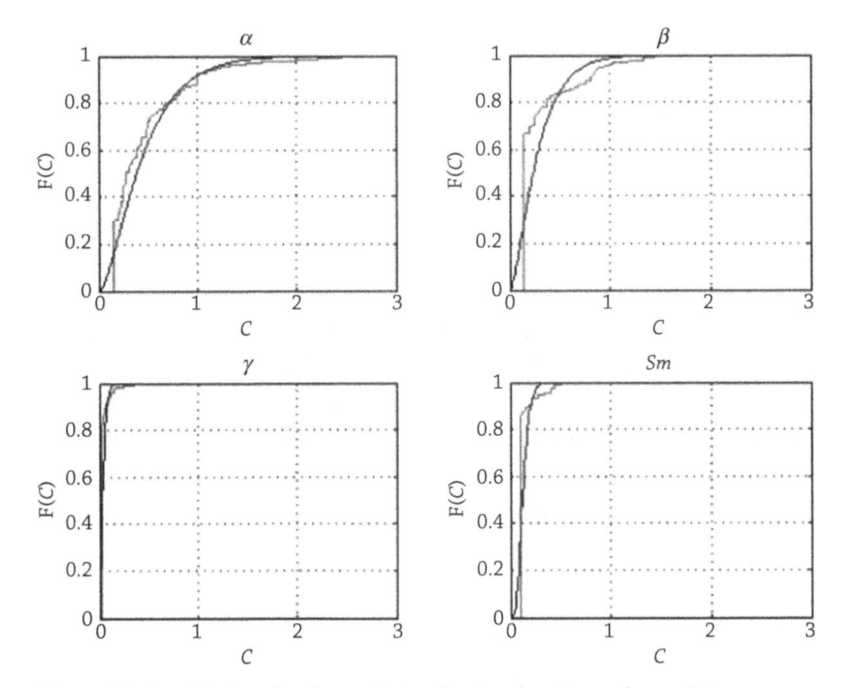

Figure 9.9 Empirical vs. fitted cumulative distributions for α, β, γ and Sm copulations.

share plus the most relevant other strategy share. The relevant pairs are $\{\alpha,\beta\}$, $\{\beta,\alpha\}$, $\{\gamma,\beta\}$, and $\{Sm,\beta\}$ (i.e., α is the relevant other strategy for β and for the other three strategies β is the most relevant). (Preliminary OLS regressions, not shown, established the relevant pairs, e.g., the fact that α barely interacts with γ.)

Table 9.2 collects the estimated coefficients of equation (9.4) using Markov Chain Monte Carlo (MCMC) regressions. To visualize the implications, Figure 9.10 collects graphs of the implied mean fitness ($\mu_j = \tilde{\kappa}_j\tilde{\theta}_j$) of each strategy j as a function of harem size and the share of the related strategy i. (Even in a 3-d plot, we are unable to display the effect of the other shares, so we hold them constant at their median level, e.g., s_α at 0.13.)

Panel A of Figure 9.10 indicates that α males gain fitness slightly from the interaction of shares of β and from harem size. However, Panel B shows that the gains to β males from other β males in interaction with harem size are much steeper, while Panel C shows that gains to γ are least. This observation gives insight into why α males tolerate β males, and why the strategy β can persist.

Another interesting implication is that beta males actually gain higher fitness than alphas in harems larger than about 100 females; see Panel D of Figure 9.10. In contrast, γ males (Figure 9.10E,F) and Sm (not shown) lose to α and β males across all harem sizes.

Table 9.2 MCMC Estimates of Structural Parameters [and 95% confidence intervals]

Strategy	τ_0	τ_1	τ_2	τ_3	a	b
α	0.965	0.755	0.681	0.496	0.391	0.0006
	[0.99,0.93]	[0.96,0.54]	[0.93,0.43]	[0.77,0.22]	[0.43,0.35]	[0.0012,0.0001]
β	0.966	0.780	0.790	0.55	0.128	0.002
	[0.99,0.93]	[0.98,0.57]	[0.97,0.60]	[0.83,0.27]	[0.15,0.10]	[0.003,0.001]
γ	0.817	0.735	0.495	0.52	0.026	0.0002
	[0.98,0.64]	[0.94,0.52]	[0.78,0.20]	[0.80,0.24]	[0.03, 0.02]	[0.0004,−0.0001]
Sm	0.98	0.86	0.78	0.51	0.083	0.0005
	[0.98,0.96]	[1.00,0.73]	[0.98,0.59]	[0.81,0.22]	[0.095, 0.0711]	[0.0007,0.0003]

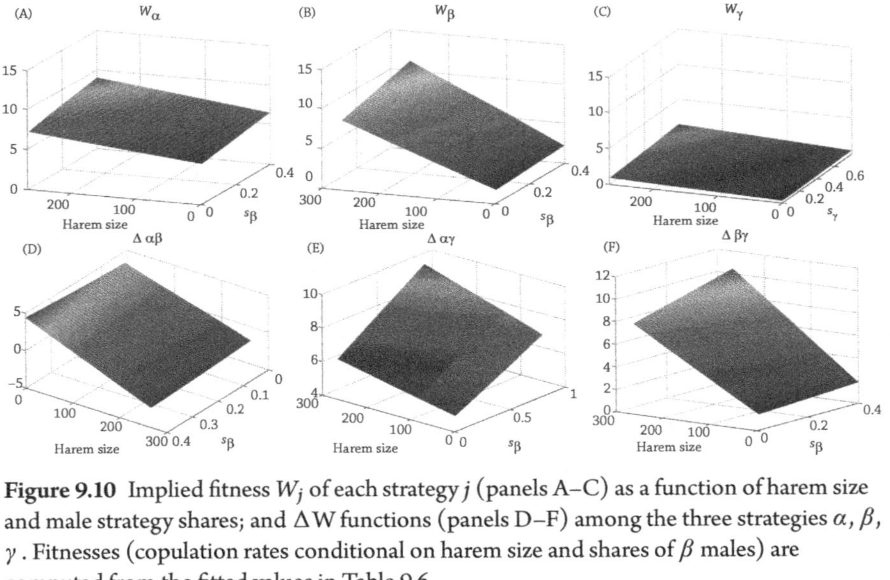

Figure 9.10 Implied fitness W_j of each strategy j (panels A–C) as a function of harem size and male strategy shares; and ΔW functions (panels D–F) among the three strategies α, β, γ. Fitnesses (copulation rates conditional on harem size and shares of β males) are computed from the fitted values in Table 9.6.

So why doesn't the apparently dominated strategy γ disappear? Recall that the computations so far are just for a single season. Perhaps γ males will look better if we consider life-cycle fitness. Indeed, from that perspective, they have three advantages: (1) the longer time to maturity implies that a smaller discount factor must be applied to α (maturation at 9–12 years) and β males (maturation at 8–10 years) than to γ (maturation at 4–5 years), (2) γ and β males can survive at much higher rates into subsequent years and, most important in terms of plasticity, (3) γ males can also transition to either α or β later in life.

To elaborate a bit on these points, it is clear that success as an alpha is very unlikely without years of building body mass, feeding out at sea, so alphas mature later. Also, a season on the beach seriously depletes body mass—these animals live off their blubber and neither eat nor drink while onshore—making it more difficult to achieve higher rank the next season.

We are now prepared to quantify these life-cycle considerations. To simplify the data analysis (and to acknowledge limitations in the data) the rest of the analysis ignores surf monsters. We lump them with gammas because they earn similar payoffs according to Table 9.6 and because most of them are older and larger gamma males or (as discussed further below) recently demoted beta or alpha males.

We will work with a modified Leftkovitch matrix $\mathbf{L}$ so that stage dynamics are of the form $\mathbf{N}_{t+1} = \mathbf{L}\mathbf{N}_t$. To spell it out,

$$\begin{bmatrix} N_{\omega,t+1} \\ N_{\tau,t+1} \\ N_{\gamma,t+1} \\ N_{\beta,t+1} \\ N_{\alpha,t+1} \end{bmatrix} = \begin{bmatrix} P_{\omega,\omega} & 0 & F_\gamma & F_\beta & F_\alpha \\ P_{\omega,\tau} & P_{\tau,\tau} & 0 & 0 & 0 \\ 0 & P_{\tau,\gamma} & P_{\gamma,\gamma} & P_{\beta,\gamma} & P_{\alpha,\gamma} \\ 0 & 0 & P_{\gamma,\beta} & P_{\beta,\beta} & P_{\alpha,\beta} \\ 0 & 0 & P_{\gamma,\alpha} & P_{\beta,\alpha} & P_{\alpha,\alpha} \end{bmatrix} \begin{bmatrix} N_{\omega,t} \\ N_{\tau,t} \\ N_{\gamma,t} \\ N_{\beta,t} \\ N_{\alpha,t} \end{bmatrix}. \tag{9.5}$$

Note that the modified Leftkovitch matrix $\mathbf{L}$ is standard in allowing weaners (ω) to advance to teen (τ) and thence to γ, but is nonstandard in allowing αs and βs to regress. This modification reflects observed reality among these animals, and allows for plasticity.

UCSC biologists, beginning with Burney LeBeouf, have gathered a lot of longitudinal data on male elephant seals (LeBeouf et al. 1983, 1994). It can be summarized as follows:

$$\mathbf{L} = \begin{bmatrix} .05 & 0 & F_g & F_b & F_a \\ .2 & .35 & 0 & 0 & 0 \\ 0 & .1 & .35 & .3 & .39 \\ 0 & .1 & .2 & .2 & .1 \\ 0 & 0 & .05 & .3 & .01 \end{bmatrix}. \tag{9.6}$$

Among the more intriguing implications of these numbers is that the α strategy is basically "one and done": hardly any (1%) α males return to that strategy the

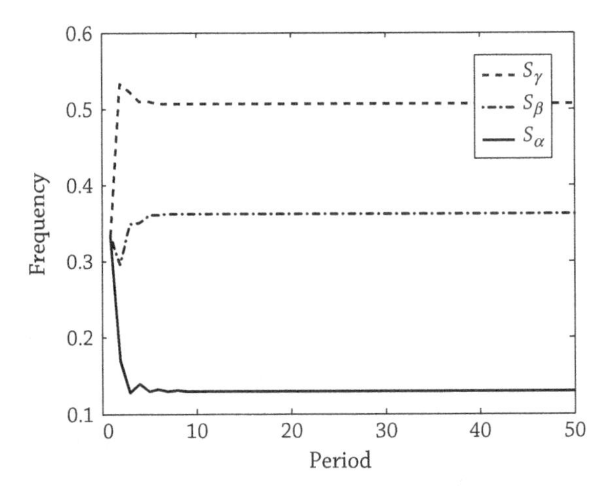

Figure 9.11 Shares of each strategy over the years.

following year; almost 40% fall to γ status, 10% come back as β males, and the rest (50%) are not seen again. By contrast, promotion of β to α occurs at the same rate (30%) as demotion to γ , and their standstill rate matches their disappearance rate (20%). While γ males often return as γ (35%), they have a modest transition to β status (20%), which can thereafter transition to α status in subsequent years (30%), and occasionally (5%) they transition directly to α status in the next season.

We estimate fecundities F_j over the season for the three male strategies as follows. Each week during the breeding season, take the conditional copulation rates described earlier, and the actual harem sizes and strategy shares, to obtain mean per capita copulation rates for each strategy that week. Add these up over the breeding season, and renormalize to obtain copulation shares. The results are $F_\alpha = 0.607, F_\beta = 0.331$, and $F_\gamma = 0.060$. Finally, multiply the copulation shares by an estimate of the number of breeding age females to obtain the fecundities used in the modified Leftkovitch matrix.

What is the long-run behavior implied by the fitted **L**? One could numerically compute the leading eigenvalue and associated eigenvector, but it is more visually appealing to simply simulate the dynamics. Beginning with initial population vector $\mathbf{N}_0 = (22, 22, 22, 22, 22)$, the population shares soon converge as seen in Figure 9.11.

The simulation yields a population growth rate and strategy shares that rapidly stabilize, thus equilibrating the strategies in terms of lifetime fitness. The steady-state growth rate is at 3% (very close to observed rates from 1975 to the present) while, as can be seen in Figure 9.11, the share of alphas reaches a steady state quickly and stabilizes at around 13% of the adult male population. Similarly, the share of beta males reaches a steady state at around 36% while the gamma share is about 50%.

The simulation clearly illustrates how a strategy like gamma (and presumably also surf monster) that is dominated each season can survive as part of a life cycle. Indeed, gamma claims the largest share, because it is far less costly than alpha and offers better prospects for future years. The data also show plasticity even within a single season—as harems shrink when females leave the beach, males accept demotion from alpha and beta to gamma and surf monster.

Our empirical analysis is ongoing, and much modeling and data collection remains to be done. The Leftkovitch matrix neglects some potentially important nuances, for example, each male who currently is a gamma is treated the same way, independent of his own personal history. Thus, we are dealing so far with random life cycles, and not exploiting their full structure nor clearly identifying the costs and benefits of plasticity. With additional data on individual transitions over the lifetime of a male strategy we could further refine the consequences of transitions within a given season (e.g., from α and β to Sm). In addition, this analysis has ignored the contribution of females to the male strategies, but these could be decisive. For example, females might boost the odds that a weaner eventually becomes an α male by providing a larger amount of milk during lactation.

9.7 DISCUSSION

Life-cycle strategic analysis is not yet mature, and opportunities abound for theorists as well as for applied researchers.

For theorists, the combination of plasticity and multi-stage payoffs is particularly intriguing. Even the simple question of when to exercise an irreversible option (e.g., when a male elephant seal should first commit to a season on the beach, or when a company should launch its new product) was not solved in any generality until the optimal stopping literature of the 1960s. Since then, there has been considerable work by mathematical economists on real options, which covers nonstrategic aspects of the problem. Two early references are Jones and Ostroy (1984), and Dixit and Pindyck (1994). There are theoretical articles on games of timing with special structure (e.g., Park and Smith 2008), but a general strategic theory of life-cycle games has not yet been developed.

Our suggested approach, perfect lifetime equilibrium (PLE), raises as many questions as it answers. In particular, what can be said about dynamic stability? Is there some simple way to characterize PLE that are stable under replicator dynamics, for example? What can be said about imitation processes that privilege some stages of the life cycle? Is there an appropriate way to (re)define fitness so that imitation dynamics converge to PLE, or to some other sort of steady state that can be characterized nicely?

In this book, theory is important to the extent that it can guide applied work. The featured application in the present chapter is elephant seals, and we believe that it will help readers develop similar biological applications. Probably it is less clear how to develop social and virtual applications, which in our view are equally promising. For that reason, we close the chapter by sketching an economic application with policy implications, based on a 2015 term paper by UCSC students Jijian Fan, Hugo Lhuillier, and Parameswaran Raman.

Is promoting education an effective way to increase social mobility and to reduce inequality? To gain insight into that question, consider an evolutionary model with two interacting populations, each with a two-stage life cycle. One population consists of highly educated people qualified for skilled jobs (as well as unskilled jobs), and the other consists of less educated people qualified only for unskilled jobs. The second stage is adult, in which the person earns a wage (the payoff) and decides on the educational attainment of her progeny, who are in the first stage. Investment in education is costly for the adult, who must pay tuition and incur other costs to support the progeny.

Wages of skilled and unskilled work are determined by supply and demand; other things equal, a shift of families from low to high investment in education will reduce the payoff (wage) premium for skilled work. (The same shift might also increase the cost of investing in education, but we can ignore that possibility for present purposes and regard the cost as constant.) The life-cycle payoff for an individual is the utility of her own wage plus some (discounted) fraction of the wage of her progeny.

It is reasonable to assume some sort of noisy best-response dynamics. The supply effect just noted means that the strategy to invest in education is apostatic—the more adults who choose that strategy, the lower its life-cycle payoff. Thus, for reasonable parameter configurations, the game is a nonlinear Hawk-Dove when lumping both populations together. When keeping the populations separate, mobility must be tracked: some fraction $p \in [0, 1]$ of the highly skilled population doesn't invest in eduction, and some fraction $q \in [0, 1]$ of the unskilled does invest.

A key insight is that (in utility terms) investing in the kids' education is more costly for low wage adults than for high wage adults. The main reason is that marginal utility of expenditure is decreasing in income, so high income adults have to make less painful sacrifices than low income adults to cover the same monetary cost of education. Thus, when the two populations are considered separately, there will be persistent inequality in this model. In equilibrium, at least one of the populations is completely specialized—either $p = 1$ and all skilled workers always invest in education, or $q = 0$ and no unskilled workers ever invest in education, or both.

In this simple model, the best way to achieve real mobility is to subsidize education for low income families by just the right amount so that in equilibrium those families can break even (in utility terms) when they invest in education. This policy also increases the equilibrium wage for low skilled workers and depresses it for high skilled workers.

APPENDIX

Formalizing a perfect lifetime equilibrium

Section 9.5 offered a verbal sketch of strategic interdependence in a life-cycle setting. Here we are more mathematically explicit.

Each individual i earns fitness W_i^t in stages $t = 1,2,...,T$, as follows. In stage t she chooses an action a_i^t in the feasible subset $A_i^t \subset A$, with three consequences:

- she receives current payoff $W_i^t = g^t(a_i^t, s^t, a_o^t)$, where s^t is the current state of play (a point in the $|A|$-simplex representing the distribution of action choices), and $a_o^t \in A_o$ is "Nature's move" (i.e., the current state of the environment);
- she survives to continue to stage $t+1$ with probability (and time discounting) $d_i^t = d^t(a_i^t, s^t)$,
- at which time her feasible set is $A_i^{t+1} = h^t(a_i^t, a_o^t)$.

Lifetime fitness is therefore $W_i^L = \sum_{t=1}^T \delta_i^t W_i^t$ where $\delta_i^t = d_i^1 d_i^2 \cdots d_i^{t-1}$.

The last two consequences can be combined. For example, d_i^t can be regarded as the probability that a_o^t is in the subset of A_o that allows a nontrivial A_i^{t+1}. However, it seems notationally useful to keep the two consequences separate.

A *strategy* α_i for player i specifies a feasible action $a_i^t \in A_i^t$ for each stage t. The strategy is *rigid* if the specified actions are independent of a_o and s. The strategy is *plastic* if, for some t, the choice a_i^t is a nonconstant function of $(a_o^1,...,a_o^t)$ and/or $(s^1,...,s^{t-1})$. A plastic strategy is *optimally payoff contingent* if no other action in A_i^t yields higher remaining lifetime fitness than the choice a_i^t that it specifies.

The relevant equilibrium concept—call it a *perfect lifetime equilibrium*, PLE—is the population analogue of perfect Bayes equilibrium. A PLE is a distribution of strategies α and a shared belief β about future states such that, at each stage $t = 1,...,T$,

a. every action a_i^t specified by a strategy in the support of α maximizes expected remaining lifetime fitness

$$W_i^{tL} = g^t(a_i^t, s^t, a_o^t) + (\delta_i^t)^{-1} \sum_{\tau=t+1}^{T} \delta_i^\tau E(W_i^\tau | \beta), \tag{9.7}$$

where s^t is the state induced by α ; and

b. The conditional expectations $E(W_i^\tau | \beta)$ for $\tau = t+1,...,T$ are consistent with Bayes theorem given α.

Example: A simplified elephant seal game

Figure 9.12 writes out explicitly a two-stage simplification of the elephant seal game. Observed behaviors are classified as

α. try to exclude other males from a stretch of beach with as many females as possible,

β. hang out at the edge of an α's territory without directly challenging him, but exclude other males from the vicinity and copulate with nearby females when the local α is busy, or

γ. wander the beach looking for available females and back off when challenged by an α or β male.

A male elephant seal's first move (node 1a) is whether to enter the beach scene when young, or to wait until next season. If he chooses e(nter), then Nature (which here includes the rest of the population) chooses the state x_1 of the beach scene that year—the number and distribution of females, the number of other males of different types, and so on. At this stage, the seal can choose rigid strategy $a = \beta$ or γ, or the plastic strategy $a = \rho$. The plastic strategy incurs a fitness cost (impounded in the fitness function u_1) and allows him to exhibit either the β or γ behavior, whichever yields higher fitness given the realization of x_1. The fitness gained at this stage is denoted $u_1(a|x_1)$.

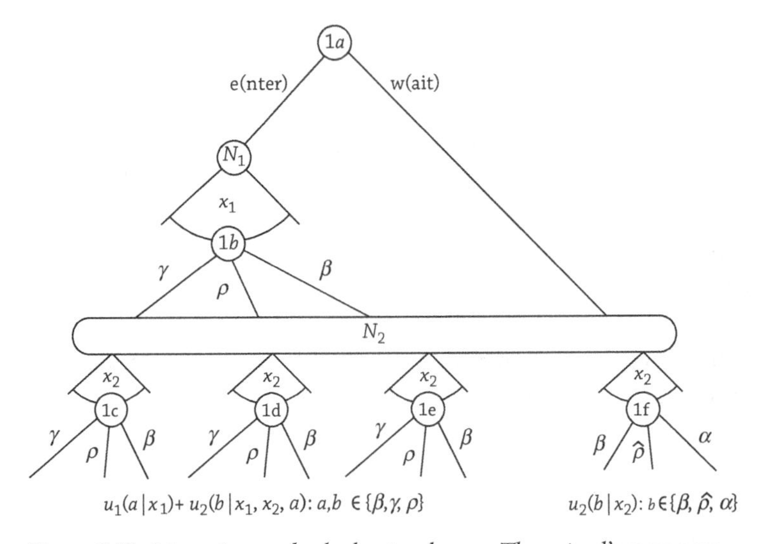

Figure 9.12 A two-stage male elephant seal game. The animal's moves are denoted by nodes 1a–f, and stochastic "Nature" moves by information sets N_{1-2}. Cumulative payoff (fitness) functions appear at the terminal nodes at the bottom of the figure.

Nature's state x_2 in the second stage is a random variable independent of a, hence the information set connecting all of Nature's second stage nodes. An animal who choses to enter at the first stage then chooses second stage strategy b from the same three alternatives as before and gains fitness increment $u_2(b|x_2,x_1,a)$. An animal who choses to wait at the first stage chooses in the second stage $b \in \{\alpha,\beta,\hat{\rho}\}$, where the plastic strategy $\hat{\rho}$ allows him to exhibit either β or α behavior, whichever yields higher fitness given the realization of x_2. The resulting fitness payoffs $u_2(b|x_2)$ are generally higher than the corresponding payoffs for animals that chose to enter at the first stage, due to the extra time feeding and avoiding fights when young.

EXERCISES

1. It is claimed in Section 9.2 that in equilibrium the plastic strategists will all choose h in the high state and will all choose d in the low state. Verify (or refute!) this claim, and any other non-obvious claims used to establish the existence and uniqueness of the given equilibrium shares for $a \in [0, \frac{1}{4}]$. Hint: Suppose that a P-strategist deviates in one of the two states. Show that the deviator makes the same choices as a rigid strategist and hence obtains lower net payoffs than that rigid strategist.
2. For the same example, establish local stability by showing that the eigenvalues of the Jacobian matrix have negative real parts at the given interior equilibrium.
3. Write down a model for the last example in Section 9.2 for two different lines of work. Verify the assertions that (a) moderate plasticity costs smooth out the pay difference across the two lines of work, and (b) the pay difference is zero in long-run equilibrium when plasticity cost is zero.
4. Verify the claim made in Section 9.4 that the nontrivial life cycle effects are not needed to analyze the case when random events z in life stage 2 are correlated with strategy choices made in life stage 1. Write down an example where an early choice affects the odds of later outcomes.
5. The following exercise doesn't involve life-cycle considerations, nor is it explicitly plastic, but does help connect the extensive form games introduced in this chapter to evolutionary models. It is inspired by Rowell et al. (2006).

   a. Members of one population (e.g., male frogs) signal their status (e.g., large or small) to a second population (e.g., female frogs in the neighborhood) but the signal may be intercepted by members of a third population (e.g., birds or snakes that eat frogs). With probability $p \in [0,1]$ the signal is accurate and with probability $1-p$ it is a lie (e.g., a small frog mimics the croak of a large frog). With probability $q \in [0,1]$ the first population believes the signal, and with probability $r \in [0,1]$ the second population believes it. Draw the corresponding extensive form game tree.

b. Draw the state space as a cube.
c. Write out parametrized payoff functions for this three-population game.
d. Simulate or solve the game for chosen parameter values.

NOTES

This chapter leaves to one side the interesting topic of "bet hedging." Bergstrom (2014) is nice recent paper on that topic, with references to previous literature.

Payoff-contingent plastic strategies are explicit in two papers by philosophers Zollman and Smead (2009, 2010) that recently came to our attention. Their main application is to the emergence of language, as formalized in signaling games. Payoff-contingent plastic strategies are implicit in Alonzo and Sinervo (2001, 2007), who considered mate choice ("plasticity" in a two-population game (male RPS, female r-K strategists) where females choose males (and/or produce the sex ratio) that will produce the highest fitness progeny in the next generation for side-blotched lizards. Following up on this form of mate choice plasticity, Sinervo et al. (2007) theorized a similar female strategy for European common lizards and Fitze et al. (2014) empirically demonstrated such mate choice that is contingent on RPS cycle frequency. See also Lively (1986) for a specialized model that includes an atemporal plastic strategy.

There is also a recent literature, for example, Leimar (2005, [et al.] 2006), Svardal et al. (2011) that compares plastic to rigid strategies intertemporally but not game theoretically. The issue is framed as genetic versus environmental morph determination, with the answer hinging on the reliability of cues.

From a longer-run perspective, payoff contingent plastic strategies are hard to distinguish from rapid adaptation, genetic or memetic. Of course, there is a vast literature on fast versus slow adaptation; Chapter 14 will discuss a large and influential strand called adaptive dynamics.

The "appealing answer" to the menopause puzzle is called the grandmother hypothesis. It remains controversial, but other serious candidates to resolve the puzzle also involve life-cycle considerations.

Given an $n \times n$ matrix M, the *characteristic equation* is $0 = |M - \lambda I|$, a polynomial of degree n in λ. Here I is the $n \times n$ identity matrix, and $|\cdot|$ denotes the determinant. The roots of the characteristic equation are called *eigenvalues* of M. See any textbook on linear algebra for details and explanations.

To be consistent, all the payoffs in Figure 9.8A should have δ^2 applied to them since they are received in the second period. However, since the same positive factor appears in all payoffs, it has no effect on the analysis and can be omitted.

Game trees are a classic way of specifying games. This specification is called the extensive form and is contrasted to the more succinct strategic or normal form, in which the game is specified by its payoff function. See any standard game theory text, for example, Watson (2013) or Harrington (2009).

BIBLIOGRAPHY

Alonzo, Suzanne H., and Barry Sinervo. "Mate choice games, context-dependent good genes, and genetic cycles in the side-blotched lizard, *Uta stansburiana*." *Behavioral Ecology and Sociobiology* 49, no. 2–3 (2001): 176–186.

Alonzo, Suzanne H., and Barry Sinervo. "The effect of sexually antagonistic selection on adaptive sex ratio allocation." *Evolutionary Ecology Research* 9, no. 7 (2007): 1097.

Bergstrom, Theodore C. "On the evolution of hoarding, risk-taking, and wealth distribution in nonhuman and human populations." *Proceedings of the National Academy of Sciences* 111, Supplement 3 (2014): 10860–10867.

Broom, Mark, Roger M. Luther, and Graeme D. Ruxton. "Resistance is useless? Extensions to the game theory of kleptoparasitism." *Bulletin of Mathematical Biology* 66, no. 6 (2004): 1645–1658.

Charlesworth, Brian. *Evolution in age-structured populations.* Cambridge University Press, 1980.

De Roos, A. M., O. Diekmann, and J. A. J. Metz. "Studying the dynamics of structured population models: A versatile technique applied to *Daphnia*." *The American Naturalist* 139 (1992): 123–147.

Dixit, Avinash K., and Robert S. Pindyck. *Investment under uncertainty*. Princeton University Press, 1994.

Harrington, Joseph E. *Games, Strategies and Decision Making*. Worth Publishers, 2009.

Hawkes, Kristen, James F. O'Connell, and Nicholas G. Blurton Jones. "Hadza women's time allocation, offspring provisioning, and the evolution of long postmenopausal life spans." *Current Anthropology* 38, no. 4 (1997): 551–577.

Jones, Robert A., and Joseph M. Ostroy. "Flexibility and uncertainty." *The Review of Economic Studies* 51, no. 1 (1984): 13–32.

Kaplan, Robert H., and William S. Cooper. "The evolution of developmental plasticity in reproductive characteristics: An application of the 'adaptive coin-flipping' principle." *American Naturalist* 123, no. 3 (1984): 393–410.

LeBeouf, B. J., and R. S. Condit. "The high cost of living on the beach." *Pacific Discovery* 36, no. 3 (1983): 12–14.

LeBeouf, Burney J., and Richard M. Laws. *Elephant seals: Population ecology, behavior, and physiology*. University of California Press, 1994.

Leimar, Olof. "The evolution of phenotypic polymorphism: Randomized strategies versus evolutionary branching." *The American Naturalist* 165, no. 6 (2005): 669–681.

Leimar, Olof, Peter Hammerstein, and Tom JM Van Dooren. "A new perspective on developmental plasticity and the principles of adaptive morph determination." *The American Naturalist* 167, no. 3 (2006): 367–376.

Lively, Curtis M. "Canalization versus developmental conversion in a spatially variable environment." *American Naturalist* 128, no. 4 (1986): 561–572.

Park, Andreas, and Lones Smith. "Caller Number Five and related timing games." *Theoretical Economics* 3, no. 2 (2008): 231–256.

Pfennig, David W. "Polyphenism in spadefoot toad tadpoles as a logically adjusted evolutionarily stable strategy." *Evolution* (1992): 1408–1420.

Qvarnström, A. "Context-dependent genetic benefits from mate choice." *Trends in Ecology and Evolution* 16, no. 1 (2001), 5–7.

Rowell, Jonathan T., Stephen P. Ellner, and H. Kern Reeve. "Why animals lie: How dishonesty and belief can coexist in a signaling system." *The American Naturalist* 168, no. 6 (2006): E180–E204.

Sinervo, Barry. "Runaway social games, genetic cycles driven by alternative male and female strategies, and the origin of morphs." In *Microevolution Rate, Pattern, Process*, A. Hendry, and M. Kinnison (eds) *Genetica* 112, no. 1 (2001): 417–434.

Sinervo, Barry, Benoit Heulin, Yann Surget-Groba, Jean Clobert, Donald B. Miles, Ammon Corl, Alexis Chaine, and Alison Davis. "Models of density-dependent genic selection and a new rock-paper-scissors social system." *The American Naturalist* 170, no. 5 (2007): 663–680.

Smead, Rory, and Kevin JS Zollman. "The stability of strategic plasticity." Unpublished working paper, Carnegie Mellon University Research Showcase @ CMU Philosophy Department (2009).

Svardal, Hannes, Claus Rueffler, and Joachim Hermisson. "Comparing environmental and genetic variance as adaptive response to fluctuating selection." *Evolution* 65, no. 9 (2011): 2492–2513.

Warner, Robert R., and Stephen E. Swearer. "Social control of sex change in the bluehead wrasse, *Thalassoma bifasciatum* (Pisces: Labridae)." *The Biological Bulletin* 181, no. 2 (1991): 199–204.

Watson, Joel. *Strategy: An Introduction to Game Theory*, Third Edition. W. W. Norton, 2013.

Williams, George C. "Pleiotropy, natural selection, and the evolution of senescence." *Evolution* 11, no. 4 (1957): 398–411.

Zollman, Kevin JS, and Rory Smead. "Plasticity and language: An example of the Baldwin effect?" *Philosophical Studies* 147, no. 1 (2010): 7–21.

10

The Blessing and the Curse of the Multiplicative Updates

MANFRED K. WARMUTH

Online learning is a major branch in the computer science field of machine learning. On the surface, online learning seems completely unrelated to evolution, but this chapter will show that there are in fact deep analogies. Our goal is to exploit these analogies to provide new insights into both evolution and online learning.

The basic task in online learning is to process a discrete stream of data. Each piece of the data stream (example) consists of an instance and a label for the instance. The instances are vectors of a fixed dimension and the labels are real valued. Each trial processes one instance vector and its label. The algorithm first receives the instance vector. It then must produce a label for the instance vector. Typically this is done by maintaining one weight per dimension of the instance vectors and predicting the label using the dot product of the instance vector and the current weight vector. After predicting, the online learning algorithm receives the "true" label of the instance vector.

For example, the instance vector might be the predictions of five experts on whether it will rain tomorrow (where 1 stands for rain and 0 for no rain). The algorithm maintains five weights where the ith weight is the current belief that the ith expert is the best expert. The maintained weight vector is a probability vector, and the algorithm's own prediction for rain might be the weighted average (dot product) of the predictions of the experts. The true label is 1 or 0, depending whether it actually rained or not on the next day. The algorithm is constructed to predict the labels "accurately" for each instance as it arrives, and to improve accuracy as more instances accumulate.

The main issue is how to update the weights after receiving the true label for the current instance vector. One choice is to update the weights by a multiplicative update (i.e., each weight is multiplied by a nonnegative factor and then the weights are renormalized). The algebra turns out to be equivalent to the discrete replicator equation seen in earlier chapters. The weights are analogous to meme or gene strategy shares and the multiplicative update factor of each gene is analogous to its fitness.

Multiplicative updates are motivated by a Minimum Relative Entropy principle and we will argue that the simplest such update is Bayes' rule for updating priors. We will give an in vitro selection algorithm for RNA strands that implements Bayes' rule in the test tube where each RNA strand represents a different model. In one liter of the RNA soup there are approximately 10^{15} different kinds of strands and therefore this is a rather high-dimensional implementation of Bayes' rule. It is also a simple haploid evolutionary process.

In machine learning multiplicative updates are investigated for the purpose of learning online while processing a stream of data. The "blessing" of these updates is that they learn very fast in the short term because the good weights grow exponentially. However, their "curse" is that they learn too fast and wipe out other weights too quickly. This loss of variety can have a negative effect in the long term because adaptivity is required when there is a change in the process generating the data stream.

We describe a number of methods developed in machine learning that ameliorate the curse of the multiplicative updates. The methods make the algorithm robust against data that changes over time and prevent the currently good weights from taking over completely. We also discuss how the curse is circumvented by Nature. Surprisingly, some of Nature's methods parallel the ones developed in machine learning, but Nature also has some additional tricks.

Out of this research a complementary view of evolution is emerging. The short-term view deals with mutation and selection (i.e., some sort of multiplicative update). However, the long-term stability of any evolutionary process requires a mechanism for preventing the quick convergence to the currently fittest strategy or gene. We think that the online updates studied in machine learning may have something to say about Nature's evolutionary processes.

10.1 DEMONSTRATING THE BLESSING AND THE CURSE

We begin by giving a high-level definition of *multiplicative updates*. Consider algorithms that maintain a weight vector. In each trial an instance vector is processed and the weights (one per dimension) are updated. Just like strategy shares, the individual weights are non-negative and they are updated by multiplying each weight by a non-negative factor. In the most common case the weights are then rescaled to sum to one. There is a systematic way of deriving such updates by trading off a relative entropy between the new and old weight vector with a loss function that measures how well the weights did on the current instance vector (Kivinen and Warmuth, 1997). See also Appendix B of Chapter 1.

Let us look at one common scenario in more detail. We have a set of *experts* that predict something on a trial by trial basis. These experts can be human or algorithmic, and the instance vector consists of their predictions. The "master algorithm" keeps one weight, $s_i \geq 0$, per expert: s_i represents the "belief" in the ith expert at trial t. At the end of the trial these beliefs are updated as follows:

$$\tilde{s}_i = \frac{s_i \, e^{-\eta \, \mathrm{loss}_i}}{Z}, \tag{10.1}$$

where $\eta > 0$ is a learning rate and $Z = \sum_j s_j \, e^{-\eta \, \mathrm{loss}_j}$ normalizes the weights to 1. Here loss_i is the loss incurred by expert i in the current trial (i.e., some measure of inaccuracy) and the $\tilde{s}_i$ weights are the new or updated weights to be used in trial $t+1$.

The update of weight vector $\boldsymbol{s}$ is the solution to the following minimization problem:

$$\tilde{\boldsymbol{s}} := \operatorname*{argmin}_{\boldsymbol{r} \in \boldsymbol{R}^n : \sum_i r_i = 1} \underbrace{\sum_i r_i \log \frac{r_i}{s_i}}_{\text{relative entropy between } \boldsymbol{r} \text{ and } \boldsymbol{s}} + \eta \underbrace{\sum_i r_i \, \mathrm{loss}_i}_{\text{loss of weight vector } \boldsymbol{r}}. \tag{10.2}$$

The relative entropy term keeps the new share vector $\tilde{\boldsymbol{s}}$ close to the old share vector $\boldsymbol{s}$ and the loss term makes $\tilde{\boldsymbol{s}}$ increase the shares of minimum loss. The non-negative learning rate η is the tradeoff parameter.

The multiplicative update process is exactly the same as discrete replicator dynamics in Chapter 1, where the factor $W_i = \exp(-\eta \, \mathrm{loss}_i)$ is the fitness of strategy i. Thus loss_i is strategy i's fitness rate scaled by minus the learning rate η. Another important connection, discussed at length in Harper (2009), is that Bayes' rule is special case of this update when the loss is the expected log loss and $\eta = 1$.

From a machine learning perspective, some basic questions would be: Why are multiplicative updates so prevalent in Nature? Why aren't additive updates more common (i.e., updates that subtract η times the gradient of the loss from the weight vector?)

As we shall soon see these multiplicative updates converge quickly (their "blessing") to the best expert (i.e., to the dominant strategy) when the data stream generating process is static. This means that all but one weight will go to zero. This is a problem when the data process changes with time. In that case the previously selected expert might perform badly on the new data, whereas one of the experts with current weight zero might predict well on the new data. Clearly wiping out the weights of all but one expert is not advantageous when the data changes over time. We call this the "curse" of the multiplicative updates. In terms of populations this means that we have a loss of variety. In the next section we will discuss a number of mechanisms developed in machine learning for preventing this loss of variety and discuss related mechanisms used in Nature.

Next we introduce in-vitro selection of RNA strands as a simple but instructive evolutionary process that connects nicely to online learning. We first show that in-vitro selection can be interpreted as an implementation of the standard Bayes' rule for updating priors to posteriors. We then use this evolutionary process to exemplify the curse of the multiplicative updates.

For example, the goal of the in-vitro selection algorithm might be the very practical problem of finding RNA strands that bind to a particular protein. Binding is a key step in drug design because the drug needs to attach to the surface of the target cell, which is identified by a surface protein. Assume we have a sheet of the target protein available, and maintain an active tube filled with dissolved RNA strands. The algorithm repeats the following three steps twenty times:

1. Dip protein sheet into the active tube.
2. Pull out and wash off RNA that stuck to surface. Discard remaining contents of the active tube.
3. Multiply washed-off RNA back to original amount and use it to fill the active tube used in the next repetition (normalization).

As a practical matter, this experiment can't be simulated with a computer because there are too many different kinds of RNA strands to track (on the order of 10^{15} strands in one liter). The basic scheme of in-vitro selection is always the following: Start with unit amount of random RNA and then in a loop, do a functional separation into "good" and "bad" RNA, and finally re-amplify the good RNA to the unit amount. The bad RNA is discarded.

The amplification in step 3 is done with the Polymerase Chain Reaction (PCR) which was invented by Kary Mullis in 1985. Roughly speaking, PCR proceeds as follows. First the single-stranded RNA is translated to double-stranded DNA. Then the following protocol is iterated for a number of steps: The substrate is heated so that the double-stranded DNA comes apart; then it is slowly cooled so that short primers can hybridize to the ends of the single-stranded DNA, and finally an enzyme (the Taq Polymerase) runs along DNA strand and complements the bases, resulting again in double-stranded DNA. Ideally in each iteration, all DNA strands are multiplied by factor of 2. After a sufficient number of iterations, the double-stranded DNA is translated back into single-stranded RNA.

We next give a mathematical description of the in-vitro selection iteration. Assume the *unknown* kinds of RNA strands in our liter of RNA strands are numbered from 1 to n where $n \approx 10^{15}$. We let $s_i \geq 0$ denote the "share" or fraction of RNA i. The contents of the entire tube are represented as a share vector $\boldsymbol{s} = (s_1, s_2, \ldots, s_n)$. When the tube has a unit amount of strands, then $s_1 + s_2 + \cdots + s_n = 1$. We let $W_i \in [0,1]$ be the fraction of one unit of RNA i that is "good" (i.e., W_i is the fitness of strand i for attaching to protein). So the target protein is represented by a second vector, the fitness vector $\boldsymbol{W} = (W_1, W_2, \ldots, W_n)$. Note that the two vectors $\boldsymbol{s}, \boldsymbol{W}$ are unknown to the researcher; clearly, there is no time to sequence all the strands and determine the fraction of attachment for each strand i. In that sense, the iterations of in-vitro selection represent some sort of "blind computation".

In the definition of the fitness vector we made a strong assumption: we assumed that the fitness W_i is independent of the share vector $\boldsymbol{s}$ (i.e., no frequency dependance). This is of course unrealistic because similar strands might compete for the same limited locations. Therefore W_i may be lower when the share s_i of strand i and the shares of similar strands s_j are high. In biological jargon, there may be some apostatic effects. However, at low concentrations they can be safely ignored, and the independence assumption might be a very good approximation.

To continue with the mathematical description, note that the amount of good RNA in tube $\boldsymbol{s}$ is $s_1 W_1 + s_2 W_2 + \cdots + s_n W_n = \boldsymbol{s} \cdot \boldsymbol{W}$. Similarly there is $s_1(1-W_1) + s_2(1-W_2) + \cdots + s_n(1-W_n) = 1 - \boldsymbol{s} \cdot \boldsymbol{W}$ of bad RNA (which is discarded). During the amplification phase, the good share of RNA strand i (i.e., $s_i W_i$), is multiplied by a factor F. If the amplification is precise, then all good RNA is multiplied by the same factor F (i.e., the factor F does not vary with strand i). At the end of the iteration, the amount of RNA in the tube is again 1. Therefore

$F\,\boldsymbol{s}\cdot\boldsymbol{W}=1$ and $F=\frac{1}{\boldsymbol{s}\cdot\boldsymbol{W}}$. The entire update of the share vector in an iteration is summarized as $\tilde{s}_i=\frac{s_i W_i}{\boldsymbol{s}\cdot\boldsymbol{W}}$.

This update can be seen as an implementation of Bayes' rule: Interpret share s_i of strand i as a prior $P(i)$ for strand i. The fitness $W_i \in [0..1]$ is the probability $P(Good|i)$ and $\boldsymbol{s}\cdot\boldsymbol{W}$ the probability $P(Good)$. In summary, the multiplicative update

$$\tilde{s}_i=\frac{s_i W_i}{\boldsymbol{s}\cdot\boldsymbol{W}} \quad \text{becomes Bayes' rule:} \quad \underbrace{P(i|Good)}_{\text{posterior}}=\frac{\overbrace{P(i)}^{\text{prior}}\ \overbrace{P(Good|i)}^{\text{data likelihoods}}}{P(Good)}.$$

Note the above Bayes' rule is a special case of update (**10.1**) and the relative entropy minimization problem (**10.2**) with learning rate $\eta=1$ and $\text{loss}_i=-\ln P(Good|i)$.

In-vitro selection corresponds to iterating Bayes' rule with the same data likelihoods. Let $i^*=\text{argmax}_i W_i$ be the strand (or model or expert) with the highest fitness factor. This strand will gather all the weight. That is, its share s_{i^*} converges to one while the remaining shares converge to zero, as in Figure 10.1. Note that the environment is static in that the data likelihoods remain unchanged. In this case, bringing up the model with highest fitness with a pure multiplicative update (Bayes' rule) increases the average fitness very quickly. That, again, is the blessing.

The standard view of evolution concerns itself with inheritance, mutation, and selection for the fittest with some sort of multiplicative update. However, multiplicative updates are "brittle" because the gradients of the losses appear in the exponents of the update factors. This is problematic when there is noise in the data and the data-generating process changes over time. The problem is compounded when the weights can only be computed up to a constant precision, since that means that weights can go to precisely zero after a few dozen periods, and the corresponding strategies are irretrievably lost. In Nature the curse is extinction—a loss of variety at the level of genes or species. In such cases, the blessing becomes a wicked curse. Some sort of mechanism is needed to ameliorate the curse of the multiplicative update: preventing weights from becoming too large or being set to zero.

10.2 DISPELLING THE CURSE

Of necessity, both Nature and Machine Learning have developed mechanisms that, to some degree, lift or at least mitigate the curse. We begin with a mechanism

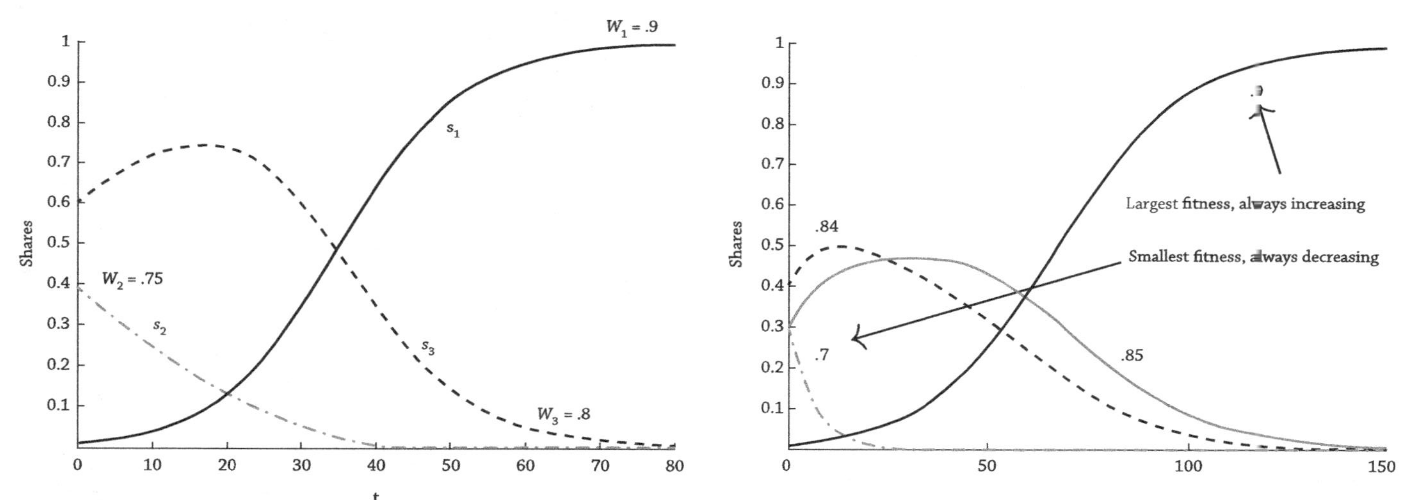

Figure 10.1 Panel A gives the shares of three models as a function of the iteration number t. The initial share vector is $s_0 = (.01, 39, .6)$, which are the starting values of the solid, dot dashed and dashed curves (respectively) at the y-axis when $t = 0$. The fitness vector for the three curves is $\mathbf{W} = (.9, .75, .8)$. The shares are updated by iterating Bayes' rule: $\tilde{s}_{t,i} = \frac{s_{0,i} W_i{}^t}{\text{normalization}}$, where t is the iteration number. Panel B gives another example. In the early iterations, the share with fitness .85 (grey curve) beats the share with fitness .84 (dashed curve), but then they are both beaten by the share with highest fitness .9 (black curve), which has a very low share initially.

from Nature, which is illustrated in two experiments. In the first, fast-mutating bacteria species are added to a tube of nutrient solution (Rainey and Travisano 1998). The bacteria diversify, and after a couple of days typically three variants survive: one near the surface (in the oxygen-rich zone), one on the side wall (light-rich but little oxygen) and one in the muck on the bottom of the tube.

This is because three different environments (niches) emerged as the nutrients were used up. Competition is most intense within each niche, and the single fittest variety survives in each. Now if the same experiment is done while agitating the tube, then only one variant survives. The agitation prevents the creation of separate niches, and there is only one homogeneous niche. The competition for the best via the multiplicative update occurs within each niche and the niche boundaries protect diversity. When the boundaries are taken away (e.g., the agitated tube is unified into a single niche), then one variant quickly out-competes the rest via the multiplicative update.

In a second experiment mentioned in Chapter 7, researchers (Kerr et al. 2002) found three species of bacteria that play an RPS game. When started on a Petri dish, colonies of each species develop that slowly chase each other around the dish: R invades S's colonies, S invades P, and P invades R. If all three species are put in a tube with liquid nutrient solution that is continuously shaken, then again only one species survives. So here the local cyclical variation on the gel created by the species themselves makes it possible to maintain the variety. This can't happen in the tube because there is too much mixing. If the nutrient gel is exhausted in the dish experiment, then the local colonies can be transferred to a new dish by pressing a sterile cloth first on the old and then on the new dish. The colonies continue to chase each other on the new dish from where they started. However, if during the transfer the cloth is imprinted twice on the new dish while rotating the cloth between imprints, then this mixes the colonies enough so that only one species survives.

We attribute this first mechanism to Nature:

Nature 1: Niche boundaries help prevent the curse.

It seems that to some extent a loss of variety due to mixing also happens for memes. Humans build ever more powerful modes of connecting the world: roads, airplanes travel, the web. The "roads" allow a mixing of memes over larger areas, and the multiplicative update of the meme shares causes a loss of variety.

We now pose a problem that we claim cannot be solved by a multiplicative update alone.

Key Problem 1. There are three strands of RNA in a tube and the goal is to amplify the mixture while keeping the percentages of the strands unchanged. This is to be done without sequencing the strands and determining their concentrations.

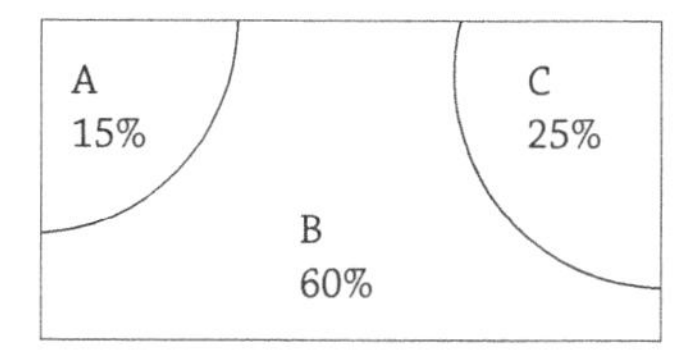

Figure 10.2 How to make more of a mixture for three strands?

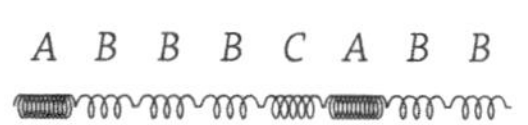

Figure 10.3 Forming long strands with approximately the right percentages.

An example is given in Figure 10.2: We want to make more of a mixture with strand A at 15%, B at 60% and C at 25%. If we simply translate to DNA, apply PCR once to the whole mixture, and translate back to RNA, then this ideally would amplify all three strands by a factor of 2. However in practice, the amplification will be less than 2 and slightly different for each strand. Iterative application of the multiplicative update will favor the strand that has the highest fitness for being replicated and we end up with a large amount of one of the strands.

There is a surprisingly elegant solution to this problem, using standard DNA techniques, that requires no detailed knowledge of the mixture's composition. The solution also tells us something about how to dispel the curse, so it is worth a quick sketch.

Translate to (double stranded) DNA and using an enzyme, add a specific short "end strand" to both ends of all strands in the tube. These end strands function as connectors between strands and make it possible to randomly ligate many strands together into long strands. Now separate out one long strand. With the help of an enzyme, add "primer strands" to both ends of that long strand. Apply PCR iteratively starting with the long selected strand, always making complete copies of the same original long strand that is located between the primers. Stop when you have the target amount of DNA. Now divide the long strands into their constituents by cutting them at the specific end strand that functioned as the connector. Finally, remove all short primer and end strands and convert back to RNA.

Note that for various reasons, the fractions of A, B, and C on the selected long strand will not agree exactly with the fractions in the original tube. However, the initial errors will not be amplified by PCR because the primers function as brackets and assure that only entire copies of the long strands get amplified. The final tube will contain the three strands at the same percentages as they appear on the initial long strand.

We just ran into the second method of Nature for preventing the curse. The long strand functions as a "chromosome." Free floating genes in the nuclei of cells would compete. Some genes would get copied multiple times during meiosis and some would die out (loss of variety). By coupling the genes into a chromosome and by assuring that the entire chromosome is copied each time, Nature prevents the curse.

Nature 2: Coupling preserves diversity.

As haploids evolve, the genes on a chromosome are selected for together (i.e., there is one fitness factor per long strand). We are selecting for the best combination, and in some sense the genes must "cooperate" for the sake of the most efficient copying into the next generation, as shown in Figure 10.4.

From a Machine Learning perspective there are already several open questions. What updates to the share vector s can be implemented with in-vitro selection? What updates are possible with "blind computation" (i.e., without sequencing individual strands)? The share vectors here are concentrations and are therefore always non-negative. However, can evolutionary processes use negative weights? Negative weights allow you to express inhibitive effects of certain features, which is very useful.

In machine learning we typically process examples $(\boldsymbol{x},y)$ where $\boldsymbol{x} \in \boldsymbol{R}^n$ are the instance vectors that have the same dimension n as the weight or share vectors s, and $y \in \boldsymbol{R}$ is a label. The loss associated with weight vector $\boldsymbol{s}$ on example $(\boldsymbol{x},y)$ often has the form $L(\boldsymbol{s}\cdot\boldsymbol{x},y)$. The simplest example is the square loss: $L(\boldsymbol{s}\cdot\boldsymbol{x},y) = (\boldsymbol{s}\cdot\boldsymbol{x}-y)^2$.

Now assume that you restrict yourself to multiplicative updates, which means non-negative weights. Nevertheless, researchers in machine learning circumvented the non-negative weight restriction using the $EG^{\pm}$ algorithm of Kivinen and Warmuth (1997). The algorithm maintains two weight vectors $\boldsymbol{s}^+$ and $\boldsymbol{s}^-$ of dimension n. In the simplest setting it predicts the label of the instance vector $\boldsymbol{x}$ as $(\boldsymbol{s}^+ - \boldsymbol{s}^-)\cdot\boldsymbol{x}$ and incurs the loss $L((\boldsymbol{s}^+ - \boldsymbol{s}^-)\cdot\boldsymbol{x},y) = ((\boldsymbol{s}^+ - \boldsymbol{s}^-)\cdot\boldsymbol{x}-y)^2$. There are now two weights associated with component i of the instance vectors: s_i^+ and s_i^-. The derivative of the loss for s_i^+ and s_i^- are the same except for a sign change. These derivatives appear in the exponents of the fitness factors for s_i^+ and s_i^-. Therefore, if s_i^+ has fitness W_i, then s_i^- has fitness $\frac{1}{W_i}$. So there are two versions of species i, and their fitnesses are reciprocal.

Thus the open question: Is this $EG^{\pm}$ trick used by Nature? Also can such an update be implemented with in-vitro selection?

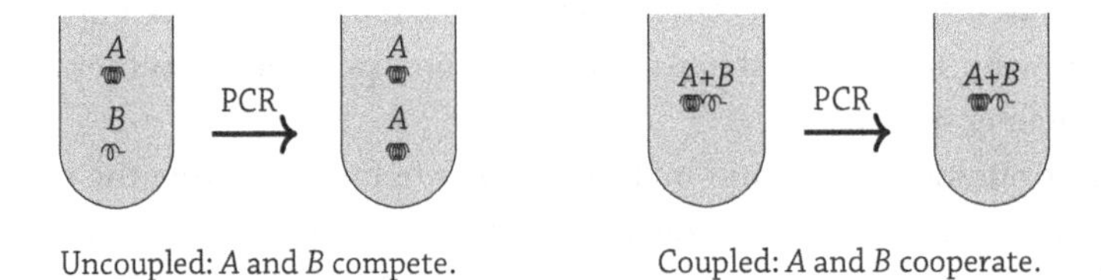

Uncoupled: *A* and *B* compete. Coupled: *A* and *B* cooperate.

Figure 10.4 Schematic depiction of coupled versus uncoupled.

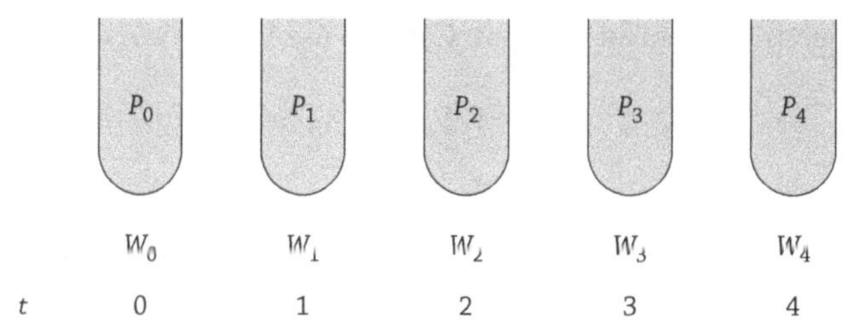

Figure 10.5 The goal is to find a set of strands that attach to $k = 5$ different proteins where each protein P_j has a different fitness vector $\boldsymbol{W}_j$.

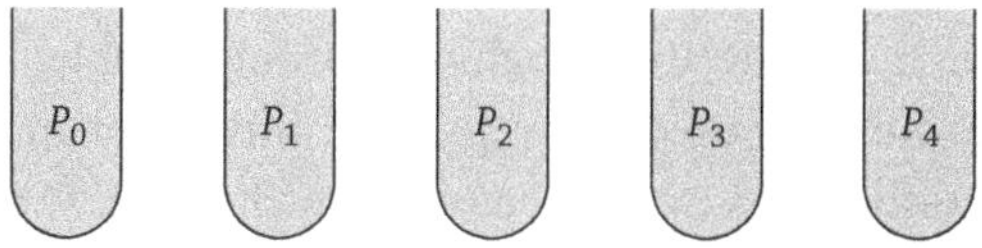

	P_0	P_1	P_2	P_3	P_4
	W_0	W_1	W_2	W_3	W_4
$W_{i,1}$	**.9**	**.8**	**.96**	.2	.04
$W_{i,2}$	.1	.01	**.8**	**.9**	**.8**
$u \cdot W_i$	.5	.405	.88	.55	.42

Figure 10.6 The fraction of attachment of strand 1, strand 2, and a tube $\boldsymbol{u}$ with 50% strand 1 and 50% strand 2. High attachment values are marked in **bold**.

We next pose a second problem that cannot be solved by a multiplicative update alone. Informally, the problem is: find a small set of RNA strands that together can bind to q different protein sheets. The previous in-vitro selection discussion assumed that $q = 1$, but $q > 1$ is a much tougher problem.

To see the difficulty, begin with the obvious approach: cycle through the q proteins and do an in-vitro selection step in turn, as in Figure 10.5. Assume protein P_j, for $j = 0, \ldots, q-1$ is characterized by fitness vector W_j. Thus in trial $t = 0, 1, \ldots$ we would do an in-vitro selection step using the fitness vector $\boldsymbol{W}_{j \bmod q}$. We assume here that no one strand exists in the tube that can bind to all the proteins and a set of strands is really needed. See, for example, Figure 10.6: Strand 1 has high attachment rate (marked in bold) on the first three proteins and strand 2 for the last three. So a tube with about 50% of strand 1 and 50% of strand 2 "covers" all five proteins in the sense that it achieves high attachment on all of them. Note that the two strands cooperatively solve the problem and this is related to the disjunction of two variables. More formally, we pose

Key Problem 2. Assume that among all strands in a 1-liter tube of RNA, there is a particular set of two strands such that for each of q proteins, at least one of the two has a high fraction of attachment. Can you use PCR to arrive at a tube

$$(\approx 0.5, \approx 0.5, \approx 0, \ldots, \approx 0)$$

which has high attachment on all q proteins?

Note that initially the two good strands have concentration roughly 10^{-15}. They are indexed by 1 and 2 in Figure 10.6. If we over-train with P_0 and P_1, then the tube will converge to $s_1 \approx 1$ and $s_2 \approx 0$. Similarly, if we over-train with P_3 and P_4, then the tube converges to $s_1 \approx 0$ and $s_2 \approx 1$. We claim this problem cannot be solved with blind computation: when the fitness vectors $\boldsymbol{W}_j$ and the initial share vector are not known, no sequencing of strands is done and the protein structure is not known. Clearly we need some kind of feedback in each trial.

Let us switch to a related machine learning problem and see what additional mechanism was used there to solve this problem. As discussed before, the tube corresponds a share vector $\boldsymbol{s}$ which is an n dimensional probability vector. Proteins correspond to example vectors $\boldsymbol{W}_j$ of the same dimension. In the machine learning setting, the components of the example vectors are binary (i.e., $\boldsymbol{W}_j \in \{0,1\}^N$). Assume there exists a probability vector $\boldsymbol{u} = (0,0,\frac{1}{k},0,0,\frac{1}{k},0,0,\frac{1}{k},0,0,0)$ with k non-zero components of value $\frac{1}{k}$ such that $\forall j : \boldsymbol{u} \cdot \boldsymbol{W}_j \geq \frac{1}{k}$. Our goal is to find a share vector $\boldsymbol{s}$ such that $\forall j : \boldsymbol{s} \cdot \boldsymbol{W}_j \geq \frac{1}{2k}$. (In Key Problem 2 and Figure 10.6, k equals 2.)

Note the we are essentially trying to "learn" a union or disjunction of k variables with a linear threshold function. Typically we are given positive and negative examples of the disjunction. In the above simplification we only have positive examples: that is, all examples need to produce a dot product that lies above the threshold of $\frac{1}{2k}$. An algorithm that achieves this is called Winnow (Littlestone 1988). In normalized form, it goes as follows.

Normalized Winnow algorithm. Let $\boldsymbol{W}_j$ be the current example and $\boldsymbol{s}$ be the current share vector. Then the updated share vector $\tilde{\boldsymbol{s}}$ is:

$$\text{If } \underbrace{\boldsymbol{s} \cdot \boldsymbol{W}_j}_{\text{good side}} \geq \frac{1}{2k} \text{then} \quad \tilde{\boldsymbol{s}} = \boldsymbol{s}, \qquad \text{"conservative update"}$$

$$\text{else} \quad \tilde{s}_i = \begin{cases} s_i & \text{if } W_{j,i} = 0 \\ \alpha s_i & \text{if } W_{j,i} = 1, \text{ where } \alpha > 1 \end{cases} \tag{10.3}$$

and re-normalize the share vector $\tilde{\boldsymbol{s}}$.

Note that if the good RNA $\boldsymbol{s} \cdot \boldsymbol{W}_j$ is at least as large as the threshold, then the multiplicative update is not executed. Linear threshold updates that are vacuous when the prediction is on the "right side" of the threshold are called "conservative updates". For a batch of examples, Normalized Winnow does passes over all examples until no more multiplicative update steps (10.3) occurred in the last pass.

Does Winnow solve our Problem 2? Yes it does, but it used an additional mechanism discovered in the context of Machine Learning.

Machine Learning 1: Prevent over-training by making the update conservative.

Here the curse of the multiplicative update is avoided by "conservatively" only executing the amplification step when the dot product $s \cdot W_j$ (i.e., the amount of "good" RNA in the functional separation) is too low. So by measuring the concentration of RNA strands on the good side (there are several ways to do this), Normalized Winnow can be implemented with blind computation.

Note that if the dot product is at least as large as the threshold $\frac{1}{2k}$, then the share vector is not updated and $\tilde{s} = s$. In the context of in-vitro selection this means that RNA strands that attached to the protein sheet are simply recombined into the tube without the PCR amplification step. However, when the dot product is too low, then the multiplicative update is executed and PCR is needed to do the normalization (**10.3**).

One can show that the amplification step occurs at most $O(k \log \frac{N}{k})$ times if there is a consistent k-literal disjunction and this is information theoretically optimal. More general upper bounds on the number of multiplicative update steps are proved for the case when there is no k-literal disjunction that is consistent, and when there are positive and negative examples. The crucial part of these bounds is that they grow logarithmically with the number of variables n. In in-vitro selection, n is the number of different strands (approximately 10^{15}). Nevertheless, the logarithmic dependence on the dimension n makes this a feasible algorithm.

The conservative update would not be necessary if we could make k-fold combinations of RNA strands and then simply select for the best combination of k by doing a multiplicative update *in each* trial. However this "coupling of strands" requires $\binom{n}{k}$ combinations which is too large even for moderate values of k. The brilliant insight of Winnow is that coupling (Nature's mechanism 2) can be replaced by thresholding and only executing the multiplicative update when you are on the wrong side of the threshold.

The next mechanism we will discuss is to cap the shares/weight from above. Again this will ameliorate the curse of the multiplicative update and in some sense preserve variety. We start with Nature and then discuss how capping is used in machine learning.

In Nature capping can be achieved by a super (apex) predator. Predators need learn how to hunt any particular species. This learning process costs considerable effort and it therefore is advantageous to always specialize on the species that is currently the most frequent. In some sense the super predator nibbles away

Figure 10.7 The lion pride hunts the most frequent species in the Serengeti, keeping all large grass eating species in check and preserving variety.

at the highest bar of the histogram of possible prey species (see Figure 10.7). Many examples across the animal kingdom are known where removing the super predator causes some species to take over, greatly reducing prey diversity. Sometimes disease can have a similar effect. The more frequent a species, the more opportunity for a disease to spread and this can keep a species from taking over.

Nature 3: A super predator preserves variety.

Perhaps surprising, super-predators loosely correspond to a similar mechanism of machine learning for battling the curse.

Machine Learning 2: Cap the weights from above for the purpose of learning a set of experts.

Assume the current weight vector s is a capped probability vector (i.e., its components lie in the range $[0, c]$, where $c < 1$). There will now be two update steps. The first one applies a vanilla multiplicative update to $\boldsymbol{s}$: $s_i^m = \frac{s_i\, e^{-\eta\, Loss_i}}{Z}$, where Z normalizes the weights to one. However, because of the normalization, the resulting intermediate weight vector $\boldsymbol{s}^m$ might have some weights larger than c. In a second capping step, all weights that exceed c are reduced to c and the total weight gained is redistributed among the remaining components that lie below c so that their ratios are preserved and the total weight still sums to one. The redistribution of the weights above c might have to be repeated a number of times. However, there are efficient implementations of this capping step. Call the resulting weight vector $\tilde{\boldsymbol{s}}$. Recall that multiplicative updates are motivated

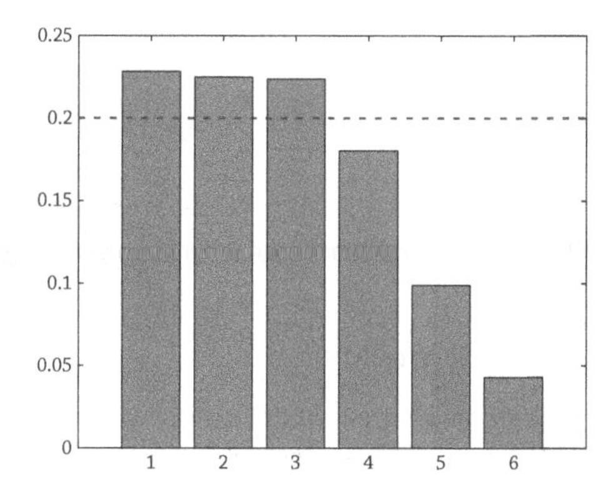

Figure 10.8 Cap and re-scale rest

by using a relative entropy as an inertia term. Similarly, the above method of capping solves a constrained optimization problem. Subject to the capping and total weight constraints, $\tilde{\boldsymbol{s}}$ minimizes the weighted sum of η times the expected loss and the entropy relative to the current weight vector $\boldsymbol{s}$.

What is capping used for in machine learning? Sets of size k from our set $\{1,2,\ldots,n\}$ can be encoded as n dimensional probability vectors: $(0,\frac{1}{k},0,0,\frac{1}{k},0,\frac{1}{k})$ with k components at $\frac{1}{k}$ and $n-k$ components at 0. We call these encodings of sets *k-corners*. Note that there are $\binom{n}{k}$ such k-corners. The capped probability simplex we seek is just the convex hull of the corners, $\{\boldsymbol{s} : s_i \in [0,\frac{1}{k}], \sum_i s_i = 1\}$, where we set $c = 1/k$. Since a share vector in the interior of the capped simplex is a convex combination of k-corners, the online algorithm maintains its uncertainty over which set is best by such mixtures.

The curse of the standard multiplicative update is that the weight vector converges to a corner of the simplex, a vector $(0,\ldots,0,1,0,\ldots,0)$ where the 1 is the position of the best expert. If, however, we run the multiplicative update together with capping at $c=\frac{1}{k}$, then the combined update converges to a k-corner of the capped simplex (i.e., to the best set of k experts). Figure 10.9 depicts the 3- and 4-dimensional simplexes plus their capped versions with $c=\frac{1}{2}$.

Note that the n-dimensional probability simplex has n 1-corners which are the n unit vectors and the capped n-dimensional simplex with $c=\frac{1}{k}$ has $\binom{n}{k}$ different k-corners, an integer much larger than n when $k \geq 2$. The combined update is computationally efficient in that it essentially lets us learn a k-variety with only n weights instead of $\binom{n}{k}$ weights. In the machine learning applications, using n weights is tolerable, but using $\binom{n}{k}$ weights is prohibitive.

An extension of this reasoning may interest technically minded readers. There is the matrix version of the multiplicative update: share vectors are generalized to density matrices which can be viewed as a probability vector of eigenvalues plus the corresponding unit eigenvectors. The "generalized multiplicative update"

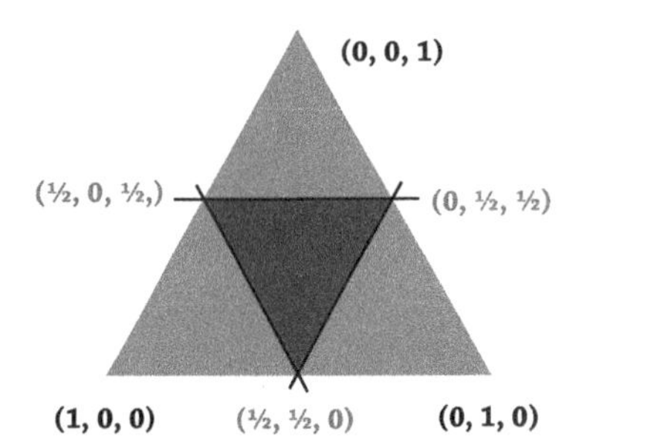

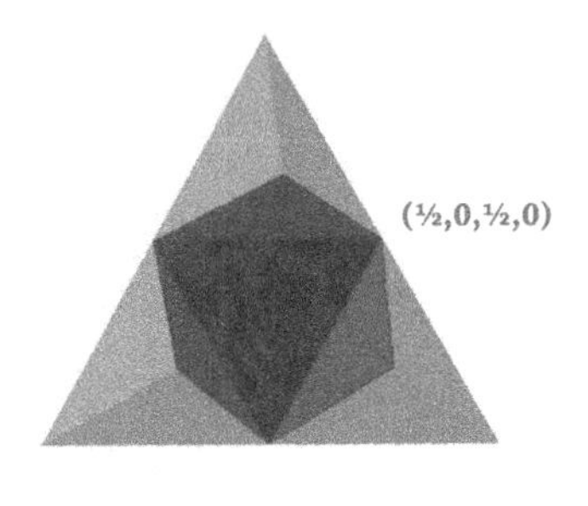

Figure 10.9 The 3-simplex can be depicted as a triangle where the three corners correspond to the single experts. The 2-corners of the capped simplex with $c = \frac{1}{2}$ correspond to the $\binom{3}{2} = 3$ different pairs of experts from a panel of three experts. The 4-simplex is a tetrahedron and the capped simplex with $c = \frac{1}{2}$ has $\binom{4}{2} = 6$ different 2-corners corresponding to the six different pairs of experts chosen from four candidates.

involves the computation of matrix logs and matrix exponentials (Tsuda et al. 2005). These updates are motivated with the quantum relative entropy instead of the regular relative entropy; which is the core concept in coding theory. When the matrix version of the multiplicative update is combined with a second capping update on the eigenvalues of the density matrix, then we arrive at an online algorithm for principal component analysis: the sets of k experts generalize to a k-dimensional subspace and capped density matrices are convex combinations of k-dimensional subspaces.

Many questions arise in this context: Is there a way to implement capping in the in-vitro selection setting by somehow modifying PCR? How does Nature avoid the $\binom{n}{k}$ combinatorial blowup when it needs to learn a conjunction or disjunction of genes? Is the matrix version of the multiplicative update used somewhere in Nature?

There is a related mechanism for battling the curse that we still need to discuss: lower bounding the shares. After doing a number of iterations of in-vitro selection, some "selfish" strand often manages to dominate without exactly achieving the wanted function. A standard trick is keep a batch of the initial mixture in reserve (which contains the initial variety of approximately 10^{15} different strands). Whenever the current mixture has become too uniform, then a little bit of the initial rich mixture is mixed into the current tube. This essentially amounts to lower bounding the shares.

A good illustration of this mechanism arises from a practical application of online learning: the *disk spindown problem*. It takes energy to shut down and restart a laptop. So if the laptop is likely to be used again very soon, then keeping it running is the best strategy for conserving energy. However, if it is likely to be idle for a long time, then it is better to shut it down immediately. In practice this dilemma is resolved with a fixed timeout: If there was no action within a fixed time τ, then the laptop simply shuts down and powers back up when the next request comes in.

A more clever way is to find a timeout that adapts to the usage pattern of the user by running a multiplicative update on a set of timeouts (Helmbold et al. 2000). So for this setting a suitably spaced set of n fixed timeouts $\tau_1, \tau_2, \ldots, \tau_n$ form our set of *experts*. As before, the *master algorithm* maintains a set of weights or shares $s_1, s_2, \ldots, s_n$ for the experts and applies a multiplicative update after each idle time:

$$\tilde{s}_i = \frac{s_i \, e^{-\eta \, \text{energy usage of timeout } i}}{Z}, \quad \text{where } Z \text{ normalizes the shares to 1.}$$

Ideally the shares of the master algorithm will concentrate on the best timeout.

The curse strikes when the data (characterized by the typical length of the idle times) change over time, as in Figure 10.10. Roughly, each typical length of the idle time favors a certain timeout value. The multiplicative update will quickly bring up the shares of the timeout values that use the least energy for the current typical idle time length and wipe out the shares of the remaining timeouts. However, this typical length will occasionally change, for example, due to a shift in workload. Unfortunately the newly needed timeouts are no longer available.

There is a simple mechanism that prevents this problem with a second update step:

Machine Learning 3: Mix in a little bit of the uniform vector.

$$\boldsymbol{s}^m = \frac{s_i \, e^{-\eta \, \text{energy usage of timeout } i}}{Z}$$

$$\tilde{\boldsymbol{s}} = (1-\alpha)\boldsymbol{s}^m + \alpha \left(\frac{1}{N}, \frac{1}{N}, \ldots, \frac{1}{N}\right), \quad \text{where } \alpha > 0 \text{ is small.} \qquad \textbf{(10.4)}$$

In Nature we have a similar situation. The multiplicative update quickly selects for the individuals most fit for the current environment. However, those individuals might not be adapted well for the environment of tomorrow and a mechanism is needed for keeping genes around that might become useful later. At a rough level, mutations function as a mechanism for lower bounding the shares.

Figure 10.10 Idle times are depicted as black line segments. Bursts of short idle times (before ↑) favor long timeouts τ_i. For long idle times (after ↑) small timeouts are better. However, at time ↑, the multiplicative update has wiped out all short timeouts.

Nature 4: Mutations keep a base variety.

We conclude with the most interesting mechanism discovered in Machine Learning for battling the curse. This mechanism works well when the data shifts once in a while, and some of the shifts are returns back to a type of data seen previously. Natural data often has this form: It shifts between a small number of different "modes": a breakfast mode, a lunch mode, and a dinner mode, and once in a while a holiday mode.

A good online update has to do two things: First, it must have the capability to bring up shares quickly, and second, it must remember experts that did well in the past because they might be needed again. The first is achieved with a multiplicative update which has good-short term properties (i.e., the shares of the good experts are brought up quickly). The mechanism needed for the second part has been dubbed "sleeping" (Adamskiy et al. 2012) because previously good models are essentially put asleep so that they can be woken up quickly when needed again.

Machine Learning 4: Use sleeping to realize a long-term memory.

A simple way to do this is to keep track of the average share vector $\boldsymbol{r}$. Instead of mixing in an α fraction of the uniform weight vector, as done in the second step of (**10.4**), mix in an α fraction of $\boldsymbol{r}$. Finally, at the end of each trial, update the average vector $\boldsymbol{r}$ so that it includes the current share vector. Note that an expert that did well in the past will have a large enough share in the average share vector $\boldsymbol{r}$ and this helps in the recovery when it is needed again.

This two-step update can be implemented within in-vitro selection paradigm (i.e., with blind computation): Maintain two tubes, one for the current share vector $\boldsymbol{s}$ and one for the average share vector $\boldsymbol{r}$. It is easy to implement the $\tilde{\boldsymbol{s}} = (1 - \alpha)\boldsymbol{s} + \alpha\boldsymbol{r}$ update by mixing the appropriate fractions of the tube. Similarly, one can update of the average tube at the end of trial t by mixing tubes: $\tilde{\boldsymbol{r}} = \frac{t-1}{t}\boldsymbol{r} + \frac{1}{t}\tilde{\boldsymbol{s}}$. However the totals of the tubes are not preserved in this two-step update: For example, an α fraction of the $\boldsymbol{r}$ tube is moved to the $\boldsymbol{s}$ tube, and now there is less than unit amount in the $\boldsymbol{r}$ tube.

This can be fixed with additional PCR amplification steps, but there is a more elegant way to implement sleeping with in-vitro selection, as described in Figure 10.11. It is essentially a Markov network with two tracks (tubes). The left (awake) tube represents the share vector $\boldsymbol{s}$ and the right (asleep) tube the share vector $\boldsymbol{r}$. Both are initialized to be uniform, with amounts γ and $1-\gamma$, respectively. In each trial, $\boldsymbol{s}$ is updated to $\boldsymbol{s}^m$ via a multiplicative update. The normalization assures that $\boldsymbol{s}$ and $\boldsymbol{s}^m$ have the same total weight. On the asleep side, nothing happens (i.e., $\boldsymbol{r}^m = \boldsymbol{r}$). In the second step, there is an exchange: both tubes send a fraction of

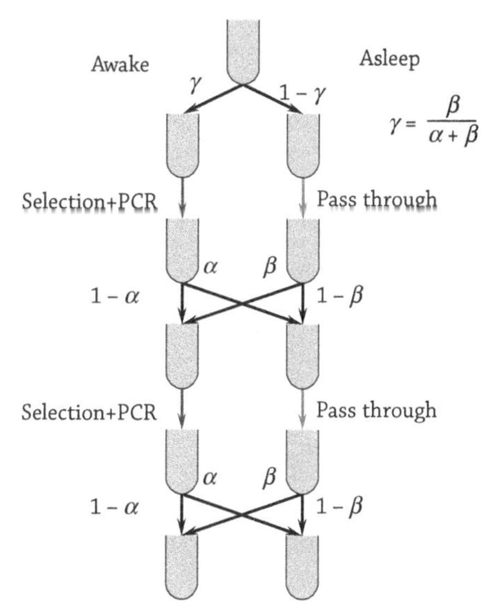

Figure 10.11 2-track method: We start with a unit amount of a rich mixture of RNA strands. A fraction of γ goes to the awake side and $1-\gamma$ to the asleep side. The awake tube participates in the selection process and the asleep tube is just passed through. After that the tubes exchange a small fraction of their content: α from awake to asleep and β from asleep to awake.

their content to the other side:

$$\tilde{\boldsymbol{s}} = (1-\alpha)\boldsymbol{s}^m + \beta \boldsymbol{r}^m$$
$$\tilde{\boldsymbol{r}} = \alpha \boldsymbol{s}^m + (1-\beta)\boldsymbol{r}^m,$$

where the exchange probabilities α and β are small numbers that need to be tuned. If the initial fraction γ of the awake side is $\frac{\beta}{\alpha+\beta}$, then the exchange keeps the fraction on the awake side at γ and the asleep side at $1-\gamma$.

A sample run of this algorithm and some of its competitors is given in Figure 10.12. The repeating segments may be seen as "modes" of the data: There is an A-mode, a D-mode, and a G-mode. Note that the two-track algorithm recovers more quickly in segments where a previously best expert is best again, that is, we return to a previously seen mode (see Figure 10.13). In Figure 10.14 we see that all previously good experts are remembered on the sleeping side (long-term memory).

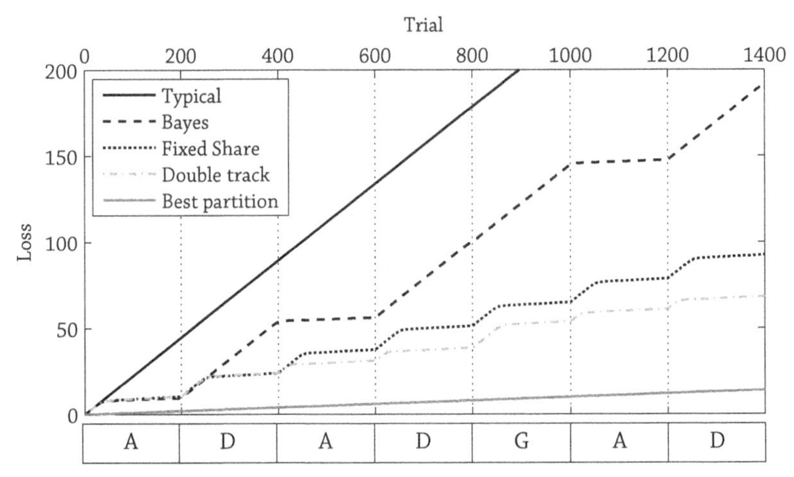

Figure 10.12 We plot the total loss of a number of algorithms when the data shifts recurrently: 1400 trials, 2000 experts, the data shifts every 200 trials. The best experts in the seven segments are A, D, A, D, G, A, D. The grey "Best partition" curve is the total loss with foreknowledge, using the best expert immediately in each segment. The loss of all other experts increase as shown in the black "Typical" curve. The dashed "Bayes" curve is just the multiplicative update: It learns expert A in the first segment and does not adjust in the later segments. However, it has the optimal slope in all later segments when A is best again. The dotted "Fixed Share" update (10.4) mixes in a bit of the uniform distribution in each trial. It learns the best expert in each segment (fixed size bump for each segment). The dash dotted "two-track" algorithm of Figure 10.11 has a long-term memory (i.e., it recovers quicker in segments where the best expert is an expert that was best in a previous segment).

Curiously enough, there is a Bayesian interpretation of the above two methods of mixing in a little bit of the average share vector or the two-track method of Figure 10.11 (Adamskiy et al. 2012): Models are either awake or asleep. For a suitably chosen set of models and prior of the models the two methods can be explained with a Bayesian update with the following caveat: the asleep models predict with the Bayesian predictive distribution (the normalizer of the Bayesian update). This means that Bayes' update for asleep models is vacuous (i.e., the prior equals the posterior). In other words, the asleep models "abstain" while being asleep and their prior is unchanged while they are asleep.

10.3 DISCUSSION

Multiplicative updates converge quickly, which is their blessing. However, multiplicative updates wipe out diversity, which is their curse. Changing conditions

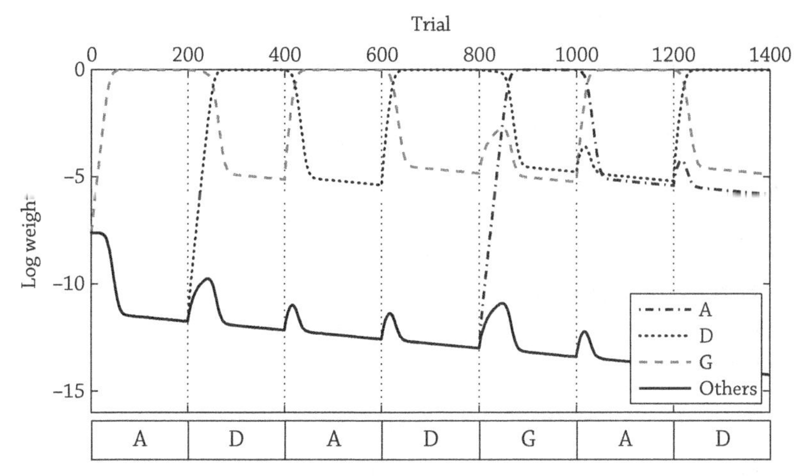

Figure 10.13 Here we plot the log weights of the *awake side* of the "two-track" algorithm. Note that the weights of the best experts are brought up quickly at the beginning of each segment. Most importantly, when an expert is best again, then it is brought up more quickly.

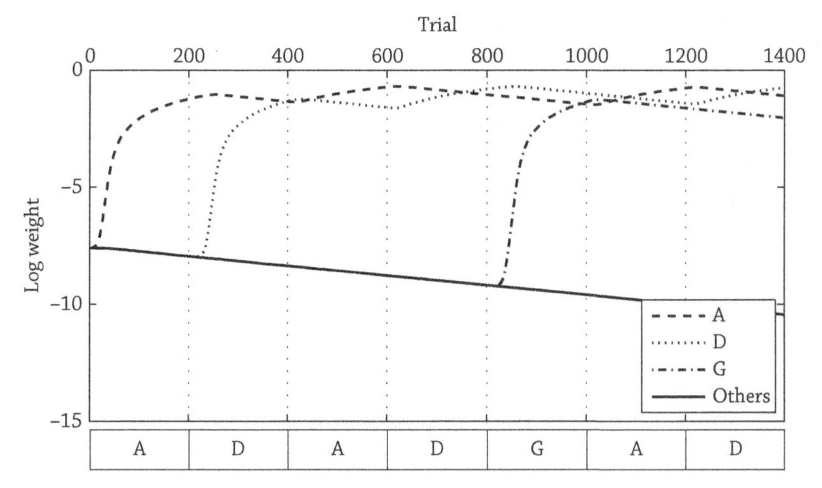

Figure 10.14 Here we plot the log weights of the asleep side of the "two-track" algorithm. Note that the weights of previously good experts decay only slowly. At the end (trial 1400) the weights of the experts A, D, and G are elevated (long-term memory).

require reuse of previously learned knowledge/alternatives and diversity is a requirement for success.

A mechanism is needed to ameliorate the curse and machine learning and Nature each have developed a number of them.

Machine Learning mechanisms:

1) conservative update for learning multiple goals
2) upper bounding weights for learning multiple goals
3) lower bounding weights for robustness against change
4) sleeping for realizing a long-term memory.

Nature's mechanisms:

1) coupling for preserving variety
2) boundaries for preserving variety
3) super-predators for preserving variety
4) mutations for keeping a base variety.

We are convinced that studying the long-term stability of evolutionary processes is an important research topic and we believe that important insights can be gained by contrasting Nature's stabilizing mechanisms with the ones developed in machine learning for multiplicative updates.

Regarding the sleeping mechanism, it seems that the two-track algorithm models evolution at different scales. The awake track represents the short-term evolution and the asleep track the long-term evolution. An interesting question is whether a third track would be useful. It is reasonable to expect that Nature makes use of the sleeping mechanism as well. At a macro level if you sow a handful of wild seed, then not all sprout immediately. Some lay dormant, and this guards against changing weather conditions. However, the deep question is how is sleeping realized at the genetic level. Does junk DNA or sex play a role?

NOTES

We use the term "multiplicative updates" specifically for updating the weights or shares by non-negative scalar factors. Note that the mutation update on a population represented by a share vector is described by a different type of multiplication: In this case the share vector is multiplied with a stochastic matrix.

Some early work on online learning can be found in Vovk (1990); Littlestone and Warmuth (1989, 1994). For the motivation of multiplicative updates that uses the relative entropy as a divergence measure see Kivinen and Warmuth (1997). A survey of in-vitro selection is given in S. Wilson and Szostak (1999).

For the observation that Bayes' rule is related to the discrete replicator equation see, for example, Harper (2009).

The relationship between various disturbance regimes and diversity has been discussed extensively in the literature. The "intermediate disturbance hypothesis" proposes that diversity will be highest at intermediate levels of disturbance (see Miller et al., 2011 for an overview).

Problems related to the replication of a mixture of strands while preserving their relative frequencies (Key Problem 1) are often discussed in the context of the evolution of early life (see, for example, Zintzaras et al. 2002; Vasas et al. 2012).

The original Winnow algorithm appeared in Littlestone (1988). The normalized version presented here was analyzed in Helmbold et al. (2002). The stabilizing effect of a super/apex predator has been well studied. See for example J. A. Estes et al. 2011 and F. Sergio et al. 2008 and the references therein, as well as the predator/prey model on the prism developed in Chapter 7.

The multiplicative updates together with capping and their application to learning principal components online were developed in Warmuth and Kuzmin (2008). The optimality of these algorithms has been shown in Jiazhong et al. (2013). The matrix versions of the multiplicative updates were developed in Tsuda et al. (2005) and Warmuth and Kuzmin (2006, 2011).

The method of making an online algorithm robust against time changing data by mixing in a bit of the uniform distribution was first analyzed in Herbster and Warmuth (1998). The idea of mixing in a bit of the average share vector was introduced in Bousquet and Warmuth (2002). The long-term memory properties of this method were also analyzed in that paper.

Finally, the Markov type two-track update given in Figure 10.11 was introduced in Adamskiy et al. (2012). Also, a Bayesian interpretation of the long-term memory updates based on the mechanism of sleeping was given in that paper.

BIBLIOGRAPHY

Adamskiy, D., M. K. Warmuth, and W. M. Koolen. "Putting Bayes to sleep." In F. Pereira, C. J. C. Burges, L. Bottou, and K. Q. Weinberger, eds., *Advances in Neural Information Processing Systems 25 (NIPS '12)*, pp. 135–143. Curran Associates, Inc., 2012.

Bousquet, O., and M. K. Warmuth. "Tracking a small set of experts by mixing past posteriors." *Journal of Machine Learning Research* 3 (2002): 363–396.

Estes, J. A. et al. "Trophic downgrading of planet earth." *Science* 333, no. 6040 (2011): 301–306.

Harper, M. "The replicator equation as an inference dynamic." arXiv:0911.1763 [math.DS], November 2009.

Helmbold, D. P., D. D. E. Long, T. L. Sconyers, and B. Sherrod. "Adaptive disk spin-down for mobile computers." *ACM/Baltzer Mobile Networks and Applications (MONET)*, pp. 285–297, 2000.

Helmbold, D. P., S. Panizza, and M. K. Warmuth. "Direct and indirect algorithms for on-line learning of disjunctions." *Theoretical Computer Science* 284, no. 1 (2002): 109–142.

Herbster, M., and M. K. Warmuth. "Tracking the best expert." *Machine Learning* 32 (1998): 151–178.

Jiazhong, N., W. Kotłowski, and M. W. Warmuth. "On-line PCA with optimal regrets." In *Proceedings of the 11th International Conference on Algorithmic Learning Theory (ALT 24)* 8139 (2013): 98–112. Lecture Notes in Artificial Intelligence, Springer-Verlag, Berlin.

Kerr, B., Margaret A. Riley, Marcus W. Feldman, and Brendan J. M. Bohannan. "Local dispersal promotes biodiversity in a real-life game of rock-paper-scissors." *Letters to Nature* 418 (2002): 171–174.

Kivinen, J., and M. K. Warmuth. "Additive versus exponentiated gradient updates for linear prediction." *Information and Computation* 132, no. 1 (1997): 1–64.

Littlestone, N. "Learning quickly when irrelevant attributes abound: A new linear-threshold algorithm." *Machine Learning* 2, no. 4 (1988): 285–318.

Littlestone, N., and M. K. Warmuth. "The weighted majority algorithm." In *Proceedings of the 30th IEEE Symposium on Foundations of Computer Science*, 1989.

Littlestone, N., and M. K. Warmuth. "The weighted majority algorithm." *Information and Computation* 108, no. 2 (1994): 212–261.

Miller, A. D., S. H. Roxburgh, and K. Shea. "How frequency and intensity shape diversity-disturbance relationships." *Proceedings of the National Academy of Sciences USA* 108, no. 14 (2011).

Rainey, P. B., and M. Travisano. "Adaptive radiation in a heterogeneous environment." *Nature* 394 (1998): 69–72.

Sergio, F. et al. "Top predators as conservation tools: Ecological rationale, assumptions, and efficacy." *Annual Review of Ecology, Evolution, and Systematics* 39 (2008): 1–19.

Tsuda, K., G. Rätsch, and M. K. Warmuth. "Matrix exponentiated gradient updates for on-line learning and Bregman projections." *Journal of Machine Learning Research* 6 (2005): 995–1018.

Vasas, V., C. Fernando, M. Santos, S. Kauffman, and E. Szathmary. "Evolution before genes." *Journal of Biology Direct* 7, no. 1 (2012): 1–14.

Vovk, V. "Aggregating strategies." In *Proceedings of the Third Annual Workshop on Computational Learning Theory* (1990): 371–383.. Morgan Kaufmann.

Warmuth, M. K., and D. Kuzmin. "Online variance minimization." In *Proceedings of the 19th Annual Conference on Learning Theory (COLT '06)* (2006). Pittsburg, Springer-Verlag.

Warmuth, M. K., and D. Kuzmin. "Randomized PCA algorithms with regret bounds that are logarithmic in the dimension." *Journal of Machine Learning Research* 9 (2008): 2217–2250.

Warmuth, M. K., and D. Kuzmin. "Online variance minimization." *Journal of Machine Learning* 87, no. 1 (2011): 1–32.

Wilson, D. S., and J. W. Szostak. "In vitro selection of functional nucleic acids." *Annual Review of Biochemistry* 68 (1999): 611–647.

Zintzaras, E., M. Santos, and E. Szathmary. "'Living' under the challenge of information decay: The stochastic corrector model vs. hypercycles." *Journal of Theoretical Biology* 217, no. 2 (2002): 167–181.

11

Traffic Games

JOHN MUSACCHIO

So far in this book, we have considered games in which either natural selection or conscious selection causes agents to play strategies with higher payoff with greater frequency. This behavior is how the mix of strategies played by the population varies over time. In this chapter, we study a related class of games—traffic games, or congestion games. In these games, as in most of the other games we study in this book, it is common to assume that there are a large number of players—so many that each player represents a negligible fraction of the population. When we assume this in the context of traffic games, we call the traffic "non-atomic," meaning that there are no indivisible chunks in the traffic.

Players choose among several possible routes to a destination based on the delays associated with each route. In turn, the delays are a function of what fraction of the players take each route. Unlike other parts of this book, our focus will be on studying the equilibria of such games rather than the dynamics that lead to equilibrium.

The equilibrium analysis turns up several surprises and several new insights. Congestion games are examples of potential games, an important class of games with special structure. Optimization techniques are especially powerful in analyzing such games, and this chapter will introduce some of the more useful techniques. We cover the Braess paradox, which demonstrates the possibility that constructing new roads can be counterproductive, even at zero construction cost.

Pigovian taxes appear first in this chapter, and will reappear in Chapter 12, as an approach to dealing with external costs and benefits. The chapter also illustrates how population games (for traffic) can combine with classic noncooperative games with only a few players (who own traffic routes). Finally, the chapter draws a precise analogy between these combined games and electric circuit analysis, and exploits the analogy to obtain some general results on "the price of anarchy." All these ideas will be explained carefully so that readers can redeploy them in new applications.

11.1 SIMPLE NON-ATOMIC TRAFFIC GAMES

We start with a simple example. Suppose a population of drivers needs to select from two alternative routes from a common origin to a common destination. Each route has a delay that grows linearly with the fraction of the drivers that select that route. The situation is illustrated on the left panel of Figure 11.1. For now, we suppose that both linear functions increase at the same rate. Moreover, drivers would like to minimize the delays they incur in getting to their destination. Clearly, each driver's delay depends on not only his own choice of route, but also on the selections of the other players. If more drivers take the upper route, the delays on the lower route will be less, and some of those drivers on the upper route will want to switch. The reverse is true if more drivers were taking the lower route. The only split that results in no drivers "wanting" to switch routes is to have half the drivers take the upper route and half the lower route. This leads to the delays on both routes being equal.

Now suppose that a third route is added to the situation. This route has a delay that is fixed, that is, it does not change with respect to the traffic taking it. Moreover, suppose that the delay is larger than the equilibrium delay when there were only the two original routes. This new situation is depicted on the right panel of Figure 11.1. Since this new road has a larger delay than on the other routes, none of the drivers want to switch to this route. Thus having half the traffic taking each of the original routes, and none on the new route is an equilibrium. A specification of what fraction of traffic takes each of the possible routes corresponds to the

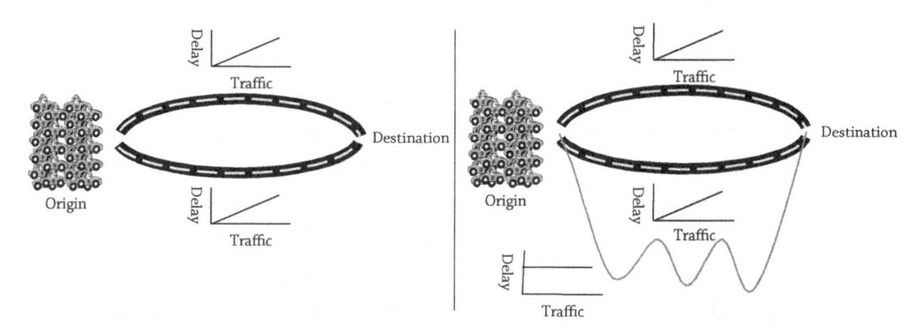

Figure 11.1 A simple traffic game. Left panel: two-route game. Right panel: three-route game.

notion of *state* that we have been using in this book. However, in the context of traffic games, it is more customary to call it a **traffic assignment**.

We can generalize from these two examples two properties that must be satisfied for a non-atomic traffic game to be in equilibrium. First, any two routes that are being used to get between a source-destination pair must have the same delay. Second, any other route between the two points not being used must have a delay that is not less than the delay of the routes being used. If these properties are satisfied, no driver can improve her delay by changing routes, hence the game is in NE. Road traffic analyst John Wardrop first observed these principles, so this notion of equilibrium in non-atomic traffic games is called a **Wardrop equilibrium**. Note that an equivalent way to describe a Wardrop equilibrium is that no player can improve his payoff by switching strategies (routes), and that the payoff (delay) of a route is unchanged by a single player changing strategies. Therefore a Wardrop equilibrium is defined exactly the same way as a NE in an evolutionary game.

A natural question is whether there always exists a Wardrop equilibrium. It turns out that there will indeed always be a Wardrop equilibrium. The key idea in proving this is to show that a traffic assignment is a Wardrop equilibrium if and only if it solves an optimization problem that has been "cooked up" in a way so that its optimality conditions match the conditions for Wardrop equilibrium.

Roughly speaking, if we can devise an objective function that gets minimized in the NE of a game, we call that function a potential function, and call the game a potential game. To make this more precise, let's for a moment consider a game with finitely many players, say N. A potential function $V(s_1,\ldots,s_N)$ is a function that maps each strategy profile $[s_1,\ldots,s_n]$ to a real number. Moreover, this function must have the property that a unilateral change of strategy of any player produces the same change in the potential function as it would in that player's individual cost function $C_i(\cdot)$. Specifically

$$V(s_i,s_{-i}) - V(s_i',s_{-i}) = C_i(s_i,s_{-i}) - C_i(s_i',s_{-i}) \tag{11.1}$$

for all players i, all strategies s_i, and opponent profiles s_{-i}. If there is a potential function, in some sense it is like the game is equivalent to a game in which all the players share a common objective. As one might expect, not all games are potential games; an example called Matching Pennies is covered in the exercises.

Since in potential games we can think of all the players trying to minimize a common objective function, a potential function in games with continuous strategy spaces should be such that the conditions for optimizing the potential function exactly match the conditions for each player to be playing a best response. Turning back to the traffic problem, let us consider the following optimization problem

$$\max_{\{f_e,\, e\in E\}} \sum_{e\in E} \int_0^{f_e} l_e(u)du \tag{11.2}$$

Subject to: Flow out of source adds to 1,

Flow into destination adds to 1,

Net flow into other nodes adds to 0,

$f_e \geq 0$ for all e.

Here the notation E refers to the set of all edges in the network, e is one of the edges, f_e is the flow on edge e, and $l_e(\cdot)$ is the delay or latency on edge e as a function of flow. Note we have assumed a single source and destination here for simplicity, even though it is possible to generalize this to multiple sources and destinations.

This is a constrained optimization problem and the constraints are listed after the "Subject to" statement. The constraints say that that flow can only be created or destroyed at the source and destination nodes respectively, and each flow must be non-negative. In an unconstrained problem, we could take the derivative of the objective with respect to each optimization variable and set each derivative to be zero to find the bottom of the "trough" in each dimension. However, there are constraints and we need to ensure that the constraints are met in our solution. The standard way to deal with this is to use the technique of Lagrange multipliers. To visualize how this works, suppose the optimal solution of the constrained problem ends up being right on a constraint. Then it is probably the case that if we solved the problem with that constraint not enforced, the optimal solution would lie beyond the constraint and actually violate it. Now suppose instead of strictly enforcing the constraint, we charge a "price" for moving the flow vector in the direction of the constraint. If this price were high we could push the optimal solution back into compliance with the constraint. The price that brings the solution just into compliance is called the Lagrange multiplier for that constraint. Also if the optimum solution is not against a constraint, we ought not charge any price for the flow vector moving toward the constraint, since there is no need to "distort" the solution to comply with a constraint that is already being met. This idea of having a zero Lagrange multiplier (price) for constraints the optimal solution is not "pushing" against is called complementary slackness in the optimization literature. Moreover, we do not know the values of the Lagrange multipliers (prices) when we start to solve the problem, but we can leave them as undetermined variables that we can find for later using the conditions that we have just described. These conditions when stated formally are known as the Karush-Kuhn-Tucker (KKT) conditions named after the discovers of them. A formal treatment of the subject can be found in Boyd and Vandenberghe (2004).

In our problem, we assign each node constraint a Lagrange multiplier λ_i, and each positivity constraint $f_e \geq 0$ a Lagrange multiplier μ_e. The objective after we add the terms that price the solution moving toward each constraint is called the Lagrangian. In our problem it is

$$\sum_{e\in E}\int_0^{f_e} l_e(u) + \sum_{\text{nodes}} \lambda_i \left(\sum_{\text{links out of } i} f_e - \sum_{\text{links into } i} f_e - \begin{cases} 1 & \text{if } i = \text{source} \\ -1 & \text{if } i = \text{destination} \\ 0 & \text{otherwise} \end{cases} \right) + \sum_{e\in E} -\mu_e f_e. \tag{11.3}$$

To proceed we differentiate with respect to each f_e and set each derivative to 0. We also observe $\mu_e \geq 0$ since we should pay a non-negative price μ_e for moving toward violating the constraint that each flow f_e is non-negative. Finally, the complementary slackness idea allows us to say that if $f_e > 0$, $\mu_e = 0$ since we are not "pushing" on that constraint and thus have no need to charge a price for moving in the direction that might eventually violate it. Using all of these ideas we can show

$$\begin{aligned} l_e(f_e) &= \lambda_i - \lambda_j \quad \text{if } f_e > 0 \text{ and edge } e \text{ connects node } i \text{ to } j, \\ l_e(f_e) &\geq \lambda_i - \lambda_j \quad \text{if } f_e = 0 \text{ and edge } e \text{ connects node } i \text{ to } j. \end{aligned}$$

Moreover, by adding the above equations for multiple edges, we can verify that the condition says that all used routes connecting the source to the destination have the same delay, and the unused routes have greater delay. Thus this condition is exactly the same as the Wardrop equilibrium condition. As long as the latency functions are nondecreasing, the optimization problem is convex and basic ideas from optimization theory ensure there is a solution.

So the only way one can have multiple Wardrop equilibria is if different traffic assignments lead to the same delays (e.g., a network with two routes that have fixed delays). Another interesting finding from this is that the players in the game end up minimizing a different objective than the total delay in the network, since that would have a form $\sum_{e\in E} f_e l_e(f_e)$. (This form can be found by adding the delay in each link times the amount of drivers that suffer that delay.)

11.2 BRAESS'S PARADOX

Consider the slightly more complex example shown on the left panel of Figure 11.2. There are two routes to the destination. Each consists of two edges—one edge with a fixed delay of 1, and the other edge with a variable delay growing linearly with the volume of traffic. The slope of the linear function is such that if all of the

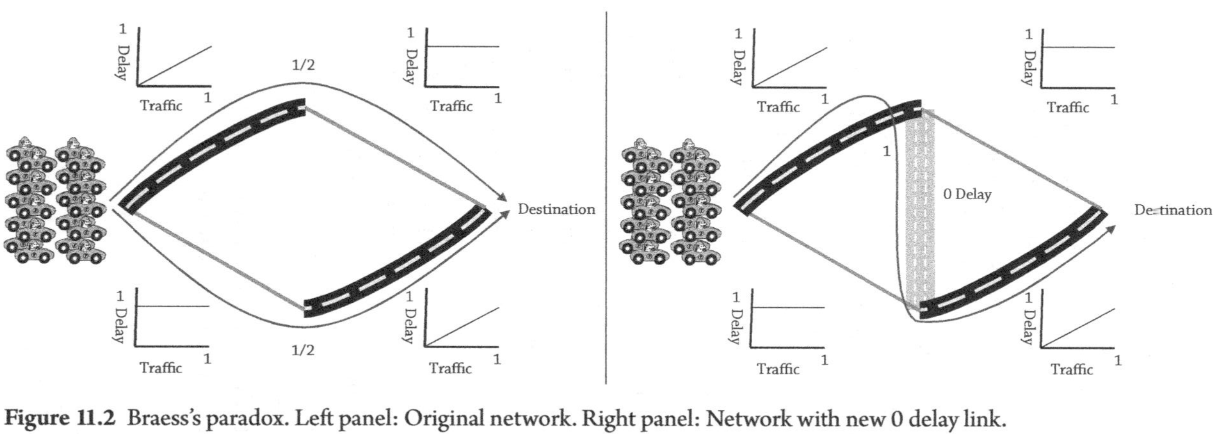

Figure 11.2 Braess's paradox. Left panel: Original network. Right panel: Network with new 0 delay link.

traffic took that edge, the delay would be 1. From the symmetry of the network, it is not hard to guess that the Wardrop equilibrium is for half the traffic to take each of the two routes. This results in a delay of $3/2$ for all the traffic – since the variable delay roads have delay $1/2$ and the fixed delay roads have delay 1.

Now suppose we alter the road network by adding a new road as depicted in the right panel of Figure 11.2. A lone driver looking to deviate from the previous traffic assignment of half on the upper route and half on the lower route discovers that she can now get to the destination with a total delay of 1 by taking variable delay road, new road, and then variable delay road. Other drivers will want to copy this, and as they do the variable delay roads traffic volume grows closer to 1. Ultimately, the Wardrop equilibrium is for all the traffic to take the same path: variable delay road, new road, and then variable delay road, with a resulting delay of 2. Paradoxically, the equilibrium delay of the road network increased by adding a new road! This example is due to mathematician Dietrich Braess, and it is thus widely known as Braess's paradox.

If a benevolent dictator or "social planner" could force the traffic on the new network to follow different routes in order to minimize the total delay in the system, it turns out the best solution is to not use the new road at all, and split the traffic between the upper and lower routes. This results in a delay of $3/2$ for all the traffic, just as before the new road was built. In traffic games, as well as other kinds of games, it is interesting to compare how much worse off selfish agents in NE compared to what could have been possible if a perfect social planner could make everyone do the optimal thing. In economics, it is typical to look at the efficiency loss—the percentage of total utility, or social welfare, that is lost in NE versus. social optimum. In the computer science community, it is a common convention to look at the ratio of NE welfare to social optimum welfare, often called the **price of anarchy**. So far in this chapter the games are formulated with costs rather than with payoffs. For such games, the price of anarchy is just the ratio of the total cost in NE to the total cost in social optimum. In the Braess's paradox example, the price of anarchy is $2/\frac{3}{2} = \frac{4}{3}$.

One question is whether $\frac{4}{3}$ is as bad as it can get with regard to the price of anarchy, or are there examples for which it is even worse?

11.3 THE PRICE OF ANARCHY WITH NONLINEAR LATENCY FUNCTIONS

We now attempt to construct an example with an even worse price of anarchy. In our example, road 1 has a fixed delay of 1, while road 2 has a very low delay when the traffic is light but increases steeply to 1 as the traffic using the road

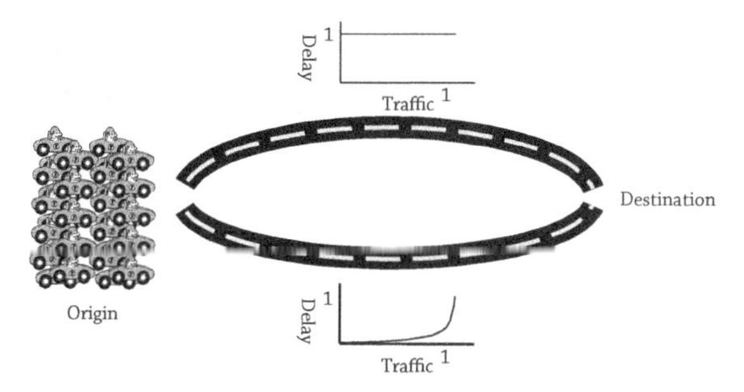

Figure 11.3 A traffic game with nonlinear latency cost.

reaches 1. The situation is illustrated in Figure 11.3. The intuition is that a social planner would try to keep the congestion level of the second road below the point it grows steeply, by "forcing" some of the traffic to road 1 instead. In contrast, users will always seek out the lowest-delay road, which is road 2 when it carries anything less than all the traffic, and when it carries all the traffic its delay is tied with that of road 1. Consequently, the only Wardrop equilibrium is that all the traffic takes road 2 and incurs a delay of 1. To make the example precise we specify that $l_1(f_1) = 1$ and $l_2(f_2) = f_2^n$ where $n \geq 1$ is a parameter. The social optimum routing can be found explicitly by optimizing the total delay, which has the form $f_1 \times 1 + f_2 \times f_2^n = (1 - f_2) + f_2 \times f_2^n$. From analyzing the first order optimality condition with respect to f_2, one can show that the optimum flow 2, f_2^*, has the form

$$f_2^* = e^{-\frac{\log(n+1)}{n}}. \tag{11.4}$$

From this expression, we see that as n gets larger (which corresponds to a steeper and steeper latency function) the fraction of traffic taking road 2 approaches but never reaches 1. In other words, the social planner puts all but a small fraction of the traffic on road 2, and when n is large the delay this traffic incurs is close to 0, while the small fraction on road 1 incurs a delay of 1. However, since almost all the traffic is on road 2 when n is large, the average delay for all the traffic is near 0, and indeed one can show that it gets arbitrarily close to 0 as n grows to infinity. On the other hand, the Wardrop equilibrium has an average delay of 1, hence the price of anarchy grows arbitrarily large as n goes to infinity.

The nonlinearity of the latency function of road 2—specifically the big difference in the slope of the function when it is heavily loaded versus when it is not—is what made it possible to have such a big difference between the Wardrop equilibrium average delay and social optimum. A natural question is how bad the price of anarchy can be if we disallow such nonlinear latency functions. This question was recently answered in the literature. It turns out that if one considers a network with affine latency functions (affine meaning the latency function of each link e takes the form $l_e(f_e) = a_e f_e + b_e$) the worst case price of anarchy is $\frac{4}{3}$, the price of anarchy we observe in the Braess's paradox example (Roughgarden 2002).

11.4 PIGOVIAN TAXES

The average delay in Wardrop equilibrium can be significantly higher than for the social optimum traffic assignment because individual users only consider their own delay and not how their route choice affects the delay of the other users in the network. When an agent's choice can impose costs (or benefits) on other parties and that agent does not consider this effect in his or her objective, it is called an *externality*. A natural approach to "fix" externalities and make users consider their effects on others is to make them pay a fine (or receive a subsidy) in proportion to the cost (or benefit) they impose on others. Such a scheme is usually called a Pigovian tax, after the English economist Arthur C. Pigou, who studied the idea in the early 20th century.

Focusing on the affine latency function case, let us consider the problem of optimizing the total delay as a social planner sees it. This problem takes the form

$$\sum_{e \in E} f_e (a_e f_e + b_e) \tag{11.5}$$

Subject to: Flow out of source adds to 1,

Flow into destination adds to 1,

Net flow into other nodes adds to 0,

$f_e \geq 0$ for all e.

We can solve this problem by constructing the Lagrangian, differentiating with respect to each flow rate, and noting the complementary slackness requirements of the Lagrange multipliers (only multipliers for tight constraints may be nonzero). Doing this we get

$$\begin{aligned} 2a_e f_e + b_e &= \lambda_i - \lambda_j \quad \text{if } f_e > 0 \text{ and edge } e \text{ connects node } i \text{ to } j, \\ 2a_e f_e + b_e &\geq \lambda_i - \lambda_j \quad \text{if } f_e = 0 \text{ and edge } e \text{ connects node } i \text{ to } j, \end{aligned}$$

where each λ_i is the Lagrange multiplier associated with the flow conservation constraint on node i. In words, the above expression says that a social planner should assign flow to links on the basis of two times the variable delay (the $a_e f_e$ piece) plus one times the fixed delay. When choosing which of two alternative routes to add a small amount of new flow, the social planner should add the flow to the one with the lower value of two times the "variable" delay plus the "fixed" delay. As a result, when the social planner is done, all the routes he assigns flow to will have the same score with respect to this cost metric, and unused routes will have a higher cost.

The conditions that describe the outcome of selfish agents—the Wardrop equilibrium—are similar, except that all the routes that are used score the same with respect to delay, which is just one times variable delay plus one times fixed delay. Hence the social planner should in a sense penalize the variable delay terms more than individual users. The intuition for that is that these variable delays are indicative of the externality—how much an additional unit of flow on a path will increase the delay of the flow already using the path. The social planner needs to take that externality into account, where an individual would not consider how his choice adds to the delay of the other traffic. The difference between how the social planner sees link cost versus an individual is just $a_e f_e$, one times the variable delay cost.

To apply the Pigovian tax idea, we need to make the individual objective match what the social planner sees. To do this, consider adding a "tax" to each link e of the amount $a_e f_e$. So far $a_e f_e$ is in units of delay, so to impose this tax we need to assume that we can find an amount of money to charge a user that is as costly to him as a time delay of $a_e f_e$. We will suppose we can choose our units of time and money to make the conversion factor between the two be equal to 1, and will not bother including that factor in our formulas. Of course that sidesteps some tricky issues, such as how such a conversion factor might be found, or how to handle the likely case that people vary in how much they value their time. These issues are important, but we will not try to address them here.

With the taxes charged on each link e set to be $a_e f_e$, users will evaluate each link by the sum of its delay $a_e f_e + b_e$ plus the tax $a_e f_e$. Observing this, one can surmise that the equilibrium reached by such users will just be the Wardrop equilibrium of a modified network for which the link delays are changed to be $2a_e f_e + b_e$ and no taxes are charged. Moreover, from our analysis of the structure of the optimal traffic assignment of the original network, we see it also corresponds to the Wardrop equilibrium solution of this same modified network with delays $2a_e f_e + b_e$. We can conclude that charging the tax of $a_e f_e$ will result in a selfish (Wardrop) equilibrium that is the same as the social optimum traffic assignment of the original network.

11.5 SELFISH PRICING

In the previous section we explored how a social planner would set link prices to steer the Wardrop equilibrium to correspond to a social optimum traffic assignment. Now we consider what would happen if each link had an owner that set the link price with the goal of maximizing the revenue from the traffic. In the model, link owners set the prices of their links, and then the traffic distributes itself across the links to achieve Wardrop equilibrium.

Since the price of one link will influence the flow that ends up on other links, this situation is a game between the link owners. With only a few link owners, the conventional notion of NE in finite player games with continuous strategy spaces is what applies. In this game, links that are parallel to each other (connecting the same pair of nodes) are competing with each other for the traffic "market" while links connected serially (back to back) complement each other since an increase in traffic on one of them results in an increase of traffic on the other. Since this class of model can capture aspects of competition, complementarity, and congestion, one can imagine many applications of this model. These include competing Internet access networks and competing "stacks" of companies for which the services of all the companies in a stack are needed to have a usable service.

Up to now, we have assumed that the amount of traffic that wants to go from a single source to a single destination is fixed. Retaining this assumption when link owners set prices presents problems. For instance, consider a network consisting of a single link. The provider here can charge an arbitrarily high price and earn an arbitrarily high amount of money. In economics terminology, this case corresponds to a monopolist with a market in which demand is completely inelastic. In practice, if the price gets high enough, eventually some of the users constituting the traffic will decide that it is not worth the price—which correspond to having demand elasticity. Moreover, in the framework we established when we investigated Pigovian taxes, users respond to a combination of delay and price, and we can convert one or the other to end up in common units. We give the name *disutility* for this sum of delay and price.

To model demand elasticity in this setting, the natural approach is to suppose that traffic decreases as disutility increases. Or conversely, as traffic increases disutility decreases—just like in a market for a good selling more of that good requires having a lower price for it. Thus there is a decreasing function $U(f)$ that gives the disutility d an amount of flow f "tolerates." Alternatively, we can work in terms of its inverse $U^{-1}(d)$, which gives the flow willing to go from source to destination at a level of disutility d.

In the earlier models, we used the notion of the price of anarchy to evaluate how much selfish users degrade the performance of the network compared to what can be achieved by a social planner. In our new setting, we have additional players, namely the link owners. The payoff of these players should just be their profit. However, up to now we have considered the *disutility* of the users constituting the traffic. Ultimately, we want to combine the payoffs of all the players to get a metric for social welfare so that we can compare the efficiency of the NE to the social optimum.

To evaluate social welfare we want to define a notion of payoff (in a positive sense) for the traffic. This notion ought to take into account the benefit the traffic gets from getting to go from the source to the destination minus the delay and monetary costs it incurs. The disutility function $U(\cdot)$ we defined gives us the means to do this, since it characterizes the willingness to pay (combined money and time cost) of all the traffic. If in equilibrium, the total flow is f and Wardrop equilibrium disutility is d, then we can use the function $U(\cdot)$ to determine the difference between the willingness to pay and what is actually paid by each unit of traffic—a notion called the consumer surplus. If the flow is f, we can consider each infinitesimal portion of traffic $[0,\epsilon]$, $[\epsilon,2\epsilon]$, ... $[f-\epsilon,f]$ where ϵ is an arbitrarily small number. The first of these portions has a willingness to pay of $U(\epsilon)$—in other words, if the disutility were as high as $U(\epsilon)$, the first ϵ mass of traffic would have been willing to go from source to destination. Since this traffic has to pay a disutility of d, the surplus from this traffic is the difference times the size of this portion of traffic, which comes to $(U(\epsilon)-d)\epsilon$. Similarly, the next of these portions has a willingness to pay of $U(2\epsilon)$ and gets surplus $(U(2\epsilon)-d)\epsilon$. Adding up these surpluses gives us a total consumer surplus of $\int_0^f (U(x)-d)dx$. We will use this expression for consumer surplus plus the profits of the link owners to quantify social welfare.

Recall that when we studied our earlier models without pricing, the price of anarchy was the ratio of the NE cost to social optimum cost. Now that we have switched to a utility model as opposed to a cost model, the customary way to take the price of anarchy ratio is the social optimum welfare divided by the NE welfare. However, at this point we have not established whether a NE of this game exists in all cases, nor whether if it exists is it unique. In cases that it exists but is not unique, the price of anarchy is usually defined using the NE with the lowest welfare.

An interesting question to ask is what is the worst price of anarchy across all possible networks and disutility (demand) functions, or if not all, a broad class of networks and disutility functions. It turns out that the price of anarchy can be infinite, if we use the customary definition of using the NE of lowest welfare when there are multiple NE. To see this, consider an network consisting of just two links connected in series so that the only possible route from the single source to destination is to cross both links. Also suppose the disutility function (demand) is chosen so that if prices are at or above some sufficiently large price p_h, the traffic falls to zero. Consider what happens when both providers charge this high price p_h. In this case, traffic is zero, profits are zero, and consumer welfare is zero since there are no users getting any benefit from the network. Hence, social welfare is zero. Moreover, neither provider can unilaterally deviate to get a positive profit since even dropping the price of one link all the way to zero will not overcome the fact that the other link needed in the route is priced so high that no customers will accept it. Hence both providers choosing p_h is a NE. As long as the social optimum prices allow some positive amount of traffic, which would always be the case unless the links had no capacity at all, the price of anarchy is a positive number divided by zero—meaning it is infinite.

One could argue that such equilibria are not realistic outcomes since it seems that if both providers were carrying no traffic, both providers would lower prices to attract some flow. Doing so would carry no risk to a provider since the payoff resulting from a price reduction could be no worse than the zero payoff the provider is getting before the deviation. With that in mind, we can "rule out" such equilibria from consideration and consider the price of anarchy amongst the remaining equilibria. To do so, we define a restricted class of equilibria that we call *zero-flow zero-price equilibria*. In a zero-flow zero-price equilibrium, if the flow through a link is zero, its price must be zero; the intuition being that a provider carrying no flow would lower his price to try to attract some flow.

In order to find the worst price of anarchy for a broad class of networks and disutility functions, we need to find a framework that describes NE and social optimum outcomes for a broad class of cases. To do that, we are going to make an analogy between the network and electric circuit, so that we can make use of some of the tools and intuition that come from circuit analysis.

11.6 CIRCUIT ANALOGY

In electrical circuits, electric current flows from points of high voltage to points of lower voltage by, loosely speaking, looking for the "easiest" path or paths. When between two points there are multiple paths of varying resistance, current distributes itself across the paths so that the voltage drops across all the paths are the same—much in the same way that in Wardrop equilibrium traffic distributes itself, across links so as to equalize the disutility between paths being used. To make an analogy between the two domains, we can think of a linear latency function as being analogous to a resistor in an electrical circuit with its resistance equal to the latency function's slope. Moreover, if the latency function is affine, with some positive latency when there is zero traffic—we can make an analogy between that and a voltage source. One might think of a voltage source as a source of energy, but in this analogy it is actually sinking energy. This is a bit like charging a battery—when current is forced through the battery in the opposite direction than the battery would release current when used as an energy source, the battery absorbs rather than releases power. In a electric circuit, if one attempts to charge a battery with a voltage lower than the battery voltage, current will flow in the opposite direction and the battery will release power. In our analogy this is like having negative traffic from source to destination, which does not make sense. To keep our analogy intact, we will add a circuit element to each path called a diode, which only allows current to flow in the direction of the destination.

The price a link owner charges is also analogous to a voltage source. Moreover the power, equal to voltage drop times current, is analogous to price times flow and hence revenue. Thus in our analogy, we can think of each link owner as choosing the voltage level of voltage source or battery in a circuit with the goal of absorbing the most power possible. Choose a voltage that is very low, and the battery will absorb a small amount of power since power is voltage times current. Choose a voltage that is very high, and some or all of the current will redistribute itself to other paths.

To complete our analogy, we need to cast the traffic demand model in terms of a circuit. If we modeled it as a "perfect" voltage source—a source that maintains constant voltage no matter how much current is drawn—it would be like an unlimited source of traffic with a willingness to tolerate disutility exactly equal to the voltage of this perfect source. In principle, this source could pump infinite power (revenue) into the system. Of course, this would not be a reasonable model. If we modeled it as a perfect current source—a source that maintains constant current no matter what—it would be like having a fixed amount of traffic that goes from source to destination no matter how high the price and disutility (i.e. completely inelastic demand). This again is not realistic, since this source too could pump infinite power into the system. Instead, we model demand as imperfect current source, a bit like a real battery. When the current drawn from the source is zero, the voltage it provides is at its peak level. As the current drawn increases, the voltage decreases. Analogously, to draw more traffic (current) the disutility (voltage drop) of the journey must be reduced. Combining all of these elements gives us the complete circuit analogy illustrated in Figure 11.4.

Now that we have constructed this circuit analogy, we can consider what tools and ideas we can borrow from the circuit literature into our problem. The following ideas will turn out to be useful.

- **Resistance aggregation formulas:** In circuits, resistances in series can be combined by adding them whereas the resistance of a parallel combination is found by adding the reciprocals of the resistances and then taking the reciprocal of the result. By analogy, these relations hold for combining the variable (the af part) of latency functions connected in a network.
- **Superposition:** In linear circuits, the effect of multiple stimuli like voltage sources connected to various places in the network can be computed by

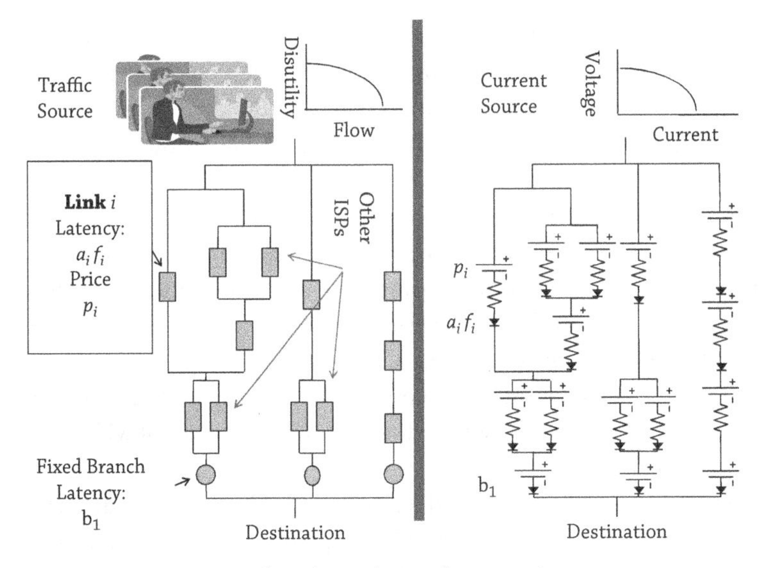

Figure 11.4 A circuit analogy for analyzing the network pricing game.

finding the effect of each stimulus individually and then adding up the effects.

- **Thévenin equivalence:** As a consequence of linearity, the behavior of a circuit in response to a voltage applied to a pair of nodes (called a "two-port" in the circuit literature) should be affine with respect to the voltage applied. Consequently, we can find the slope and offset of this affine response. We term these the Thévenin equivalent resistance and voltage, respectively, in honor of the discoverer of this property. The Thévenin equivalent resistance can be found by inspecting the network, pretending all voltage sources are closed circuits (wires), and combing the resistances using the resistance aggregation formulas.

We will use the Thévenin equivalent tool to describe how each link provider "sees" the rest of the network from the point of view of where it is connected. Crucially, the Thévenin equivalent resistance describes how much flow is lost by a link provider when he raises his price a small amount. For example, consider a network of N parallel providers with overall demand described by a disutility function with slope s—which can be modeled as a voltage source in series with a resistor of resistance s connected at the top of circuit and bottom of the circuit as in Figure 11.4. The Thévenin equivalent provider 1 sees, which we denote δ_1, can be computed by using the parallel resistance formulas as $\frac{1}{\frac{1}{a_2}+\cdots+\frac{1}{a_N}+\frac{1}{s}}$. Note that the slope of the disutility function appears as another resistance in parallel, since when we replace all the voltage sources by wires, the resistor of resistance s is in parallel topologically with the other resistors. From the principle of superposition, a small price increase of ϵ by provider 1 should result in a $\epsilon/(\delta_1 + a_1)$ reduction in flow. This is because we can think of the price increase of ϵ as voltage acting on the circuit in isolation, which would push ϵ volts against the Thévenin equivalent resistance δ_1 in series a_1 with the resistance modeling the latency function of provider 1. Moreover, this is in the negative direction since it passes through provider 1 bottom-up and not top-down. Adding this to the original flow, we get that the flow should be reduced by $\epsilon/(\delta_1 + a_1)$. This of course could be derived directly from the equations of the model themselves, but it is useful to have the tools of resistance aggregation, superposition, and Thévenin equivalence at our disposal to write relations like this, without having to re-derive them.

To eventually model NE, we need to understand under what conditions link providers find themselves not wanting to change price. Building on the previous discussion, if provider i increases price by ϵ then his flow should decrease by ϵ/δ_i where δ_i is the Thévenin equivalent resistance seen by provider i. If the disutility function is nonlinear, we can describe it locally linear – and thus equivalent to some voltage source connected to a resistor with slope s. However, as we get further from the point of linearization, the actual flow will be less than predicted by the linear model because we have assumed concavity. Thus, the actual flow will be a bit less than predicted by the linearization, and that "bit" less should be of order ϵ^2 or more since anything of order ϵ should be captured in our linearization. Provider i's profit following an ϵ price increase can be found by taking the product of price and flow to find

$$(p_i+\epsilon)\left(f_i - \frac{\epsilon}{\delta_i + a_i} - o(\epsilon^2)\right) = p_i f_i + \left[-\frac{p_i}{\delta_i + a_i} + f_i\right]\epsilon - o(\epsilon^2) \qquad \textbf{(11.6)}$$

where a_i is the resistance (latency function slope) of provider i's link and f_i is the flow before provider i perturbs his price. For provider i to be best responding, it must be that any perturbation does not increase profit. We see the above expression is less than the original profit $p_i f_i$ for all nonzero ϵ if and only if

$$\frac{p_i}{f_i} = \delta_i + a_i. \qquad \textbf{(11.7)}$$

Hence, if the provider i is best responding, he is maintaining a ratio of price (voltage) to flow (current) of $\delta_i + a_i$. In circuits, the circuit element that maintains a fixed ratio of voltage to current is simply a resistor. Hence, we can apply the following trick—replace the voltage source of each provider with a resistor of size $\delta_i + a_i$. Then, solve the circuit. Since in this solution everyone is best responding to each other, it should be an NE. This is basically correct, except there are some other details that need to be taken into account. In particular, the Thévenin equivalence idea applies to linear circuits. However, recall that we included diodes in our circuit to prevent flow going backwards through providers whose disutiility (voltage) is higher than its competitors. A diode "looks" linear when it is not blocking flow, and looks like an open circuit when it is. Moreover, there's a third possibility in which a diode is just at the "edge" of blocking flow. In this case, the branch the diode is in is carrying no traffic, but if its competitors raised prices slightly it would be carrying traffic. In contrast, if competitors lowered prices slightly, they could steal no more traffic from the branch with this "edge" diode since the branch is already zero. In such cases, the presence of would-be competitors that carry no traffic may push other providers to have lower prices than they would have if that competitor did not exist—sort of like the others feel the threat of entry of the would-be competitors. A careful analysis takes all of these possibilities into account. For brevity, we will skip the details of how that can be done.

Since we seek to study the price of anarchy, we should also try to model the social optimum pricing as a circuit as well. Our discussion of Pigovian taxes makes this relatively easy. Recall that we deduced that the tax should equal $a_i f_i$ where f_i is the flow and a_i is the slope of the latency function. That same result applies here. Moreover, if we focus on the ratio of price to flow we see that $p_i/f_i = a_i$. Hence, we can employ the same trick we just introduced for studying the NE of this game. We replace the voltage sources modeling price on each link with a resistor of size a_i. Note that we are putting smaller resistors into the circuit than we did to model NE. Consequently, the equilibrium current (flow) will be higher in social optimum than in NE—which is probably what one would have expected.

Even with these tricks, there is still more work to be done. We have to write equations that describe the solution of these circuits, then write expressions for social welfare in social optimum and NE, and then show the ratio is bounded by some number or expression. This can all be done for topologies we call simple parallel-serial topology consisting of an arbitrary number of parallel branches, each composed of a number of serial links. More general combinations of parallel and serial links can be analyzed as well, but the results, so far, are weaker for these topologies. The analysis works by describing the system with matrices and vectors, and using some linear algebra tricks.

It turns out that the bounds are found to be a function of a parameter y we term the *conductance ratio*. The conductance ratio is the conductance of the most conductive branch divided by the conductance of network as a whole. The conductance ratio is therefore a measure of how concentrated the capabilities of the network are in a single branch. A conductance ratio near 1 means that most of the conductance of the system is concentrated in a single branch. The smaller the conductance ratio, the more that the overall conductance of the system is distributed across multiple branches. Thus, in a sense the conductance ratio reflects the market power or concentration of the system. As one would expect, the price of anarchy bounds that we find increase as the conductance ratio approaches 1. The following theorem comes from and is proven in Musacchio (2009).

Theorem 1. *Consider the game with a simple parallel-serial topology. Consider the following ratio*

$$y = \frac{\max_i 1/a_i}{\sum_i 1/a_i}, \tag{11.8}$$

which is the conductance of the most conductive branch divided by the overall conductance. The price of anarchy for zero-flow zero-price Nash equilibria is no more than

$$\begin{cases} \frac{1}{4}\frac{m^2+2m(1+y)+(y-1)^2}{m} & y \leq 1-m/3 \\ \frac{m^2(2-y)+m(4-y^2-y)+2(y-1)^2}{8m-6my} & y \geq 1-m/3 \end{cases} \tag{11.9}$$

where m is the maximum number of providers connected serially. Furthermore, the maximum of the above bound occurs when $y=1$, and consequently the price of anarchy is no more than

$$1+m/2. \tag{11.10}$$

The bounds given in Theorem 1 are illustrated in Figure 11.5. Note how the price of anarchy falls as the conductance ratio falls (i.e., becomes less monopolistic). Also note the increase in the price of anarchy as the number of serially

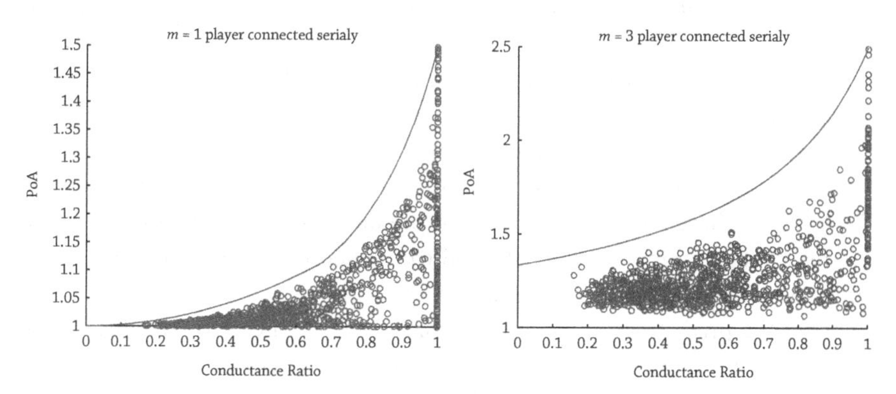

Figure 11.5 Simple parallel serial PoA bound for the cases where there are either $m=1$ or three providers connected serially. The bounds are plotted as a function of the conductance ratio—the ratio of the conductance of the most conductive branch to the conductance of the whole circuit. The points are the PoA of randomly generated example topologies.

connected providers increases. This is an example of the well known "double marginalization" effect in which serially connected providers tend to overcharge (as compared to social optimum) because they do not consider how their price increases will hurt the profits of the other serially connected providers.

The results of Theorem 1 are for simple-parallel serial topologies. For general parallel serial topologies (with arbitrary groupings of links connected in parallel and series), it turns out the same bounds hold, with an additional factor of 2. The argument is given in Musacchio (2009). We do believe that the bound with a factor of 2 is not tight, but it remains an open problem to get a tighter bound.

11.7 DISCUSSION

Using the framework of population games, we have demonstrated that selfish behavior can lead to systemwide inefficiency and also quantified that inefficiency in some cases. In addition, we have taken population game models of routing choice and combined them with classic models of games with finitely many players to analyze pricing games between link owners connected in parallel branches of series combinations. While we posed these games as routing games, they also model the situation in which several substitutable "stacks" of complementary services compete for customers, and each service provider is free to set prices.

This type of modeling exposes many interesting phenomenon, but it is also important to recognize some of its limitations. For many reasons, it is quite difficult to empirically verify how closely drivers in road networks follow the behavior predicted by congestion game models. For instance, collecting data on route choices of drivers with a particular source-destination pair in mind is a challenge, though that has in the past few years gotten somewhat easier with the proliferation of GPS-enabled mobile phones. Leveraging this new data source is currently a major area of investigation by road traffic researchers. Also, controlled experiments, to demonstrate Braess's paradox for instance, are usually not practical for such large systems. However, there are anecdotal examples of roadways being suddenly shut down and having traffic congestion not worsen. Thus we can have some confidence that the phenomena like Braess's paradox are real. Still, we cannot expect stylized game theoretic models to make exact predictions of road traffic behavior in the real world.

As for the predictions of the pricing games, we also need to take them with caution. The "double marginalization" effect that happens in our models with providers connected in series is a well-known and studied phenomenon. Often in such situations, the parties (link owners) involved could be expected to solve the problem by negotiating a contract. They might do so because they realize they can make more money by cooperating to create a bigger "pie" or combined profit. After agreeing to a split of the enlarged pie that makes both parties better off than if they were non-cooperative, they might then write into the contract punishments for breaking the contract. This is more or less the idea of the Coase theorem, due to the work of Nobel Laureate Ronald Coase in the 1950s and 1960s. This "theorem" (though not originally posited in theorem form by Coase) says that bargaining between all parties affected by an externality should result in an efficient outcome, provided transaction costs are low enough. As a practical matter, however, bargaining often breaks down, for example, when the parties do not have complete information about one another. Thus it is useful to have models like our pricing game that describe the outcome of completely uncooperative situations. It sheds light on what *could* happen, but does not perfectly say what *will* happen in the real world. It provides a benchmark from which the parties involved might bargain to achieve a more efficient outcome.

EXERCISES

1. Consider the Matching Pennies game. It has two players, who simultaneously choose between H ("Heads") and T ("Tails"). If both players play the same strategy then player 1 wins, while if they play different strategies, player 2 wins. Here "win" means have a cost of -1 (i.e., a unit gain) and lose is a cost of 1. Prove that Matching Pennies is not a potential game. Hint: Suppose there were a potential function V. Use equation (11.1) to write an inequality for each player facing each possible opponent strategy. String together the inequalities to obtain

$$V(T,T) > V(H,T) > V(H,H) > V(T,H) > V(T,T)$$

which is a contradiction that proves the theorem.

2. Consider two links connected in series. Each link has a latency that grows with slope 1 (time units per traffic unit). The series combination is fed by a source of demand that generates a maximum of 1 unit of traffic if there were no disutility (a path of zero price and zero delay), and the traffic drops linearly with slope 1 (traffic unit per unit of disutility).

   (a) What will be the flow of traffic if both link owners charge zero price? What is the social welfare (provider profits plus consumer surplus) under this scenario? Hint: Model the demand source as a voltage source of 1 volt connected in series with a 1 Ohm resistor. Note when current is drawn from the demand source, the voltage supplied from the demand source drops linearly, like the demand source we want to model. Similarly, model the links as 1 Ohm resistors. Solve for the current in the circuit. Verify that the consumer surplus is half the power dissipated in the resistor in the demand source.

(b) What flow will occur in social optimum? What is the social welfare under this scenario? Hint: modify your circuit from part (a) by adding two more resistors in your series combination. Each of these resistors models the effect of the price charged by each of the links. Write the words "price 1" and "price 2" next to the respective resistors to remind yourself of this. Recall that the optimum link price should be such that the price to flow ratio should be the same as the slope of the latency function of the link. Thus, each of these two new resistors is also 1 Ohm. Solve for the current in the circuit. Calculate the consumer surplus as before. Verify that the provider profits correspond to the power dissipated in each of the two new resistors you just added to your circuit.

(c) What flow will occur in NE? (Look only for what we called zero-flow, zero-price equilibria.) What is the social welfare under this scenario? Hint: modify your circuit for part (b) by changing the resistance of the price resistors you added in part (b) to have a resistance of 1 Ohm plus the Thévenin equivalent of the other parts of the circuit (excluding the resistors modeling prices). For each link, the "other" parts of the circuit are the resistor modeling the other link's latency and the resistor modeling demand, both in series. Thus each price resistor has a resistance of $2+1=3$. Solve for the current, from which you can then calculate social welfare.

3. Consider two links connected in parallel. This parallel combination is fed by a source of demand that generates a maximum of 1 unit of traffic if there were no disutility (a path of zero price and zero delay), and the traffic drops linearly with slope 1 (traffic unit per unit of disutility).

(a) What flow will occur in social optimum? What is the social welfare under this scenario? Hint: Draw a circuit similar to the one you drew in exercise 2. However, make 2 parallel branches after the model of the demand source. Each branch should have a 1 Ohm resistor representing latency, and another 1 Ohm resistor representing the optimal price. Solve for the overall current in the circuit, by observing that the overall resistance of the two parallel paths of 2 Ohms is

$$\frac{1}{2^{-1}+2^{-1}}=1.$$

From this you should be able to find the current in each branch, the provider profits, and the consumer surplus.

(b) What flow will occur in NE? (Look only for what we called zero-flow, zero-price equilibria.) What is the social welfare under this scenario? Hint: modify your circuit for part (a) by changing the resistance of the price resistors to have a resistance of 1 Ohm plus the Thévenin equivalent of the other parts of the circuit (excluding the resistors modeling prices). The other pars of the circuit consist of the other link's latency in parallel with the resistor modeling demand, giving a Thévenin equivalent of $\frac{1}{2}$. Consequently the price resistors should each be 1.5 Ohm.

NOTES

Most tangents and entries to the published literature are already contained in the text in this chapter. The Discussion touches on vertical relations between firms, for which there is a wide literature in the Economics field called Industrial Organization; see for example the classic textbook by Tirole (1988). There is also a string of papers beginning with Selten et al. (2007) testing traffic models in laboratory experiments.

Acemoglu and Ozdaglar (2007) is the origin of the selfish traffic model discussed in section 11.5. They also develop the first price of anarchy bounds that are generalized in this chapter. The "Price of Anarchy" expression was introduced in the Koutsoupias and Papadimitriou (1999).

BIBLIOGRAPHY

Acemoglu, D., and A. Ozdaglar. "Competition in parallel-serial networks." *IEEE Journal on Selected Areas in Communications* 25, no. 6 (2007): 1180–1192.

Body, Stephen, and Lieven Vandenberghe. *Convex Optimization*. Cambridge University Press, 2004.

Koutsoupias, E., and C. H. Papadimitriou. "Worst-case equilibria." In *Proceedings of the 16th Annual Symposium on Theoretical Aspects of Computer Science* (1999): 404–413. Trier, Germany.

Musacchio, J. "The price of anarchy in parallel–serial competition with elastic demand," Tech Report: UCSC-SOE-09-20, April 2009.

Pigou, A. C. *The economics of welfare*. Macmillan, 1920.

Roughgarden, T. "How bad is selfish routing?." *Journal of ACM* 49, no. 2 (2002): 236–259.

Selten, Reinhard, Thorsten Chmura, Thomas Pitz, Sebastian Kube, and Michael Schreckenberg. "Commuters route choice behaviour." *Games and Economic Behavior* 58, no. 2 (2007): 394–406.

Tirole, Jean. *The theory of industrial organization*. MIT Press, 1988.

Wardrop, J. "Some theoretical aspects of road traffic research". *Proceedings of the Institute of Civil Engineers, Part II* 1, no. 36 (1952): 352–362.

12

International Trade and the Environment

MATTHEW MCGINTY

How do you build evolutionary game models that speak to economic policy questions? The main purpose of this chapter is to provide some solid examples. It draws mostly on published work that deals with international trade and environmental economic policy.

Before building models, the chapter offers some general perspectives on economic applications of evolutionary game theory. The intellectual roots go back to Thomas Malthus, an 18th-cenutry moral philosopher, and to mid-20th century economist Armen Alchian. The first section touches on the connections between rationality and adaptive behavior, on economic profits as payoffs (or fitness), and on dynamic interpretations of economists' distinction between (static) long-run and short-run equilibrium.

The next section constructs firms' payoff functions from the underlying market demand and cost conditions. It continues with a discussion of externalities, defined as impacts on payoffs of third parties not directly involved in a market transaction. We saw one sort of example, congestion costs, in Chapter 11. In the present chapter we will encounter other sorts of negative externalities such as air or water pollution associated with manufacturing or power generation.

The presence of externalities generally implies that the free-market outcome is not socially optimal, so there is potentially a role for government policies to increase average payoff. The chapter includes a brief discussion of government policies including pollution taxes, subsides for green technologies, and international agreements, emphasizing their impact on the relative payoffs to different strategies and the types of evolutionary equilibria that may emerge.

After presenting models of international trade and government policy interventions, the chapter also briefly mentions some other policy-oriented economic applications of evolutionary games.

12.1 ECONOMICS AND EVOLUTIONARY GAME THEORY

Malthus' (1798) "Essay on the Principle of Population" was famous in its day and, more importantly to us, helped shape the emerging disciplines of economics and biology. The essay's main prediction, that sustainable progress is not possible in the average economic standard of living, has (so far) turned out to be wrong, but nevertheless the sorts of questions Malthus asked, and much of his approach, have proved quite durable in economics. Indeed, referring to economics as "the dismal science" goes back to Mathus' essay. Perhaps even more consequentially for biological science, Darwin (1859, Chapter 4) developed his central concept, natural selection, as an adaptation and extension of Malthus' argument.

Alchian (1950) argued that economists' standard first principles of perfect foresight and perfect rationality are not literal descriptions of behavior, but rather are good approximations of evolutionary equilibrium. Clearly individual humans cannot perfectly anticipate the future, nor are most individual consumers or firms perfectly rational in choosing actions that maximize their payoffs. Alchian proposed instead that individual humans (including managers of firms) are adaptive. They respond to payoff differences, and consequently actions that yield a payoff advantage increase in prevalence. The result, Alchian argued informally, typically is a good approximation of perfect rationality, perfect foresight, and common knowledge about rationality. Friedman (1991) mentions attempts to make the argument more rigorous and discusses how evolutionary games apply to economics.

Sugden (2001) continued this line of thought by suggesting empirical exercises aimed at determining the degree of imperfect rationality. The empirical learning models introduced in Chapter 8 are one way to develop an adaptive approach to rationality.

The present chapter mainly models the behavior of firms. As in any application of evolutionary games, we need an observable proxy of fitness, or payoff. Of course, the canonical assumption in economics is that firms' sole purpose is to maximize their profit, and this seems promising for evolutionary models as well. Alchian emphasized that strategies which generate greater profit have a higher survival rate in competitive industries. Schaffer (1989) models a fixed number of infinitely lived firms that periodically fire and hire managers. Each manager plays a fixed strategy and a firm changes strategy by firing the manager and hiring a new manager. A manager's survival depends on the relative profitability of her strategy compared to competing firms. Thus in Schaffer as well as in Alchian, relatively profitable strategies increase in prevalence. We will return to the question in the notes, but until then we will assume without further discussion that firms' payoff (or fitness) is economic profit.

Standard static economic equilibrium concepts are commonly referred to as either short run (SR) or long run (LR). The distinction is the scope of the choice set. In the SR, firms choose a quantity of output, taking as given the distribution of "types." Here a firm's type includes its internal organization and mode of production, the characteristics of the product it produces, and the production function and some of its inputs (e.g., physical capital). In the LR, firms can choose some (or perhaps all) aspects of its type.

To preserve this useful distinction in an evolutionary model, we will consider two distinct time scales. In the SR, firms can all adjust output in response to profit differentials, holding fixed their own type and the distribution of other firms' types. In the LR they adjust type in response to the profit differentials defined by SR equilibrium. Examples presented in the next few sections emphasize LR adjustment of modes of production. Technology diffuses with some inertia as existing factories outlive their useful lives and "expire," and firms and households slowly respond to policy and local conditions.

Economists define an externality as an impact on an individual who is not part of a market transaction. Externalities may be positive or negative and may be generated by the act of consumption or the act of production. Sulfur dioxide pollution from electricity production is an example of a negative production externality. Economists agree that the free-market outcome is generally not optimal for society when externalities are present.

Consequently, there can be a useful role for government policies such as a tax (for negative) or a subsidy (for positive) externalities. The Pigovian taxes discussed in Chapter 11 are a good example. More generally, government policy may alter the payoffs of all strategies, and thus alter the evolutionary equilibrium. In many economic applications of evolutionary games, then, the goal is to identify policies that may improve on free-market outcomes.

12.2 STATIC COURNOT MODEL

One of the most commonly used mathematical models in economics is also one of the oldest. In 1838 the French economist Antoine Cournot proposed a simple model where firms independently choose what quantity of output to produce to maximize their individual profit. The choices imply strategic interdependence and give rise to a best-response function which shows the profit-maximizing level of output as a function of the other firms' output choices. In fact, Cournot's solution concept of a simultaneous best-response predates the NE by 112 years. The most parsimonious example used in the literature has linear market demand and a constant cost of production for each unit. While Cournot considered only two firms, the model has been extended to allow for any number I of individual firms which can be considered a population of identical entities whose single goal is to maximize profit.

Each of the I firms produces an identical good and output for firm i is denoted q_i. Demand is assumed to be a linear function of the total amount of output Q, where $Q = \sum_{i=1}^{I} q_i$ and P is the market price

$$P = a - bQ. \tag{12.1}$$

The parameter $a > 0$ is called the choke-price and is the maximum price that the market will bear before demand is zero. If price were zero demand would be $\frac{a}{b}$, thus the potential size of the market is increasing in a. The demand slope is $-b$, which indicates how rapidly market price declines when there is an additional unit of the good. The total revenue (TR) earned by firm i is Pq_i and depends on its own output and every other firm's output via the market price. A unit increase in output by one firm reduces the market price by b and lowers the revenue for all other $I - 1$ firms. This is the source of the strategic inter-dependence between firms and is also an example of a negative externality.

There is an analogy between market demand in Economics and an open-access resource. In his famous essay "The Tragedy of the Commons," Garrett Hardin (1968) recalls that Adam Smith in 1776 put forth the idea that individuals seeking to maximize their own personal payoff will produce an outcome that is best for society as a whole. Building on the work of later economists, Hardin recognized that this proposition fails when externalities are present. When each herder decides how many cows to let graze in the commons (i.e., open-access pasture), he gains the full benefit from each additional cow, but the costs of over-grazing are spread over all herders, and so each herder bears only a fraction of the cost he imposes. Hence each individual herder has an incentive to over-graze, and the resulting outcome is not socially optimal. We saw the same logic at work in the "price of anarchy" analysis in Chapter 11.

The same logic also applies to firms in a market. When a firm sells one more unit of output it receives all the revenue, but this additional unit of output lowers the market price by b, which is a cost that is borne by all I firms. The result is "over-production" compared to the level of output that would maximize the combined profit of the I firms.

The parameter $c > 0$ represents the cost to produce a unit of the good and is initially assumed to be constant and identical across firms. The total cost (TC) for the firm thus is a linear function of output,

$$TC_i = cq_i. \tag{12.2}$$

Profit for firm i (π_i) is the difference between total revenue earned by the firm and total cost,

$$\pi_i = TR_i - TC_i = (a - bQ)\,q_i - cq_i. \tag{12.3}$$

Each firm individually chooses quantity to maximize its own profit, taking other firms' output as given. The first-order condition is

$$\frac{\partial \pi_i}{\partial q_i} = a - 2bq_i - bQ_{-i} - c = 0, \tag{12.4}$$

where the aggregate output of the $I-1$ other firms is denoted $Q_{-i} = \sum_{j \neq i} q_j$. Solving (12.4) for q_i yields the best-response function, q_i^{br}, which is the output level that maximizes profit as a function of the combined output of the other firms:

$$q_i^{br} = \frac{a - c - bQ_{-i}}{2b}. \tag{12.5}$$

Repeat this procedure for the other $I-1$ firms and recognize that symmetry implies that in equilibrium $Q_{-i} = (I-1)\,q_i$. The simultaneous solution to the I first-order conditions results in the Cournot-Nash equilibrium individual and market quantities:

$$q^{ne} = \frac{a-c}{b(I+1)}$$
$$Q^{ne} = \left(\frac{I}{I+1}\right)\frac{a-c}{b}. \tag{12.6}$$

Returning the NE quantities to equation (12.3) results in equilibrium profit

$$\pi^{ne} = \frac{1}{b}\left(\frac{a-c}{I+1}\right)^2 = b\left(q^{ne}\right)^2. \tag{12.7}$$

Next, consider the cooperative, or monopoly, outcome that would be obtained if there was a single firm, or if all I firms internalized the externality and coordinated output to maximize their combined profit

$$\pi = (a - bQ)\,Q - cQ. \tag{12.8}$$

The cooperative outcome is the choice Q^o that maximizes industry output (12.8), so

$$Q^o = \frac{a-c}{2b}. \tag{12.9}$$

A comparison of Q^{ne} and Q^o shows that the failure to internalize the external costs results in over-production (from the firms' perspective) whenever there is more than one firm. In general, government policy in the form of taxes, subsidies, or other policies can overcome this problem and align individual incentives with optimal outcomes. This is the simple static Cournot model. There have been a wide variety of extensions which include asymmetric costs, nonlinear demand, and cost functions and products that are not identical.

Two interpretive remarks may be in order. First, we are not advocating policies that maximize firms' joint profits; the impact on consumers should also be taken into account. The output level Q^o is a useful benchmark, not necessarily an ideal to be pursued. Second, unlike many biological models, payoffs in the Cournot model are not naturally interpreted as arising from random pairwise encounters. It is better thought of as accruing to firms "playing the field." In particular, a firm's profit depends on the output of all I firms via the market price. This is true for both the single and multiple population models in the applications that follow.

12.3 GREEN TECHNOLOGY DIFFUSION

Using the Cournot model as a building block, we now construct evolutionary models when there are two alternative strategies or 'types' available to a single population I of firms. The current state of the system $(s, 1-s)$ is the distribution of firms across the two types, and the payoffs are the short-run NE payoffs under Cournot competition, as in equation (12.7). Thus we focus on the long-run analysis, implicitly assuming that, for any fixed s, the short-run dynamics converge globally to the unique NE.

Following McGinty and de Vries (2008), the two firm types are called clean (c) and dirty (d) and they may differ in terms of production costs. For example, electricity can be produced either by conventional coal-fired power plants at unit cost $c_d > 0$, or by environmentally friendly technologies such as wind, solar, or biomass at unit cost $c_c \geq c_d$. Likewise, organic farmers generally have higher unit production costs than conventional farms that use pesticides. On the other hand, the government may choose to subsidize the clean type of production an amount ϕ per unit of the good, so that the effective unit production cost for the clean good is $c_c - \phi$. Hence we have the total cost functions

$$TC_c = (c_c - \phi)q_c \qquad (12.10)$$
$$TC_d = c_d q_d.$$

The two types may also differ somewhat in terms of demand. Even if the product is electricity that is indistinguishable to the end user, there are consumers that are willing to pay a higher rate (i.e., a price premium) for environmentally friendly produced electricity. For example, Central Vermont Public Service is a utility that in 2009 charged 16.5 cents per kilowatt hour for electricity produced from a renewable source (methane) compared to 12.5 cents for electricity produced from coal-fired power plants. Likewise, organic produce often sells at a price premium relative to conventionally grown produce.

Consumers regard the two types of goods as different but similar enough to be substitutable. To deal with this, we need to modify the Cournot model slightly. Demand is still linear but (as in Dixit 1979) the market price for each type of product now also depends on quantities produced of both types:

$$P_c = a_c - bQ_c - \delta Q_d \qquad (12.11)$$
$$P_d = a_d - bQ_d - \delta Q_c,$$

where P_i, $i = c,d$ is the price and $Q_c = \sum_{c=1}^{sI} q_c$ and $Q_d = \sum_{d=1}^{(1-s)I} q_d$ is the quantity of the clean and dirty goods. Given current shares there are sI clean firms that each produce the quantity q_c and $(1-s)I$ dirty firms that each produce the quantity q_d.

As in the basic Cournot model, the slope parameter $b > 0$ indicates how rapidly price declines in the amount of the own type. Without loss of generality we will set $b = 1$; this normalization arises from redefining output units appropriately, for example, in $1/b = 10$ ounce packets instead of bushels. The new parameter $\delta > 0$ represents the degree of substitutability of the two goods. If $\delta = 0$ then the goods are completely unrelated and there are two distinct markets. In this case output of one type of good does not impact the price of the other type. If $\delta = b = 1$ then consumers regard the goods as perfect substitutes. Thus $0 \leq \delta \leq 1$. To the extent that (at least some) consumers prefer the clean good we have $a_c > a_d$.

Profit for the two types of firms is the difference between total revenue and total cost $\pi_i = TR_i - TC_i$, so

$$\pi_c = P_c q_c - (c_c - \phi)q_c = [a_c - Q_c - \delta Q_d]\, q_c - (c_c - \phi)q_c \qquad (12.12)$$
$$\pi_d = P_d q_d - c_d q_d = [a_d - Q_d - \delta Q_c]\, q_d - c_d q_d.$$

Define the cost margin (potential profitability) for the two goods as $\theta_c \equiv a_c - c_c$ and $\theta_d \equiv a_d - c_d$. Each individual firm chooses quantity to maximize profit, taking all other $I - 1$ firms' output as given.

Following the same procedure as for the basic Cournot model reveals that the short-run Cournot-Nash equilibrium quantities are

$$q_c = \frac{(\theta_c + \phi)[(1-s)I + 1] - \delta\theta_d(1-s)I}{s(1-s)I^2(1-\delta^2) + I + 1} \qquad (12.13)$$

$$q_d = \frac{\theta_d(1 + sI) - \delta(\theta_c + \phi)sI}{s(1-s)I^2(1-\delta^2) + I + 1}. \qquad (12.14)$$

Note that both numerators and denominators are quadratic functions of the current state s. More substantively, note that the subsidy ϕ increases the output of a clean firm and decreases the output of a dirty firm. For given state s, the market quantity of each type is $Q_c = sIq_c$ and $Q_d = (1-s)Iq_d$. Returning these quantities to the payoff function in (12.12) results in simple expressions for the profit of each type of firm:

$$\pi_c = (q_c)^2 \tag{12.15}$$

$$\pi_d = (q_d)^2. \tag{12.16}$$

Putting together equations (12.13)–(12.16), we see that each type of firm's profit is a quotient of fourth-order polynomials in the state variable s, hence the payoff functions $w_i = \pi_i$ are nonlinear. But they turn out to be quite straightforward to analyze, as we can begin to see in Figure 12.1. The payoff difference for the given parameter values has a unique downcrossing near $s = 0.4$ for the given parameters. We know from Chapter 2 that it therefore is a generalized Hawk-Dove game, with a unique interior EE that is globally stable.

The intuition is also straightforward. When s is close to zero there are very few clean firms, and their scarcity gives them a profit advantage. Conversely when s is close to 1, most firms are clean resulting in a profit advantage for dirty types. This rare-type advantage implies a strictly decreasing Delta function and thus a HD-type game.

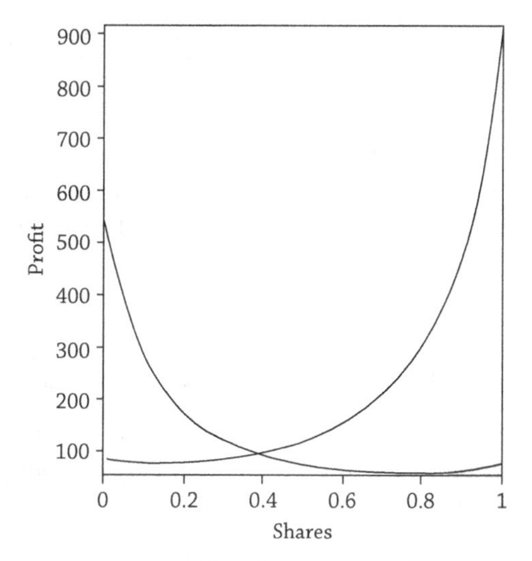

Figure 12.1 Payoffs w_i for types $i = c, d$, given parameters $I = 10, \theta_c = 96, \theta_d = 100, \delta = 0.98$, and $\phi = 0$.

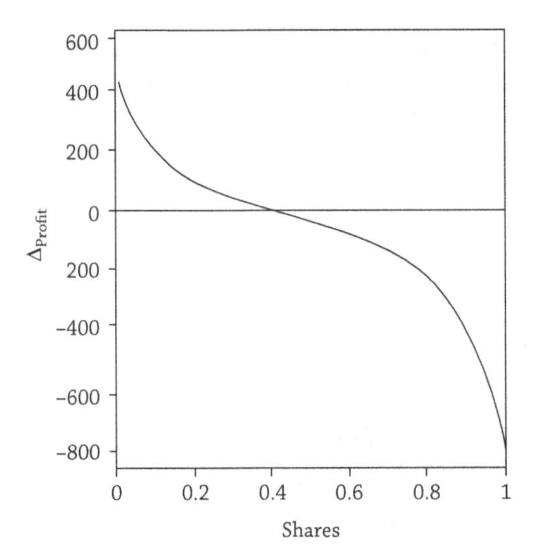

Figure 12.2 Payoff difference Δw_{cd} for same parameters.

What happens if we change the parameters? For a given state s, an increase in the subsidy to $\phi > 0$ shifts up the profit for clean firms and shifts down the profit for the dirty type. Since the (interior) EE equates profit across types, the location of EE must therefore shift to the right (i.e., the equilibrium share s^* increases).

All comparative statics results can be obtained analytically as follows. Plugging equations (12.13) and (12.12) into $\Delta w_{c-d} = pi_c - pi_d$, we obtain

$$\Delta w_{c-d} = \frac{[(\theta_c+\phi)[(1-s)I+1]-\delta\theta_d(1-s)I]^2-[\theta_d(1+sI)-\delta(\theta_c+\phi)sI]^2}{[s(1-s)I^2(1-\delta^2)+I+1]^2}. \tag{12.17}$$

McGinty and de Vries (2008) show that (12.17) defines a Hawk-Dove game with an interior equilibrium $s^* \in (0,1)$ when the degree of product differentiation is moderate, and the differences in the cost margin (θ_c relative to θ_d) is not too large. Specifically, they find an interior EE when

$$\delta \in \left[\frac{(\theta_c+\phi)(1+I)-\theta_d}{I\theta_d}, \frac{\theta_d(1+I)-(\theta_c+\phi)}{I(\theta_c+\phi)}\right]. \tag{12.18}$$

However, if the goods are highly substitutable and there is a large difference in the potential profitability then there is a dominant strategy and one good completely drives out the other in the EE. If

$$\theta_c+\phi > \theta_d \text{ and } \delta \geq \frac{\theta_d(1+I)-(\theta_c+\phi)}{I(\theta_c+\phi)} \text{ then } s^* = 1 \tag{12.19}$$

and all firms are clean. If

$$\theta_c + \phi < \theta_d \text{ and } \delta \geq \frac{(\theta_c + \phi)(1+I) - \theta_d}{I\theta_d} \text{ then } s^* = 0 \qquad \textbf{(12.20)}$$

and the dirty method of production completely drives out the clean mode. Note that there is a role for government policy via the subsidy for the clean good ϕ. The government can eliminate the dirty method of production entirely with the appropriately defined subsidy. This clearly depends on the price premium consumers are willing to pay for the clean good compared to the additional cost of the clean mode, θ_c relative to θ_d.

If the clean technology is new (an "innovation" in Econ-speak or a "mutant" in Bio-speak) then the initial condition is $s \approx 0$. Depending on the degree of substitutability, the government can offer a temporary small subsidy to speed convergence to the interior EE, or perhaps offer such a large permanent subsidy ϕ that adopting the clean technology becomes a dominant strategy. Government can also decide to do nothing ($\phi = 0$) if the degree of substitutability is small and the clean technology is very costly relative to the price premium consumers are willing to pay.

McGinty and de Vries (2008, Proposition 3) show that the subsidy has a greater effect when the goods are highly substitutable. The subsidy for the clean good reduces pollution via the substitution effect, but interior EE profit and output are equal across types, so the clean subsidy also increases dirty output per firm via the output effect and thus the per-firm profit of dirty firms. What is the net effect? It turns out that when the products are very similar, the substitution effect is more likely to dominate, in which case overall pollution will be reduced.

12.4 INTERNATIONAL TRADE

Our next economic application involves two populations: firms in two different countries that engage in international trade, as in Friedman and Fung (1996). In their model, the firms in each country again choose between two alternative production methods. Neither is especially clean or dirty; instead they differ in the "mode" or form of internal organization and so have different implications for both cost and demand.

An example the authors had in mind is automobile manufacture. In the 1980s, U.S. firms still used a traditional hierarchical form of organization, which we will refer to as mode A. It features fixed worker tasks, large buffer inventories of inputs purchased from separate suppliers, merit-based competitive promotion, and top-down decision making with many layers of middle management. The Japanese (inspired largely by American academic W. Edwards Deming) by this time had perfected a much more flexible mode, call it B, that featured flatter hierarchies, job rotation within work groups, close relationships within the supply-chain network, just-in-time delivery of the inputs needed for production, and seniority-based promotion.

Cost-side externalities produce an asymmetry between the two competing modes. Worker pay in the B mode depends primarily on seniority, while in the A mode it depends primarily on productivity. This means that the best early and mid-career workers in the B mode can be "skimmed" and hired away by A mode firms that offer them higher pay. Another cost-side externality, dubbed the "network effect,"arises from the close (often exclusive) supplier relationships that enable B mode firms to minimize costs. Type A firms, when sufficiently prevalent, will bid up input prices and disrupt exclusive (or even close) supply chain relationships. Mode B firms' unit production costs therefore are increasing in the number of type A firms, due to both network and skimming effects.

These cost side effects are captured by the parameter β. At any point in time there are sI firms using the type B mode of production and $(1-s)I$ using the type A mode. The total cost of production is

$$TC_A = c_A q_A \tag{12.21}$$
$$TC_B = (c_B - \beta s) q_b.$$

The cost of B mode firms is decreasing in their prevalence at rate β, where $c_B - \beta s > 0$ for all values of $s \in [0,1]$ implies the parameter restriction $\beta \in [0, c_B)$.

On the demand side, we again allow for the possibility that consumers perceive the goods as imperfect substitutes. Thus the prices are

$$P_A = a_A - bQ_A - \delta Q_B \tag{12.22}$$
$$P_B = a_B - bQ_B - \delta Q_A.$$

Again we normalize the own-price coefficient to $b=1$, and impose $0 \leq \delta \leq 1$. Note that increasing the share of either mode, and therefore the production quantity, tends to depress its price and thus its profitability. This so-called "glut effect" again advantages the rare strategy.

For the moment, assume that no international trade is possible. Then in each country the short-run NE quantities and profit are obtained as earlier, and have a roughly similar form as the quotient of fourth-degree polynomials in s.

Figure 12.3 points up a crucial difference. For the chosen parameter values, the profit functions intersect at $s^* = 0.5$, but the A mode has a profit *advantage* when common (at $s=0$) and so does the B mode (at $s=1$). The intuition is that the skimming and network effects overpower the glut effect, and so we have a nonlinear coordination game (CO in the notation of Chapter 2). However,

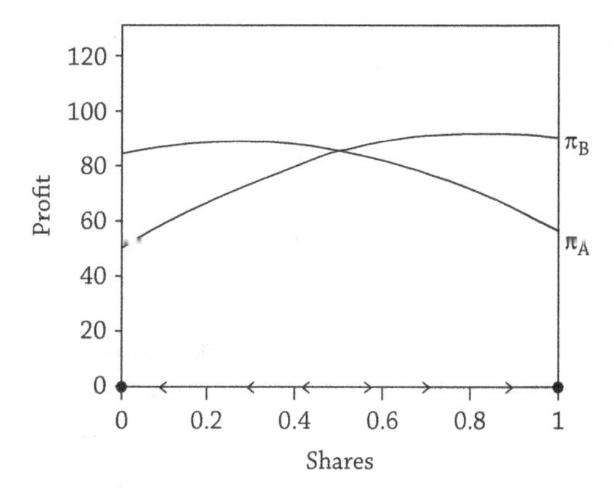

Figure 12.3 Mode A and mode B payoffs. Parameter values are $a_A = a_B = 110$, $b = 1$, $\delta = .98, c_A = 10, c_B = 14, \beta = 8, I = 10$. Redrawn from Friedman and Fung (1996), Figure 1.

when the substitutability parameter δ is reduced from 0.98 to 0.95, the interior equal profit point $s^* = 0.5$ becomes evolutionarily stable and the game becomes a nonlinear version of Hawk-Dove.

Now consider the two-population game played by firms in the two countries when we allow international trade. The trade model has two populations with state variables s_1 and s_2 representing shares of B-mode in the home country (with I_1 firms) and the foreign country (with I_2 firms). Thus the state space is the unit square, $s_1, s_2 \in [0,1]$. The current state (s_1, s_2) is fixed in the short run, but evolves according to the profit advantage in the long run.

To spell this out, each firm decides how much output to produce for the home market (denoted h) and the foreign market (denoted f). The home market demand functions become

$$P_A = a_A - b(Q_A^h + Q_A^e) - \delta(Q_B^h + Q_B^e) \tag{12.23}$$
$$P_B = a_B - b(Q_B^h + Q_B^e) - \delta(Q_A^h + Q_A^e)$$

where Q_A^h is home production and Q_A^e is foreign production of type A output that is exported into the home market. The foreign market has similar demand, but there is no reason to think that the parameters a, b, δ are the same across nations. On the cost side, we assume that the skimming and network externalities depend only on the national share of type B firms; in this model, labor and other inputs are sourced nationally. This gives us cost functions for B mode firms of the form

$$TC_B^h = (c_B^h - \beta^h s_1) q_b^h \tag{12.24}$$
$$TC_B^f = (c_B^f - \beta^f s_2) q_b^f.$$

The cost functions for the A mode firms are independent of the state and, as usual, are linear in output:

$$TC_A^h = c_A^h q_A^h \qquad (12.25)$$
$$TC_A^f = c_A^f q_A^f.$$

The short-run equilibrium has each firm choosing how much to produce for each market given its own type (A or B) and the state (distribution of types). The initial state is the autarky evolutionary equilibria from the separate games. The relative importance of the cost-side effects and the degree of product differentiation determine which of several equilibria may emerge in the long run.

Figures 12.4 and 12.5 show that the asymmetric equilibria $(s_1 = 1, s_2 = 0)$ or $(s_1 = 0, s_2 = 1)$ are typical evolutionary equilibria. Thus the model predicts that in LR equilibrium with free trade in outputs, each nation will completely specialize in a different organizational mode.

Another possibility arises when we allow for international trade in inputs as well as in goods. In this case, workers from one nation can re-locate to the other nation. Even in SR equilibrium we equate the cost of production for a given mode across nations, as the network and skimming effects are global rather than contained within a nation.

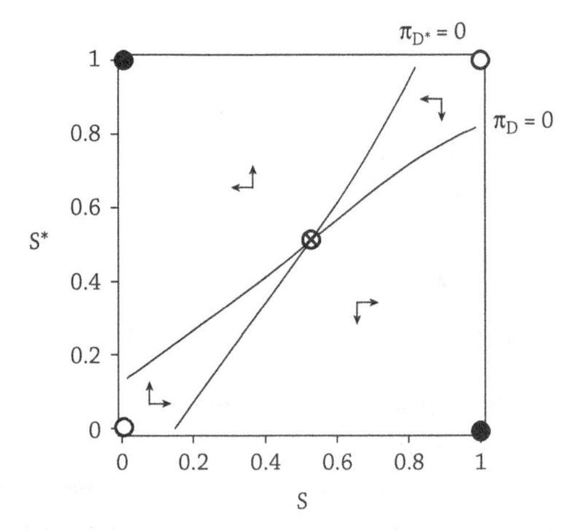

Figure 12.4 Monotone dynamics for the free trade model (12.23)–(12.25), for identical countries with parameters as in Figure 12.3 except $\delta = .95$. The labels π_D and π_{D^*} refer respectively to the home and foreign payoff differences Δw_{AB}^h and Δw_{AB}^f.

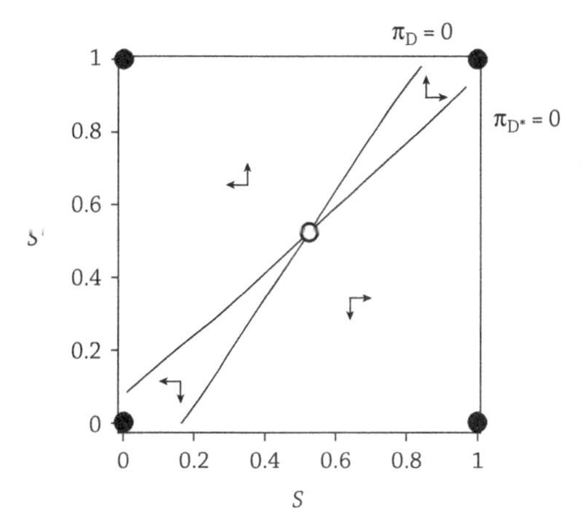

Figure 12.5 Same as 12.4 except that $\delta = .98$ (i.e., glut effects are returned to their weak baseline value), $I = 20$ and $a_A = a_B = 220$ (i.e., the home country is twice the size of the foreign country).

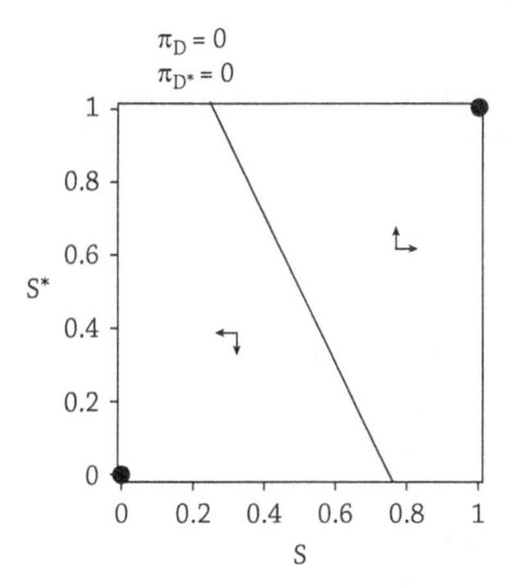

Figure 12.6 Monotone dynamics when inputs as well as outputs are traded internationally. Parameter values are as in Figure 12.5. This and the previous two figures are redrawn from Friedman and Fung (1996).

This complete trade model has a single zero-profit locus separating the two basins of attraction. Figure 12.6 shows that with mobile inputs, symmetric equilibria are obtained $(s_1 = 0, s_2 = 0)$ or $(s_1 = 1, s_2 = 1)$. If the initial condition is $s_1 = 0, s_2 = 1$ in Figure 12.6, the system will evolve towards $s_1 = 0, s_2 = 0$: all firms choose the type A production method. For the parameters used here, this is the least efficient state. Surprisingly, the inefficient type A production mode can completely displace the efficient type B method in both nations.

12.5 INTERNATIONAL TRADE AND POLLUTION TAXATION

McGinty (2008) combines features of the models presented so far to analyze trade and environmental policy. It is a trade model where there is a negative production externality for one of the goods (called the dirty good). There are two sectors of each nation's economy, denoted clean (c) and dirty (d), where dirty sector production creates pollution. For example, the clean sector could be services and the dirty sector could represent manufacturing. The model considers the incentives for a pollution tax under autarky (single population) and under trade (two populations). Firms may switch sectors in response to an endogenous payoff differential. The proportion of firms in each sector is the state variable, with s proportion of firms being clean and $(1 - s)$ proportion of firms producing the dirty good.

Demand is a simplified version of (12.1) where both the slope and intercept a and b are normalized to unity and there is no substitutability across sectors:

$$
\begin{aligned}
P_c &= 1 - Q_c \\
P_d &= 1 - Q_d.
\end{aligned}
\tag{12.26}
$$

As usual, each of the I firms is assumed to have constant cost per unit of output, which can also be normalized to zero. Government may impose a tax t per unit of the dirty good. This can also be thought of as a tax per amount of pollution, given the amount of pollution created by each unit of the dirty good.

For the moment, assume autarky (no trade). Maximizing SR profit given s, as in the previous sections, yields

$$
\begin{aligned}
q_c &= \frac{1}{sI + 1} \\
q_d &= \frac{1 - t}{(1 - s)I + 1}.
\end{aligned}
\tag{12.27}
$$

Firm output is declining in the number of firms of the same type, as usual due to a glut effect, which is especially strong here due to the lack of substitutability. Note that the national government can determine the level of pollution via the tax t. The tax decreases dirty output and increases the price of the dirty good, but has no impact on the clean sector in the short-run with s fixed.

Given the demand normalizations, profit of each type of firm is $\pi_i = (q_i)^2$. Instead of the usual payoff difference Δw, it is convenient here to express the payoff advantage as a ratio

$$\Pi_D \equiv \frac{\pi_c}{\pi_d} = \left[\frac{(1-s)I+1}{(sI+1)(1-t)}\right]^2. \tag{12.28}$$

Thus a clean firm has a payoff advantage at states s where $\Pi_D > 1$.

Monotone (or sign-preserving) dynamics can be written $\dot{s} = A(s)\Pi_D(s)$, where $A(s) > 0$ for $\Pi_D(s) > 1$ and $A(s) < 0$ for $\Pi_D(s) < 1$. This generates a unique, stable interior EE at $s = \frac{I+t}{I(2-t)}$ for all tax rates $t < \frac{I}{I+1}$. Thus under autarky (no trade) and moderate pollution taxes, we have a single-population Hawk-Dove game in each country. If the pollution tax rate exceeds $\frac{I}{I+1}$, then the clean sector is a dominant strategy and the EE is $s = 1$. In the absence of a tax the EE is $s = 0.5$.

Thus it seems that the government tax policy determines the allocation of firms across sectors in a straightforward manner. One subtlety can be seen by evaluating firm profit at the EE $s = \frac{I+t}{I(2-t)}$, where

$$\pi_c^{EE} = \pi_d^{EE} = \left[\frac{2-t}{I+2}\right]^2. \tag{12.29}$$

The pollution tax t reduces the LR profits of clean as well as dirty firms, because all profitable deviations are eliminated at the EE. The tax reduces the SR profitability of dirty firms, decreasing their prevalence, which boosts their LR profitability. Clean firms see a reduction in profit at the EE from the tax as the number of clean firms increases.

Socially optimal tax policy balances the damage from pollution with the benefits that the dirty good provides. These benefits include the benefits to consumers (called consumer surplus), the profit generated by the dirty firms and the tax revenue. The optimal tax policy increases t from zero until the reduction in consumer surplus and firm profit equals the increase in benefits from tax revenue and pollution abatement. Comparing the optimal tax rate for a fixed value of s with the optimal tax at the EE reveals a surprising result that follows from the subtlety noted in the previous paragraph. If government anticipates that the tax will result in firms fleeing the taxed sector they will choose a higher tax rate. The reason is that the reduction in dirty output from the tax for each firm is partially offset by an increase due to fewer competing dirty firms.

International trade generates an additional impact on socially optimal tax rates. A fixed tax policy generates a result known as the *pollution haven hypothesis*: the

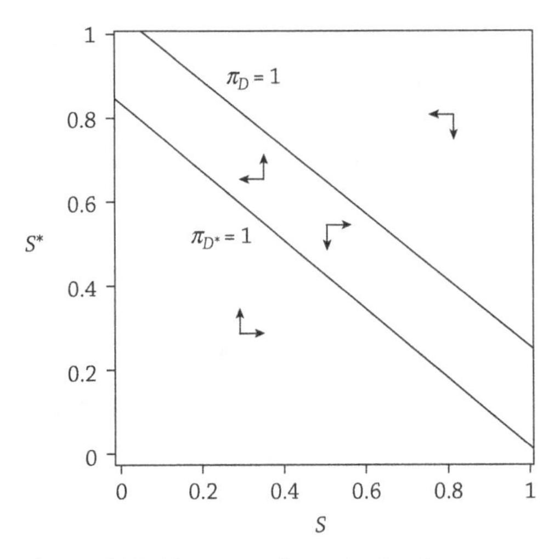

Figure 12.7 Monotone dynamics for the pollution tax model when home tax exceeds foreign tax.

nation with the lower tax rate produces the polluting good. That is, in the EE with international trade, the nation with the higher tax rate will be completely specialized in the clean good, and the nation with the lower tax rate will become a pollution haven, producing all of the dirty good. Figure 12.7 is the phase-diagram when the home nation (no asterisk) has a higher tax rate than the foreign nation (denoted with an asterisk).

The foreign nation can reverse this pattern of specialization by increasing their pollution tax sufficiently to shift the foreign locus above the home locus.

If governments are aware of this effect then they will endogenize the tax rates and recognize that the relative tax rates matter. This generates a "tax game" between governments where the tax rates are strategic complements. When pollution damage is a convex function, the optimal tax rate is increasing in dirty output. Figure 12.8 shows that with endogenous pollution policy an interior EE is obtained where both nations are incompletely specialized. However, the NE of the tax game yields an EE with higher taxes than would be jointly socially optimal. Due to the incentive to shift pollution to the other nation there is a race to the top, which ends beyond the cooperative level.

12.6 OTHER ECONOMIC APPLICATIONS

We quickly summarize a few other recent economic applications of evolutionary game theory to international and environmental economics. The purpose is not

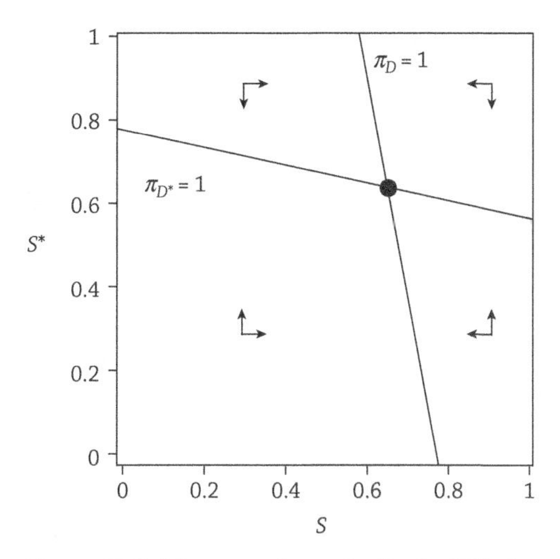

Figure 12.8 Monotone dynamics for the NE of the pollution tax game.

to make a complete survey, but rather to advertise the original publications, and encourage interested readers to develop new applications.

A clean environment is a global public good, meaning that everyone can simultaneously derive benefits without reducing the good, and no one can be prevented from receiving the benefits. Arce (2000) and (2001) uses evolutionary game theory to consider international agreements when an environmental public good has four possible aggregation technologies. These are weakest link, weaker link, best shot and better shot. For weakest-link technology the public good is determined by the minimum contribution, and with best shot it is determined by the maximum. Using a pairwise matching rule in a finite population, several interesting results emerge. Heterogenous behavior in a homogenous population may be an evolutionary steady state. Much more optimistic results are obtained when a leader can move first and match subsequent contributions by others.

Fisher and Kakkar (2004) model international trade and the evolution of comparative advantage. A nation has a comparative advantage in a good that it is relatively more efficient at producing. Individuals have payoff u (for utility) which is a Cobb-Douglas function of two goods x_1 and x_2, where $u = x_1^{\alpha} x_2^{1-\alpha}$. There is a Ricardian production function with constant unit labor input requirements a_1 and a_2. This means it takes a_1 units of labor to produce one unit of good 1 and a_2 units to produce good 2. Each individual has one unit of labor to devote to production of one of the goods. Hence, output, Y, for each individual is $Y = \left(\frac{1}{a_1}, 0\right)$ if they chose to produce good 1 and $Y = \left(0, \frac{1}{a_2}\right)$ if they choose to produce good 2. There is a continuum of individuals with an incentive to trade, since the utility and production functions imply that $u = 0$ if they do not trade.

Individuals are randomly pairwise matched and the surplus from a fertile match is divided according to the Nash bargaining solution. A fertile match occurs when individuals with different goods are matched and the bargaining rule implies an equal share of the surplus. Under autarky they obtain a single population Hawk-Dove game with $s^* = 0.5$ and social welfare is maximized. The trade model has identical preferences, but different unit input labor requirements across nations. They find that the EE reflects the pattern of comparative advantage. Perverse patterns of trade may occur when they restrict matching such that a home firm is only matched with a foreign firm. The result is that Hawk-Dove games have a stable mix of both types in a single population setting, but that occurs in a two-population game when random pairwise matches are drawn from different populations.

Dijkstra and de Vries (2006) investigate the effects of pollution on the location of firms and consumers in a spatial model. The paper shows how the location incentives depend on the policy response when there is a negative production externality (pollution). They argue for the evolutionary approach since firms and households will not instantaneously respond to payoff differences. The paper considers three policy regimes: laissez faire (i.e. no governmental policy), pollution taxation and direct compensation for pollution damages. Under laissez-faire households have an incentive to stay away from firms, but firms have no incentive to stay away from households. Second, a Pigovian pollution tax is set equal to the marginal damage, which is decreasing in the distance between firms and households. Hence, under taxation both households and firms have an incentive to stay away from each other. Third, under compensation firms directly pay households the amount of pollution damage. This makes households indifferent with respect to their location, but firms have an incentive to locate away from households as their negative pollution externality is completely internalized. They find that taxation always leads to a local welfare optimum (NE) but that compensation may lead to a global welfare optimum (EE). They stress the idea that path dependence to a Pareto inferior evolutionary equilibrium may be difficult to break.

McGinty (2010) transforms the static stage game of international environmental agreements (IEAs) from Barrett (2001) into an evolutionary game and shows that several important differences occur. The stage game is not robust to any trembles, that is, if any signatory changes their action the whole agreement collapses (called a lynchpin equilibrium in Barrett 2001). Furthermore, signatories to an agreement have a (much) lower payoff than non-signatories to an agreement, but higher than if the lynchpin is pulled and the agreement collapses. Thus "trembles" may in fact be intentional as a signatory would prefer to be outside of an agreement. By contrast, in an evolutionary equilibrium the payoffs inside and outside an agreement are equated and the EE is robust to trembles.

One conclusion is that a credible zero-sum system of transfers may eliminate the basin of attraction of the Pareto inferior EE. For the evolutionary IEA the two populations are Annex I (developed) and non-Annex I nations (developing) as defined by the Kyoto Protocol on greenhouse gas emissions. Developed nations are subject to abatement requirements under Kyoto, while developing nations are not. See Exercise 2 to get a sense of how this works.

Schaffer (1989) shows that Milton Friedman's (1953) assertion that natural selection will always favor profit maximizers need not hold. Economic natural selection will not drive non-profit maximizers out of the market when firms have market power. Due to the price externality, a non-profit-maximizer can have a larger profit than a profit-maximizer. Using a simple two-player example Schaffer shows that a profitable deviation by one player one can lead to an even larger increase in profit for the other player. Thus, the non-maximizer will have a survival advantage. Similarly, a spiteful action that reduces own payoff will increase the survival probability if the action reduces the other player's payoff by an even larger amount.

Benabou (1993) shows how a city may become segregated with high and low skilled labor endogenously choosing different locations. The evolution occurs via a state-dependent externality in the form of education. The cost of becoming either a low or high skilled worker is declining in the number of high skilled workers, however the effect is stronger on other high skilled workers. High skilled workers earn more and thus outbid lower skilled workers for housing in areas with more high skilled workers. This results in segregation. However, production requires both high and low skilled workers thus segregation may be inefficient. As Benabou puts it, "At the heart of the model lies the interplay of local and global interactions: community spillovers in education, and neoclassical complementarity in production, respectively." (620)

Possajennikov (2009) considers the evolution of conjectures, which are beliefs about how a rival will respond to a different action. The Nash conjecture is zero response since the NE implies a best response taking the other players' actions as given. A consistent conjecture is one that is correct and equal to the actual best-response slope. When the conjecture is a constant, Possajennikov shows that consistent conjectures will emerge as an EE. Players are randomly drawn from two populations and those conjectures with a payoff higher than the population average will increase in prevalence. The paper shows "...that more rationality (consistency) should lead to the same result as less rationality (evolution)..."

EXERCISES

1. Read Malthus' (1798) essay and write down a simple, single-population dynamic model for per capita standard of living that captures his argument.

Then consider how to modify the model to accommodate the empirical fact that, at least in the industrialized world over the next two centuries, per capita income rose far above subsistence level.

2. Use the static Cournot model from Section 12.2 to show Schaffer's 1989 idea of relative fitness. Suppose we begin at the zero-profit competitive equilibrium where price equals marginal cost or $P = a - bQ = c$, hence market quantity is $Q^* = \frac{a-c}{b}$. Consider a duopoly with two firms each producing half the competitive output level $q_1 = q_2 = \frac{Q^*}{2} = \frac{a-c}{2b}$.

    (a) Show that this symmetric competitive equilibrium is a 'symmetric evolutionary equilibrium' in the sense that if firm 1 continues to produce $q_1 = \frac{Q^*}{2} = \frac{a-c}{2b}$ and firm 2 deviates and chooses any other quantity that firm 1 will always have a higher profit. Do this by showing profit for both firms for $q_2 < \frac{a-c}{2b}$ and $q_2 > \frac{a-c}{2b}$.
    (b) Does this argument hold if "survival" is based on absolute rather than relative profit?
    (c) Find firm 2's best-response to $q_1 = \frac{a-c}{2b}$ and call this q_2^{br}. Hence firm 2 is a profit maximizer given $q_1 = \frac{a-c}{2b}$, but firm 1 is *not* a profit maximizer since $q_1 = \frac{a-c}{2b}$ is not a best response to q_2^{br}. By comparing profit, which firm is more likely to survive, the profit maximizer (firm 2) or the non-profit maximizer (firm 1)?

3. Use the model from Section 12.3 to determine the effects of adding another policy instrument. A subsidy ϕ per-unit of clean output requires a taxpayer expenditure of $sIq_c\phi = Q_c\phi$. Now assume that the government wishes to continue to encourage green technology diffusion, but cannot afford the subsidy program. A solution is to charge a tax t per unit of the dirty output, subject to the constraint that the tax and subsidy program is revenue neutral and thus there is no cost to taxpayers (ignore administrative or compliance issues). A revenue neutral policy implies that the subsidy cost $sIq_c\phi = \phi Q_c$ is equal to the tax revenue $(1-s)Iq_d t = tQ_d$. How do the results from the tax and subsidy policy compare to those in Section 12.3?

4. Reformulate the model in Section 12.5 as a subsidy for the clean good, rather than a tax on the dirty good.

    (a) In the evolutionary equilibrium do dirty firms have an increase in absolute or relative fitness as a result of the subsidy program?
    (b) Does the subsidy increase both consumer surplus and firm profit?

(c) Next, in a two-nation trade model suppose the governments play a "subsidy game" where they recognize that the relative subsidy rates matter. Is there a race to the top or to the bottom?

5. Consider the following social dilemma game, due to Barrett (2001). There are 100 players, $N_1 = 50$ are type 1 and $N_2 = 50$ are type 2. Each player has two possible actions, a or p. Let s_1 be the proportion of type 1 players choosing a and let s_2 be the proportion of type 2 players choosing a.
The payoff function for type 1 players is: $\pi_1^p = \frac{1}{2}(3s_1N_1 + 6s_2N_2)$ and $\pi_1^a = \pi_1^p - 100$.
The payoff function for type 2 players is: $\pi_2^p = 3s_1N_1 + 6s_2N_2$ and $\pi_2^a = \pi_2^p - 100$.

(a) Find the NE of the game where all players choose their action simultaneously.
(b) Next, suppose the following stage game. In stage 1 all players choose whether to join a coalition or not. In stage 2, those who choose to join the coalition collectively choose a or p for all the coalition's members. In stage 3 all players outside the coalition individually choose a or p. Find all NE of this stage game. Compare equilibrium payoffs for all players with those from your answer in part (a). How robust is your equilibrium to trembles by the players?
(c) Next, consider the possibility of side-payments in a four-stage game. In stage 1 type 2 players choose to join the coalition or not. In stage 2, type 2 nations in the coalition collectively choose a or p for all the coalition's members, and a side-payment T to pay to type 1 players should they join the coalition and choose a. This choice is made to maximize the sum of payoffs to the type 2 players in the coalition. In stage 3, type 1 players choose to join the coalition (playing a and receiving T) or remain outside. In stage 4 all players that did not join the coalition (independently) choose a or p. Find all NE of the stage game and the equilibrium T. Find the equilibrium payoffs for both types of players inside and outside the coalition.
(d) Reformulate this stage game as two separate single-population evolutionary games with state variables s_1 and s_2. Next, reformulate this situation as a two-population evolutionary game with a "playing the field" matching rule. What results would be obtained from random pairwise matching where a single player from population 1 was matched with a player from population 2?

NOTES

Standard biographies of Darwin (e.g., Barlow 1958, p. 120), acknowledge Malthus' crucial influence on Darwin's development of evolutionary theory.

The analysis in this chapter simply assumes that the short-run dynamics converge globally to the unique Cournot-Nash equilibrium for any fixed *s*. Although this is very plausible, we have not yet introduced techniques that would enable a formal demonstration. The technical problem is that the strategy space in the Cournot model is a continuum (non-negative output levels). Chapter 14 touches on evolutionary dynamics for continuous strategy spaces, but a serious analysis is beyond the scope of this book.

Section 4 assumes that modes A and B are mutually exclusive and are the only alternatives. Friedman and Fung (1996) cite papers that supported that assumption, which seemed quite plausible in the late 1980s. However, since then the advent of information technology, distributed computing, and other developments seems to have enabled hybrid modes to emerge. Indeed, in recent years, firms in the both the United States and Japan (and beyond) use global supply networks that minimize inventories and that feature flatter hierarchies.

BIBLIOGRAPHY

Alchian, A., (1950). "Uncertainty, evolution, and economic theory." *Journal of Political Economy*, 58(3), 211–221.

Arce, D. G. (2000). "The evolution of heterogeneity in biodiversity and environmental regimes." *Journal of Conflict Resolution*, 44(6), 753–772.

Arce, D. G. (2001). "Leadership and the aggregation of international collective action." *Oxford Economic Papers*, 53, 114–137.

Barlow, N. "The autobiography of Charles Darwin 1809–1882". Collins, (1958).

Barrett, S. (2001). "International cooperation for sale." *European Economic Review*, 45, 1835–1850.

Benabou R. (1993). "Workings of a city: Location, education and production." *Quarterly Journal of Economics*, 108(3), 619–652.

Cournot, A. (1838). "Researches into the mathematical principles of the theory of wealth," translated by N.T. Bacon, Macmillian, 1987.

Dijkstra, B. and de Vries, F. (2006). "Location choice by households and polluting firms: An evolutionary approach." *European Economic Review*, 50, 425–446.

Dixit, A. K. (1979). "A model of duopoly suggesting a theory of entry barriers." *Bell Journal of Economics* 10, 20–32.

Friedman, D. (1991), "Evolutionary games in economics." *Econometrica* 59, 637–666.

Friedman, D. and Fung, K.C. (1996). "International trade and the internal organization of firms: An evolutionary approach." *Journal of International Economics*, 41, 113–137.

Friedman, M. (1953). "The methodology of positive economics." In: *Essays in positive economics*, University of Chicago Press.

Fisher, E. and Kakkar, V. (2004). "On the evolution of comparative advantage in matching models." *Journal of International Economics*, 64, 169–193.
Hardin, G. (1968). "The tragedy of the commons." *Science*, 162, 1243–1248.
McGinty, M. (2008). "An evolutionary race to the top: Trade, oligopoly and convex pollution damage." *The B.E. Journal of Economic Analysis and Policy*, 8:1.
McGinty, M. (2010). "International environmental agreements as evolutionary games." *Environmental and Resource Economics*, 45, 251–269.
McGinty, M. and de Vries, F. (2009). "Technology diffusion, product differentiation and environmental subsidies." *The B.E. Journal of Economic Analysis and Policy*, 9:1.
Possajennikov, A. (2009). "The evolutionary stability of constant consistent conjectures." *Journal of Economic Behavior and Organization*, 72, 21–29.
Schaffer, M. E. (1989). "Are profit-maximisers the best survivors? A Darwinian model of economic natural selection." *Journal of Economic Behavior and Organization*, 12, 29–45.
Sugden, R. (2001). "The evolutionary change in game theory." *Journal of Economic Methodology*, 8:1, 113–130.

13

Evolution of Cooperation

Any treatment of how cooperation evolves, especially among humans, must tackle several big issues that have so far remained quietly in the background. These include macroevolution (or major qualitative transitions) versus microevolution (everyday adaptation and share dynamics); controversies over group selection and inclusive fitness; preferences (or utility) as distinct from fitness, and how both shape evolution; the origins of memes and their co-evolution with human genes; and subtle and second-order free riding.

Each of these big issues deserves a book (or shelf of books) in its own right, so we will need a different expositional strategy in this chapter. Instead of carefully developing a few models, we will mainly discuss how the big issues connect to each other, occasionally illustrate with simple model, and mention alternative approaches, old and new. Even so, we will need touch on some new technical points such as the Folk Theorem. The notes are therefore longer than usual, and will help the reader navigate primary sources for the technical points and the big issues.

The chapter opens with a new graphical representation of the basic conflict between self interest and group interest. It then presents two established solutions to the conflict: kin selection and bilateral reciprocity. (Biologists sometimes refer to the accounting behind the first solution as inclusive fitness and refer to the second solution as reciprocal altruism, while economists regard the second solution as a consequence of the Folk Theorem.) Although many social creatures use them to ameliorate social dilemmas, these solutions can not, by themselves, explain widespread cooperation among humans. The rest of the chapter seeks distinctively human solutions, and suggests a memetic approach grounded in human moral sentiments.

13.1 COORDINATION, COOPERATION, AND SOCIAL DILEMMAS

As we first saw in Chapter 2, the essence of a coordination game is that, although there are multiple equilibria, there is no real conflict between individual optimization and social efficiency. Once the other players coordinate on a particular pattern of behavior, it is efficient, and also in your own self-interest, for you to do the same.

For example, if everyone else drives on the right side of the road, so should you. Schools of fish such as anchovies play a coordination game. By joining a large school, a lone anchovy will on average increase its own fitness and also slightly increase the fitness of the other anchovies. (Why? Because neighbors' reactions will allow the anchovy to recognize a predator sooner, and as the neighbors dart in different directions before reconverging, the anchovy will have a better chance at escaping.) Likewise, a lone pelican saves energy by joining a flock flying in V-formation and riding the bow wave of the bird in front. At the same time its own bow wave helps the bird behind, and it even helps the bird in front by slightly reducing the drag on its wings. By growing near to each other in a forest, redwood trees keep down invasive species and increase local rainfall, benefitting themselves as well as the neighbors. In the equilibrium of such symmetric coordination games, each player helps others at the same time that she helps herself, generating what biologists refer to as mutualism.

In other games coordination may be more difficult. Suppose that you need to meet someone with no cellphone. You both want to coordinate but may have a problem without some shared understanding, or a coordination device such as the town hall clock at noon. Another difficulty is that payoffs may not be symmetric. In a territorial assignment game, for example, some territories may be better than others. More generally, in a dominance hierarchy, those at the top get the best opportunity and those lower down in the hierarchy can only pick over what is left. A feudal prince, for example, does better than a serf, but both are better off following established custom than battling each other. Even in these difficult coordination games, playing your part of an equilibrium strategy profile serves both self-interest and group interest.

Some games are even more difficult because there is a direct conflict between group interest and self interest. Such games are called social dilemmas, and here achieving efficiency takes true cooperation, not just coordination. We've seen several examples already, including prisoner's dilemma, tragedy of the commons and public goods.

Figure 13.1 helps sort things out. It takes the perspective of a player called Self, who interacts with other players, collectively called Other. Relative to non-action (or to the status quo strategy), each alternative strategy available to Self potentially increases or decreases the player's own payoff, and at the same

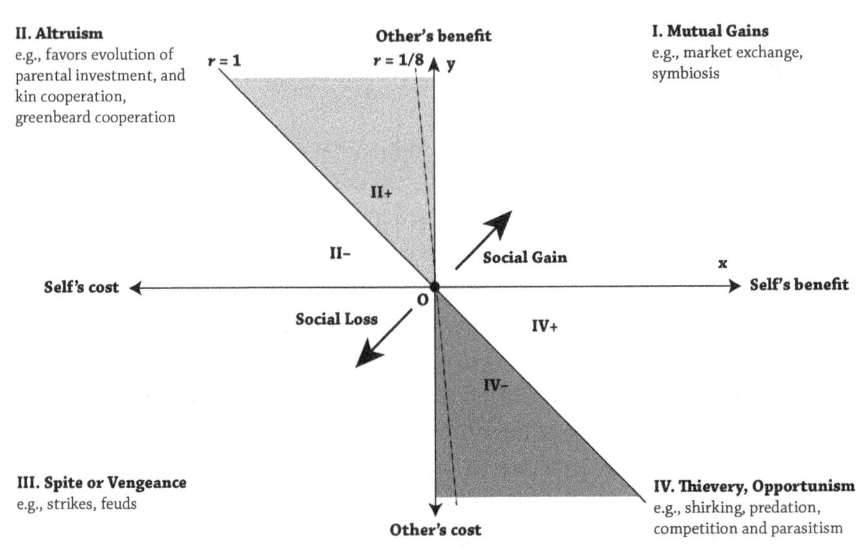

Figure 13.1 Types of social interactions. The origin $\mathbf{O} = (0,0)$ marks the status quo fitness. In quadrant I, both the actor ("Self") and conspecifics ("Other") benefit from a departure from the status quo. In quadrant II, the action helps Other at a cost to Self; the fitness sum is positive in subquadrant II+ (shaded) and negative in II–. In quadrant III, both Self and Other incur fitness costs. In IV, Self gains at the expense of Other, and the fitness sum is positive in IV+ but negative in the shaded subquadrant IV–.

time has an impact on other players' payoffs. The net impact on Other (summing across all other players) also may be positive or negative.

Evolution directly favors activities whose fitness impact on Self is positive, regardless of the impact on Other. Geometrically speaking, Self's iso-fitness lines are parallel to the vertical axis and lines further east (i.e., to the right in the diagram) represent higher fitness. Hence shares should increase most rapidly for actions whose payoffs are furthest east. We conclude that unaided evolution pushes creatures into quadrants I and IV.

By contrast, the group as a whole gains fitness when the sum (or average) of the members' payoffs is as high as possible. Social efficiency is best promoted by actions that equally weight payoffs to Other and Self. Geometrically speaking, social iso-efficiency lines are parallel to the line of slope –1 through the origin, and the group does best when actions are chosen that are farthest to the Northeast.

Actions in quadrant I serve self-interest and group interest simultaneously, as in coordination games. Economists refer to these as mutual gains, and biologists as mutualism.

However, efficient altruism (subquadrant II+) is not favored by unaided evolution, while inefficient opportunism (IV–) is favored. In these shaded subquadrants we have a direct conflict between what is good for the individual and what is good for the group.

For example, consider again the simple two-player prisoner's dilemma with payoff matrix $W = \begin{pmatrix} 1 & -1 \\ 2 & 0 \end{pmatrix}$. The status quo is for both players to choose the second action ("defect"), yielding the payoff sum $0+0=0$. A unilateral choice by Self to instead play the first action ("cooperate") yields the payoff vector $(-1,2)$ in subquadrant II+, an increase in social efficiency since the payoff sum is $-1+2=1>0$. Social efficiency is maximized if Other reciprocates, yielding payoff vector (1,1) in quadrant I and payoff sum 2. But unaided evolution increases the share of defect because it is the dominant action, and consequently the group is unable to reap substantial mutual gains.

Another classic social dilemma is the Tragedy of the Commons, mentioned in Chapter 12. Here we model it as an n-player game with payoff function $u(x_i, X_{-i}) = x_i - h(x_i + X_{-i})$, where X_{-i} is the sum of the $n-1$ other players' actions, and the harm (or social cost) function h is smooth and increasing with $h(0)=0$ and derivative $h' \in (\frac{1}{n}, 1)$. For example, x_i might be the profit a firm makes from selling its output, while h is the health cost associated with total industry output, due to air or water pollution. In the eponymous example, x_i is the number of cows (or sheep) farmer i grazes in the common pasture, and h is the social cost of overgrazing.

We have a tragedy because the personal benefit $\frac{dx_i}{dx_i} = 1$ from increasing one's action $x_i \in [0,1]$ exceeds the personal harm h' but is overshadowed by the social harm nh'. Therefore Self's action $x_i = a > 0$ appears as a point in inefficient subquadrant IV–. For example, with status quo from $x_i = 0 = X_{-i}$, the point on the graph is $(a - h(a), -(n-1)h(a))$. This is in the inefficient subquandrant since $0 < \int_0^a [1 - h'(x)]dx = a - h(a)$ and $0 > n\int_0^a [\frac{1}{n} - h'(x)]dx = a - nh(a) = [a - h(a)] + [-(n-1)h(a)]$. Thus it pays for Self to choose x_i as large as possible, but when everyone does this the payoff sum is the lowest possible (i.e., the outcome is as bad as it could possibly be).

In general, social efficiency requires seizing opportunities in shaded region II+ and preventing activities in shaded region IV–, contrary to the push from unaided evolution. How might that happen? There are several ways, as we will see in the following sections, and on the surface they seem quite different from each other. From the perspective of Figure 13.1, however, they all do the same thing—they all rotate the vertical axis (or Self's iso-fitness lines) counterclockwise. In other words, they all internalize the externalities and thereby convert social dilemmas into coordination games.

13.2 SOLUTION K: KIN SELECTION

The first way to resolve social dilemmas is to funnel the benefits to kin. The intuition goes back to Darwin (1856, Ch 7) and Haldane (1955), but was first developed formally by Hamilton (1963, 1964), and popularized in Dawkins (1976).

Here is the reasoning. Suppose that a prosocial trait, such as sharing food or defending a common-pool resource, is controlled by a gene at some particular locus. That trait will spread if its gene has higher fitness than alternatives. The gene bears the fitness cost of the trait but also enjoys the fitness benefits of prosociality if—and this is the crucial caveat—the beneficiaries carry the same gene. For example, an individual who shares food with others has lower survival probability than an individual who consumes all the food herself, but the sharer confers higher fitness on its beneficiaries. If they carry the same gene, then that gene may gain higher fitness than a gene for nonsharing. Likewise, for an antisocial trait, the gene internalizes the costs to the extent that they are borne by closely related individuals.

Here is the basic algebra; see the notes for elaboration. Suppose that, relative to status quo, an individual (Self) bears fitness cost C for some genetically controlled behavior, and other individuals $i = 1, \ldots, n$ each enjoy benefit b. Let r_i denote i's degree of relatedness to Self, defined as in Chapter 4 as the probability that i and Self are more likely than the population average to bear the same gene. Let $r = \bar{r} = \frac{1}{n}\sum_{i=1}^{n} r_i$ be the average degree of relatedness of the beneficiaries, and let $B = nb$ be the total benefit to others. Then the fitness increment (often called the *inclusive fitness*) for the prosocial gene is $\Delta w = -C + rB$. The trait will spread to the extent that $\Delta w > 0$ (i.e., that $rB > C$). Thus we have

Hamilton's rule: *A prosocial trait with personal cost C that brings total benefit B to individuals with average relatedness r will gain share if and only if it has positive inclusive fitness, so*

$$rB > C. \tag{13.1}$$

Figure 13.1 illustrates the geometry. When $r = 1$, as with two cells in your body (or as between clones), the locus $\Delta w = 0$ (i.e., $B = C$, coincides with the $-45°$ line separating the efficient from the inefficient portions of quadrants II and IV). A gene benefits exactly to the extent that the group benefits. That is, when genes are identical within the group, the vertical line (representing zero fitness increment for Self) is rotated counterclockwise 45°. Thus the externality is completely internalized and the conflict between group and self-interest evaporates.

For lesser values of r the conflict is ameliorated but not eliminated. The dashed line in Figure 13.1 is the locus $\Delta w = 0$ or $rB = C$ when $r = \frac{1}{8}$, as with first cousins. Evolution now favors efficient altruism in the narrow gray wedge east of the dashed line in subquadrant II+, and disfavors inefficient opportunism in the narrow dark wedge west of the dashed line in IV–. The line has slope $\frac{-1}{r} = -8$, and it represents counterclockwise rotation by a bit more than 7°, or about a sixth of the rotation that would completely eliminate the conflict between Self and group.

Hamilton's approach to social behavior got a major boost from the fact that ants and bees, two of our planet's most spectacularly social species, belong to the insect order Hymenoptera, whose unusual reproduction process yields exceptionally high values of r. The unusual process is called haplodiploidy: unfertilized eggs develop into haploid males and fertilized eggs develop into diploid females. One consequence is that full sisters share 100% of paternal genes and (as usual) 50% of maternal genes, hence have $r = 0.75$ instead of $r = 0.50$ as in most diploid species (including humans). In Figure 13.1, the $\Delta w = 0$ locus has slope -1 for clones or identical twins, -1.5 for sister bees and ants, and -2 for ordinary sibs.

Ants and bees exemplify the extreme division of labor, or caste system, that biologists call *eusociality*: Queens do all the reproduction and Workers collect all food, take care of young and defend the colony. This implies altruism in that Workers do not directly obtain fitness through progeny, but only indirectly by helping Queens. Eusociality has evolved repeatedly in the Hymenoptera. Many biologists (but not all—see the notes) believe that the unusually high degree of relatedness of Worker to the Queens tips the balance in favor of reproductive sacrifice on the part of the workers.

Eusociality has also evolved in termites, a completely different insect group. Termite colonies also typically have a single queen that produces all the eggs, while workers collect food and warriors defend the colony. High values of relatedness r in termite colonies come from inbreeding. For example, sib-mating can increase the probability of shared genetic material to 0.75 among their own progeny. This is only after a single generation. Repeated inbreeding events can lead to values for relatedness that rival or even exceed the 0.75 of hymenoptera.

The naked mole rat is a eusocial vertebrate. In a naked mole rat colony, a single female reproduces, while other females are workers, and the males do little work except for defense of the colony. Inbreeding in their colonies produces average values of r around 0.8, once again consistent with Hamilton's approach.

Inbreeding is not very common in general due to three fitness costs. In the short run of a few dozen generations, deleterious recessive alleles appear more often in the homozygous state than they do in an outbred population, and thus lower fitness. Eventually, however, natural selection will purge these deleterious alleles from a persistently inbred population. The second cost remains even in the long run: overall genetic variation is reduced, which lowers the "adaptability" of the population to novel environments. That long-run problem is not so bad for termites and naked mole rats because they occupy some of the world's most stable niches—they live in subterranean environments that have remained essentially unchanged for 65 million years for naked mole rats, and 100 million years for termites. The third cost relates to gains from rearing related sibs, it drops with inbreeding (Bartz 1979).

A similar consideration applies to ordinary parenting, where two unrelated individuals each combine half of their genes to create progeny with relatedness $r = 0.5$ to either parent. Why wouldn't evolution favor asexual cloning (as with amoebae) so each individual could rear identical ($r = 1.0$) copies of itself and thus double its fitness, from the gene's perspective? The best current evolutionary explanation is that this asexual strategy will be targeted by parasites once it becomes common, and/or will accumulate deleterious mutations that cannot be purged by the mechanisms of sexual reproduction and natural selection. In essence, diseases stabilize the cooperative system of sexuality.

Parental investment is another evolved social behavior illuminated by inclusive fitness. The eponymous trait of mammals is that females have mammary glands, evolved specialized structures for nursing their young. The mom incurs a fitness cost in producing milk, but it is worthwhile according to Hamilton's rule when the benefits to offspring are at least twice as large, given the usual relatedness $r = 0.5$ between mom and kids. A lioness sometimes will also nurse her sister's cubs ($r = .25$). According to Hamilton's rule, this makes evolutionary sense only when the benefit/cost ratio is twice as large as for her own cubs, for example, when they are already full.

How about male parental investment? Most male mammals are polygynous, seeking out multiple female partners, and invest little in their offspring. Birds exhibit a variety of mating systems, including biparental care, female only care, male only care, and even brood parasites that do not care for their young but deposit them in the nests of other parents (either of the same species, or an entirely different species).

With biparental care, both parents contribute nearly equally to child-rearing, a highly cooperative strategy. It is susceptible to cheating, the evolutionary "temptation" to leave the rearing of progeny to the other (now single) parent while the cheater invests in another brood of progeny with another partner. The abandoned partner is in an evolutionary bind. Also abandoning the $r = 0.5$ progeny will lessen her (or his) own fitness, but sticking with them gives the other parent's 0.5 genetic share a free ride.

The free-rider problem in birds is ameliorated by a monogamous lifestyle (an idea first posited by Emlen and Oring). When resources are distributed at great distances and it is not economical to defend all but a small resource patch, birds tend to monogamy. In the case of concentrated and abundant resources at a local scale, birds tend to polygyny.

Limited resource availability can lead to systems in which grown progeny stay behind to help parents raise their younger siblings. When the habitat is very densely settled by breeding adults, it is nearly impossible for progeny to disperse and carve out their own territory. Then the cost C of remaining with the parents is low for the young adult birds, while r is nearly 0.5 and the benefit B is high for the new progeny. In these circumstances, Hamilton's equation justifies staying home with the parents.

Humans are unusual among mammals (and especially among primates) in that both parents usually contribute to raising the young. As we will see in Section 13.5, biparental investment is part of the evolved behavioral suite that promotes cooperation among groups of humans.

Kin selection is an important solution to social dilemmas, but it has its limits. To work properly, non-kin must be largely excluded from the benefits of altruistic behavior. Otherwise the average relatedness r drops until $rB < C$ and then, by Hamilton's rule, cooperation fails. To exclude non-kin, there must be reliable kin recognition and/or limited dispersal.

The notes recount a recent controversy regarding inclusive fitness. Our take-away, which goes back to Grafen (1984), is that inclusive fitness should be thought of as an alternative accounting scheme that helps build intuition regarding indirect effects in evolution. Inclusive fitness is an appealing answer to the question, what objective function should evolved organisms appear to maximize? To the own-fitness increment arising from Self's act, inclusive fitness adds (a) the r-weighted sum of fitness increments that act bestows on all Others, but omits (b) the fitness increment for Self resulting from all Others' acts. The standard fitness accounting scheme, by contrast, omits (a) but includes (b). Both schemes are consistent and complete, but they differ in focus and implementability. We have just argued for the insights arising from inclusive fitness' focus on how Self's act affects others. However, we should acknowledge that implementing a complete bookkeeping for (a) is a real challenge when there are many alleles and genetic loci whose fitness implications interact for Others as well as Self.

13.3 SOLUTION R: BILATERAL RECIPROCITY

About the same time as Hamilton was beginning to develop his ideas, several leading game theorists, working together and separately, developed a purely rational approach to cooperation. By the early 1960s they all came to realize that the key was repeated interaction and patience.

To illustrate their reasoning in simplest form, we present a very stylized two-player game with alternating moves. In odd-numbered periods Player 1 can, at personal cost $C > 0$, provide a benefit $B > C$ to Player 2, who has the same cooperative opportunity to benefit Player 1 in even numbered periods. Suppose that periods are equally spaced and both players have one-period discount factor $\delta > 0$ (more about this below). If both players always cooperate, the (present value of the) payoff to Player 1 will be

$$\Delta w = -C + \delta B - \delta^2 C + \delta^3 B - \cdots = [-C + \delta B](1 + \delta^2 + \delta^4 + \cdots). \quad \textbf{(13.2)}$$

Suppose that each player i believes that the other player will cooperate until i chooses not to cooperate, and after that will never cooperate again—game theorists call this strategy Grim Trigger. Then i will get payoff zero following any defection, versus Δw if she always cooperates (or if she follows Grim Trigger herself). We conclude that it will be a NE for both players to follow Grim Trigger (and so always cooperate) if $\Delta w > 0$ in equation (13.2). Since the last factor in the equation must be positive, the condition for cooperation is simply $[-C+\delta B] > 0$ (i.e., $\delta B > C$).

The last inequality looks just like Hamilton's rule, with the discount factor $\delta \in [0,1]$ replacing the relatedness coefficient r in equation (13.1). The geometric interpretation is exactly the same: as δ increases from 0 to 1, the locus $\Delta w = 0$ rotates 45° counterclockwise from the vertical to the off-diagonal, shrinking the zone of conflict between Self and group until (at $\delta = 1$) it disappears.

New insight arises from the nature of δ, which represents the intensity of repeated interaction. We can write $\delta = \frac{p_c p_r}{1+\rho}$, where p_c is the probability that the interaction will continue at least one more period, p_r is the probability that the other player will reciprocate at the next opportunity, and ρ is the real interest rate or impatience rate indicating how benefits received now trade off against benefits received one period later. Frequent, reliable interaction between patient individuals implies δ just slightly less than 1. The corresponding $\Delta w = 0$ locus is very near the off-diagonal, indicating virtual elimination of the social dilemma. On the other hand, the dashed line in Figure 13.1, corresponding to $\delta = \frac{1}{8}$, will arise from a less reliable or frequent partner, or from greater impatience, and it indicates only slight amelioration of the social dilemma.

Game theorists in the 1960s and 1970s found that minor variants of the $\delta B > C$ condition enable cooperation in many more complicated games played out over time. See the notes for pointers to the Folk Theorem literature that generalizes our simple illustration.

Robert Trivers, then a Harvard graduate student in evolutionary biology, was the first to appreciate the parallels between inclusive fitness and the repeated game logic just illustrated. His 1971 article uses that logic to show that cooperation is evolutionarily viable between unrelated individuals who interact repeatedly. Referring to the interaction as "reciprocal altruism," Trivers argues that it can explain grooming and alliances in several primate species, and warning calls by sentinel birds and ground squirrels as well as some examples of interspecies cooperation such as cleaner fish. He conjectures that reciprocal altruism may also explain human social emotions, a matter we will revisit in a later section.

The best-known evidence supporting Trivers's theory is Gerald Wilkinson's (1984) study of vampire bats in Puerto Rico. The bats live in stable colonies where the average degree of relatedness is less than 1/8. A young bat signals hunger by licking its mother's wings, and she feeds the youngster by regurgitating partially digested blood directly into its mouth. But adult vampire bats sometimes also need nurture. They become weak, and their lives are in danger, if they are unsuccessful for two consecutive nights in finding a large mammal (such as a cow or pig) whose blood they can drink. Wilkinson noticed that unsuccessful adults often used the same hunger signal as youngsters, and often got fed by an unrelated bat, but only if the currently hungry bat had not previously refused to help the currently well-fed bat. Since the benefit B to a hungry bat of a gram of regurgitated blood is at least 3 times its fitness cost C to a well-fed bat, Wilkinson's observations seem to fit Trivers's theory quite nicely.

Very few other examples of reciprocal altruism have been found in the animal kingdom, except in primates. There it is quite common. Alliances between distantly related adult males are often observed among chimpanzees, as are alliances among female bonobos. Of course, backscratching is proverbial.

With primates as with vampire bats, kinship seems to provide a scaffolding that enables reciprocity to evolve. Take backscratching, for example. Most mammal moms groom their kids. That behavioral pattern could be extended to unrelated adults and become bilateral reciprocity in species, such as monkeys and apes, with the capacity to recognize individuals and recall previous interactions. It seems clear that bilateral reciprocity can arise from (and complement) kin selection.

13.4 SOCIAL PREFERENCES: A PROBLEMATIC SOLUTION

Our main interest in this chapter is cooperation among humans. We humans are especially good at exploiting once-off opportunities with a variety of different partners, so bilateral reciprocity can't be the whole story. Nor can inclusive fitness, since even in tribal groups average relatedness r is typically less than 1/8 (e.g., Smith 1985).

What other explanations might there be? Could it be that we *just like to help our friends*? Introspection tells us that there is something to this, as does neuroscience beginning with Rilling et al. (2002). Most people really do care about others, at least their friends, and are willing to make some sacrifices to benefit them. The idea is so simple and appealing that we should see how far it can take us.

Suppose that your happiness is an increasing function of $x+\theta y$, that is, you are willing to sacrifice up to θ units of personal payoff to increase others' payoff by a unit. In other words, you'd take an action with personal cost C as long as the benefit B to others satisfies $\theta B > C$. Once again, we have a variant of Hamilton's rule, now with the preference parameter θ replacing the relatedness coefficient r in equation (13.1).

The algebra and geometry must look familiar by now. Friendly preferences of the sort just described partially internalize the externality, and parameter values

$\theta \in [0,1]$ potentially can explain exactly the same range of social behavior as genetic relatedness and repeated interaction.

This explanation looks good at first glance, but hard questions lurk just beneath the surface. Unlike r or δ, the parameter θ is not a given feature of the environment, but instead is an evolved trait. So we must ask how friendly preferences can evolve: how can $\theta > 0$ arise, and how can it resist invasion?

To answer that question we first must deal with fundamental issues regarding preferences, social or otherwise. What are preferences, and what evolutionary advantages can they convey?

Economists think of preferences as summarizing contingent choice, and evolutionary economists have recently begun to see evolved preferences as Nature's way of delegating choices that must respond to local contingencies. Of course, some behavior doesn't need to be delegated and can be more or less hard-wired (e.g., to drop a hot potato). Here a finely calibrated response is unnecessary since burnt fingers tend to reduce fitness in almost all contingencies. But contingencies are crucial to some other sorts of behavior. For example, tribeswomen may gather only certain kinds of root vegetables and only in certain seasons, and prepare them only in certain ways. They and their families would lose fitness if they ate poisonous roots, or didn't cook some good roots properly.

It is logically possible to hardwire contingent responses via trial-and-error mutations over many generations, generating what biologists refer to as innate behaviors that are purely genetically determined. However, even with a moderate number of contingencies the combinatorics are daunting – one needs about 2^{100} alternative alleles to fully specify behavior with only five root species in four seasons with five preparation methods. Even worse, changes in the environment can make the hardwiring obsolete before it can become established. As Robson and Samuelson (2011, Section 2.3) put it,

> A more effective approach may then be to endow the agent with a goal, such as maximizing caloric intake or simply feeling full, along with the ability to learn which behavior is most likely to achieve this goal in a given environment. Under this approach, evolution would equip us with a utility function that would provide the goal for our behavior, along with a learning process, perhaps ranging from trial-and-error to information collection and Bayesian updating, that would help us pursue that goal.

For the rest of the chapter we will take it as given that preferences, or utility functions, drive much of human behavior, including most social behavior, and we will study how the preferences evolve. That is, we will consider models of how preference parameters evolve under learning and/or genetic selection.

Such models distinguish between fitness payoff and utility payoff. Evolution is driven purely by fitness, that is, by own material payoff x arising from one's own acts and those of others. Choice, on the other hand, is driven by preferences, that is, by utility payoff that may include other components such as θy, the joy of helping others.

It might seem at first that evolution offers no scope for such additional components. Creatures (including people) who make choices more closely aligned with fitness should, it would seem, displace creatures whose choices respond to other components. But social interactions complicate the analysis because fitness payoffs depend on others' choices as well as one's own, and one's own behavior can affect others' choices. The correct statement is sometimes called the

Principle of Indirect Evolution: *For a given sort of preferences to evolve, creatures with such preferences must receive at least as much material payoff (or fitness) as creatures with feasible alternative preferences.*

Evolution here is indirect in that it operates on the preference parameters (such as θ) which determine behavior rather than directly on behavior. Armed with this principle, we are now prepared to see when $\theta > 0$ can evolve.

To recap, an individual with $\theta = 1$ values own material payoff equally with Other's material payoff, and in that sense follows the Golden Rule. Of course, $\theta = 0$ represents a selfish Self who is indifferent to the impact his actions have on others—this might be called the Brass Rule. The group as a whole is better off the closer a member's θ is to 1, but any individual will gain higher material fitness with lower θ, because he saves on costs incurred helping others without (given the discussion so far) reducing benefits received from others. Thus the cheater problem reappears in a form that sometimes is called *subtle free riding*: those with lower $\theta > 0$ gain more material fitness than those with higher θ. Without some further support, evolutionary forces will undercut friendly preferences and they will eventually disappear, or never appear in the first place.

Vengeful preferences rescue friendly preferences. The idea here is that social preferences are state dependent: my attitude θ towards your payoffs depends on my emotional state, for example, friendly or hostile, and your behavior systematically alters my emotional state. If you help me or my friends, then my friendliness increases, as captured in the model applying a larger positive θ to your material payoff. But if you betray my trust, or hurt my friends, then I may become quite angry with you, as captured by a negative θ. In this emotional state, I am willing at some personal material cost to reduce your material payoff, that is, I seek revenge. This might be called the Silver Rule: be kind to others who are kind to you, but also seek to harm those who harm you or your friends.

The algebra of vengeance involves a double negative but is otherwise very simple. An action in Quadrant III of Figure 13.1 is represented by $x,y < 0$, that is, both Self and Other incur losses relative to the status quo. In an angry state with $\theta < 0$, punishing (imposing the loss $y < 0$ on) a culprit Other will bring Self satisfaction $\theta y > 0$, which can more than offset her personal fitness loss $x < 0$.

Everyone loses fitness from such punishment and in that sense it directly impairs efficiency, but the indirect effects can be even more important. If Others punish Self for being insufficiently friendly, then the material advantage of having a lower θ shrinks or disappears. Cheaters no longer prosper, and neither do subtle free riders. That is, although less friendly people still incur fewer costs of altruistic acts, they now incur increased costs due to punishment by vengeful group members. If vengeance is sufficiently intense, it supports a high average value of θ, and thus promotes efficient social behavior.

But we are still not done. Punishment is also costly to the avenger, so less vengeful preferences seem fitter. What then supports vengeful preferences: who guards the guardians? This is called the *second-order free rider problem*. As explained in the notes, it has provoked a considerable literature in its own right. Samuelson (2001) summarizes the early work as follows. There is no second-order free rider problem when Others can see Self's degree of vengefulness. In this case, sometimes called "transparent disposition," a high degree of vengefulness brings high material payoff because it deters free riding and no costly punishment is necessary. However, in the opposite case of "opaque disposition," in which Others can not directly tell whether Self is quite vengeful or not at all vengeful, then the second-order free rider problem is fatal: less vengeful types indeed drive out more vengeful types and cooperation fails.

In later sections we propose a solution based on group interactions. The second order free rider problem is greatly ameliorated when the costs of vengeful behavior are shared widely within a group—at small cost to each individual member, the group as a whole can impose large costs on a single culprit. The cost is even lower (though it never quite disappears) when such group sanctions deter most selfish behavior, so that vengeful episodes are rare. But group dynamics are complex and will require new considerations.

In this section we have seen that social preferences do seem to be part of the story of how humans cooperate, but there are deep problems in modeling their evolution. We must account for both friendly and hostile preferences, and for contingencies that flip them from hostile to friendly or the reverse. Interactions matter within (and perhaps across) groups, not just within pairs of individuals. Learning somehow plays an important role, and so does emotion. It seems complicated.

13.5 EARLY HUMAN NICHES

To help us keep our bearings while sorting things out, we will take a look at the world in which humans first evolved. Here we must deal with macroevolution—the major evolutionary transitions, such as those that produced multicellular organisms, or (more to the current point) the human race. Unlike microevolution—differential survival and fecundity from one generation to the next, as studied in earlier chapters—in macroevolution the main driving forces are contested by several schools of thought. The traditional **internalist** school emphasizes biological constraints and the time it takes to overcome them. For example, this school suggests that human brains take millions of years to evolve from chimp-like brains under the best of circumstances.

We will take cues from two alternative approaches. The **externalist** school emphasizes the environmental crucible. For example, human-like brains are not adaptive in most situations, but can evolve relatively quickly when they become advantageous. The **co-evolutionist** school notes that traits are adaptive (or not) only in conjunction with other traits and other evolving species. For example, human-like brains are adaptive only in conjunction with certain sorts of social behavior in certain sorts of niches.

Surely all three schools have valid points, but our account of human cooperation will de-emphasize internalist perspectives, which go back to Victorian thinkers. That is, we won't claim that creatures much like us were inevitable given enough time. Instead, we will argue that a distinctively human form of cooperation evolved in the extraordinary environment of Africa during the Pleistocene era, 50,000 to 2.5 million years ago.

How unusual was that environment? Normally, different ecosystems lie in stripes. For example, in northernmost Africa today, a narrow, fertile coastal stripe supports a rich Mediterranean suite of plants and animals. Just to the south lies a broad stripe of desert, the Sahara. South of that is the Sahel, more suitable for farming and ranching but prone to drought. Further south, we see a savanna stripe—grasslands with some trees. Beyond that are stripes (or large patches) of forest and more savanna. Stripes can move when the climate changes slowly (e.g., the Sahel/Sahara boundary has advanced and retreated over the past thousand years).

In most previous eras, weather didn't change greatly from one decade or century to the next. Figure 13.2 shows that the Pleistocene was different. The temperature chart for the last 2 or 3 million years looks a bit like a seismograph when an earthquake rumbles through. Our ancestors had to deal with increasingly violent temperature fluctuations around a general cooling trend.

This crazy weather scrambled Africa's ecosystem stripes, and our ancestors lived in an ecological kaleidoscope. When temperatures rose, the monsoons and trade winds of Africa brought heavier rainfall. Lake Chad swelled into Mega-Lake Chad, the middle Congo River overflowed its banks and became a vast inland sea, and smaller lakes speckled the central Sahara. When temperatures fell, drought swept Africa. Often, a broad swath of desert stretched from the Sahara along the Rift Valley down to the Namib and Kalahari, dividing and isolating patches of savanna. The many layers of volcanic ash in east Africa's Rift Valley tell of zones

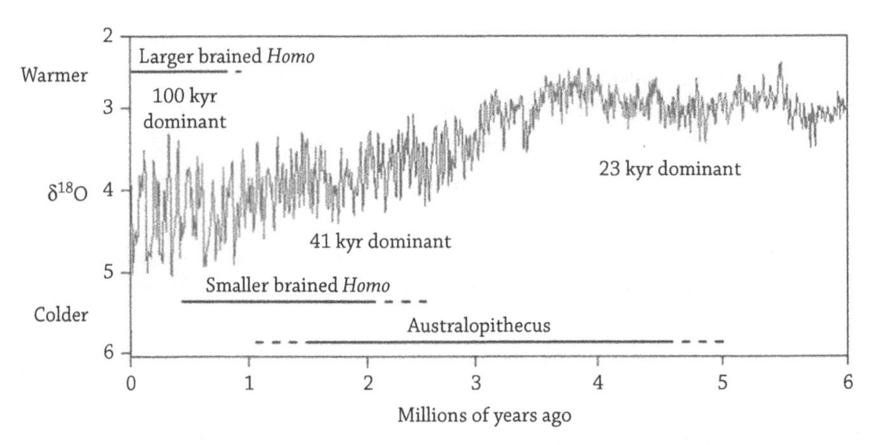

Figure 13.2 Temperature cycles in the Pleistocene era and earlier. Vertical axis is a composite record the temperature, proxied by the concentration of an oxygen isotope found in marine core samples. Horizontal axis is millions of years before the present, so the more distant past is to the right. Chart compiled from original sources by Richerson (2012), Figure 3.

of destruction and occasional winters lasting years. Plant and animal species that, in more settled times, would live in the separate biomes of desert, savanna and forest were all thrown together, and the extinction rate was quite high.

Genetic evolution wasn't fast enough to allow large animals to adapt in the Pleistocene kaleidoscope. When the environment is stable for thousands of generations, genetic evolution promotes specialists who make efficient use of niches. But the African Pleistocene favored generalists, who could do reasonably well in a broad range of environments. Even more, the kaleidoscope favored creatures who could adapt within a generation or two by learning new behavior and transmitting it via memes.

Big-brained birds (like ravens and parrots) and some mammals adapt via memes, transmitted by individual learning. Often the moms teach these memes to their kids. Primates (and elephants) are especially good at imitating parents and successful others.

Before discussing our primate ancestors, we should emphasize that relying on memetic transmission has its costs. First of all, it takes a relatively large brain to carry and transmit memes well. Mammal brains typically absorb up to 5% of adult base metabolism, reducing the resources available for digestion, armor, foot speed, and other uses. Meme-friendly creatures also seem to require longer juvenile periods, which puts a burden on parents. Delayed maturity is a life-cycle strategy that retards the rate of increase, and creates vulnerabilities. But it's worthwhile if niches come and go too quickly for genetic adaptation.

Beginning with *Homo erectus* about 2 million years ago, our primate ancestors went all-in for big brains (16% of base metabolism!), high protein diets to fuel brain development, long childhoods, and social learning. Prior to that, a key evolutionary event was the change from the ancestral polygynous mating system of the great apes, where large aggressive males vie for female harems, to one characterized by monogamy and biparental investment. This transition is evinced by the loss of traits adapted to aggressive polygyny—males much larger than females, and endowed with enlarged canine teeth. The fossil evidence indicates that *Ardipithecus ramidus* of 4.5 million years ago is the first hominid with only slight sexual dimorphism, a hallmark of monogamy (White et al. 2009).

Monogamy enables paternal investment, which in turn amplifies meme transfer. Both sexes could now participate in meme evolution, males perhaps in tool making and hunting and females in gathering and young child rearing. Biparental investment co-evolved with delayed maturity. Young chimps receive only about two years of care (from mothers only), but evidence suggests that our *Homo erectus* ancestors spent their first 8 years in their parents' care, rising to about 15 years in *Homo sapiens.* Extended biparental care allowed for an increase in the learning period and a more egalitarian division of instruction.

These lifestyle developments, and the emphasis on meme adaptation and transmission, arguably fueled the rapid evolution of the hominid neocortex. Costly brain development was supported by a more carnivorous diet, augmented by amazing technological innovations like fire and cooking. These expanded our ancestors' niches, allowing more calories to be extracted from tough plants and even from toxic tubers rendered digestible and harmless via cooking.

Meme transmission was enhanced by numerous genetic changes in neocortex development from early hominids right down to our recent ancestors; indeed Mekel-Bobrov et al. (2005) present evidence of such mutations that arose as little as 5800 years ago and soon spread across the globe. The rapid dispersal perhaps was due to the spread of agriculture with attendant gene mixing and flow, or perhaps due to skirmishes among neighboring proto villages and proto cities. The innovation of war likely fueled an even more rapid technological arms race. Our species has surely been shaped by a rapid co-evolution of memes and genetic traits that enhance meme transmission.

To summarize, humans evolved in a kaleidoscope world. That world gave an unusual advantage to generalist species, especially species which could fine-tune their behavior to respond to dangers and opportunities that appeared and disappeared within a few decades or centuries. Our ancestors adapted to that world by specializing in carrying and transmitting memes, which can evolve as rapidly as the kaleidoscope turns. Meme transmission co-evolved with a suite of physical traits including a large neocortex, and lifestyle traits like hunting and cooking, monogamy, biparental care, and prolonged childhood. We humans are pretty efficient distance walkers but pathetically slow, weak, and thin-skinned. We lack fangs, claws, and heavy-duty digestion. But our memetic capacity is unrivaled and, as elaborated in the next section, our ability to cooperate surpasses that of ants.

13.6 SOLUTION M: MORAL MEMES

Which memes make a difference? For us, the crucial memes are those that govern human social behavior. Taken together, they constitute what we shall refer to as the *moral system*. In this section we argue that early humans evolved the moral system as a third solution to social dilemmas. It relies on reciprocity and kinship yet it is distinct from them, as a vine is distinct from the trellis it climbs.

The moral system originally evolved in fairly small and stable groups of people, who knew each other intimately. Such groups have a shared understanding of proper behavior—a **moral code.** As diagrammed in Figure 13.3, the code prescribes certain sorts of behavior and prohibits others. These code elements

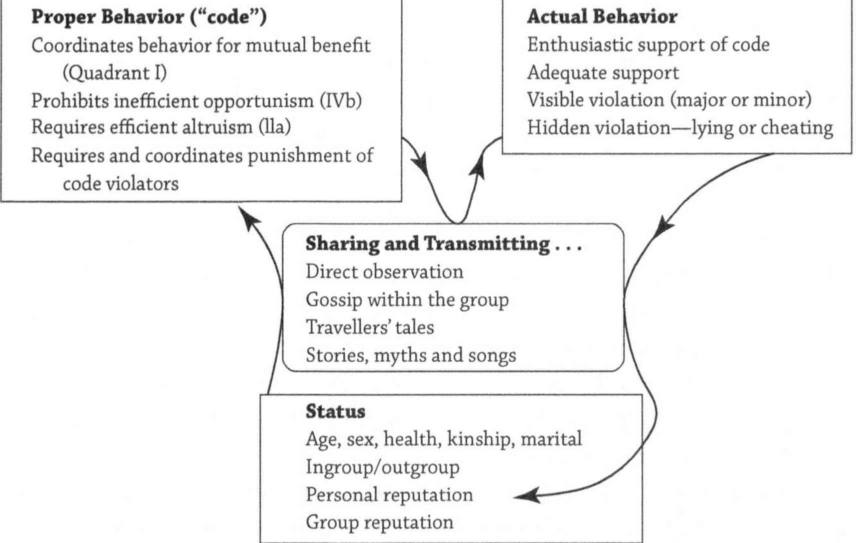

Figure 13.3 The Moral System.

(or norms, as they are sometimes called) often are contingent on age, sex, and status. For example, the moral code for a particular group of hunter/gatherers might prescribe that, among other things, (1) big game should be shared equally among all heads of households; (2) those who mildly transgress (e.g., taking too much occasionally) should be punished mainly via mild ridicule; (3) adult males should threaten any outsiders entering our valley and try to kill them if they stay overnight; and (4) insiders who fail to support that prescription should be severely punished via exile.

The code itself is a collection of memes, but the moral system as a whole relies on (and co-evolved with) a suite of genetic capacities. These include (a) the capacity to share understandings and information, via speech, memory, and cognition; this capacity enables the transmission of key memes via stories and song, and also information (often via gossip) on who is lying or cheating and who is reliable. (b) The capacity to keep track of individuals, using face recognition, memory, and cognition; this enables a code that depends delicately on status. (c) The capacity to detect intentions (to "mind read") from body language, gestures, chuckles and groans, and speech; this is a key to adjusting reputation, hence status, and to code enforcement. Most importantly, (d) the capacity for social emotions such as anger (to motivate enforcement of prescribed punishments), guilt (to deter cheating and maintain internal reputation), love (to motivate prescribed actions in quadrant II+ and to deter prohibited actions in IV–), and the joy of teamwork. The code must hook into these emotions to gain traction.

The moral system is a human universal, but contents of moral codes vary widely as they evolve rapidly in response to local conditions. Successful codes usually are those that best enable flexible cooperation in the local environment.

The Nuer and Dinka are a well-documented example (Kelly 1985). For many centuries the Dinka people maintained a pastoral lifestyle in the upper Nile basin. Beginning sometime in the early 1700s, a single mutation occurred in their moral system. One local group, which called itself the Nuer, slightly changed marriage customs in a way that created stronger bonds between the in-laws families. Although the Nuer were otherwise culturally (and genetically) identical to the Dinka, the intergroup bonds enabled the Nuer to form larger raiding parties than their neighbors. That gave them an unbeatable military edge and, decade by decade, the Nuer took territory away from their Dinka neighbors. Men either joined voluntarily, or were driven away or killed. Women willingly or unwillingly took new Nuer husbands and raised their children as Nuer. The Nuer were poised to take over the remaining Dinka territory when the British intervened around 1880. (In late 2013, as this passage was being drafted, the Nuer-Dinka rivalry resurfaced in news stories about political instability in South Sudan.)

To summarize, moral memes co-evolve with genes that supply the necessary capacities. The moral system arose among our Pleistocene ancestors who lived in small, fairly stable groups of individuals who knew each other intimately. It allowed them to cooperate, and solve the social dilemma in a far more flexible fashion than ever before.

13.7 ILLUSTRATIVE MODELS

To help crystallize ideas, we sketch a series of simple models, drawn mainly from Rabanal and Friedman (2015). We begin with the prisoner's dilemma game mentioned right after introducing Figure 13.1, and make it explicitly sequential. Self has already chosen to cooperate, and Other then decides whether to cooperate (C) or defect (D). Of course, in the absence of social preferences and a moral system, Other's best response is D, and cooperation unravels.

Now suppose that Self has the ability, at personal cost ch, to inflict a chosen amount of harm $h > 0$ on Other, where $c \in (0,1)$ can be thought of as a punishment technology parameter, and also suppose that Self may have a vengeful disposition. Specifically, suppose that in angry state triggered by Other choosing D, Self gets utility bonus of $v \ln h$ from inflicting harm h; in other emotional states, Self's utility is just the material payoff. The motivation (vengeance) parameter $v \geq 0$ captures an individual's temperament, and it evolves according to material payoffs only.

Thus, following D, Self maximizes $U = -1 - ch + v \ln h$ by choosing $h^* = v/c$. (To see this, take the first-order condition $0 = \frac{\partial U}{\partial h}$ and simplify.) Self's material payoff in this case is $W^S = 1 - ch^* = 1 - v$. The material (also utility) payoff here for Other becomes $2 - h^* = 2 - v/c$. If $v > c$, then this payoff is less than the cooperation payoff of 1, so D is no longer a best response for Other. The threat of vengeance thus rationalizes Other's cooperation and justifies Self's initial trust. The first-order free rider problem is solved!

But this happy conclusion assumed that Other can see that indeed $v > c$, that is, that Self has a transparent and quite vengeful disposition. Of course, it is in the interest of a non-vengeful ($v = 0$) Self to mimic the outward appearance of a vengeful type, but not incur the cost of actually inflicting harm. To see how this form of second-order free riding works, we assume opaque dispositions and obtain the payoff matrix

Self:	Other: C $[p]$	D $[1-p]$
$[q]$: $v > c$	1, 1	$-(1+v), 2-\frac{v}{c}$
$[1-q]$: $v = 0$	1, 1	$-1, 2$

The vengeful type of Self gets payoffs shown in first row. Facing a share p of cooperators and $(1-p)$ of defectors, the expected payoff is

$W_v^S = 2p - 1 - v(1-p)$. The unvengeful type of Self, with payoffs shown in second row, gets expected payoff $W_0^S = 2p - 1$. Hence for $p < 1$ the vengeful type is strictly dominated and $\Delta W^S = -v(1-p) < 0$. We conclude that the fraction q of vengeful types shrinks to zero.

Meanwhile, payoff to cooperative Other is $W_C^O = 1$, while defectors receive expected payoff $W_D^O = 2 - \frac{qv}{c}$. Hence $\Delta W^O = \frac{qv}{c} - 1$, which is negative once q drops below the threshold $\frac{c}{v} \in (0,1)$. It follows that ultimately p also shrinks to 0. Thus the state converges to $(p,q) = (0,0)$, whose basin of attraction includes the entire interior of the (p,q) square. We conclude that neither vengeance nor cooperation is evolutionarily viable in this model.

We now change the model so that vengeance is backed by a moral code. Code compliers gossip among themselves, usually learn the identity of any defector, and avoid interacting with them. But gossip is imperfect, as captured in the parameter $e \in (0,1)$: Self mistakenly avoids Others playing C with probability e, and with the same small probability encounters Other playing D. Using techniques from Chapter 5.2, we can summarize this gossip effect in adjustment matrix $A = \begin{pmatrix} 1-e & e \\ 1 & 1 \end{pmatrix}$.

The moral code enjoins code compliers (K) to share equally the cost v of punishing, to the utility-maximizing degree $h^* = \frac{v}{c}$, any defector they encounter. A chapter-end exercise invites you to verify that taking the the Hadamard product with the adjustment matrix and subtracting compliers' extra cost $\frac{ev(1-p)}{q}$ yields the payoff matrix

Self:	Other: C $[p]$	D $[1-p]$
$[q]$: K	$1-e-\frac{ev(1-p)}{q}, 1-e$	$-e-\frac{ev(1-p)}{q}, e(2-\frac{v}{c})$
$[1-q]$: N	$1,1$	$-1,2$

The expected payoff for non-compliers (N) coincides with that of unvengeful Self in the previous model, $W_N^S = 2p - 1$. Cost sharing boosts K's expected payoff when D is relatively rare and K is relatively common, and reducing encounters with D always helps. Thus $W_K^S = p - e(1 + \frac{v(1-p)}{q})$ is not dominated by W_N^S; the payoff difference

$$\Delta W^S = 1 - p - e\left(1 + \frac{v(1-p)}{q}\right) \tag{13.3}$$

can be positive or negative.

The moral code also shifts the payoff difference for Other. Instead of $q\frac{v}{c} - 1$, it is $\Delta W^0 = W_C^O - W_D^O = [1 - qe] - [qe(2 - \frac{v}{c}) + 2(1-q)]$. Simplifying slightly,

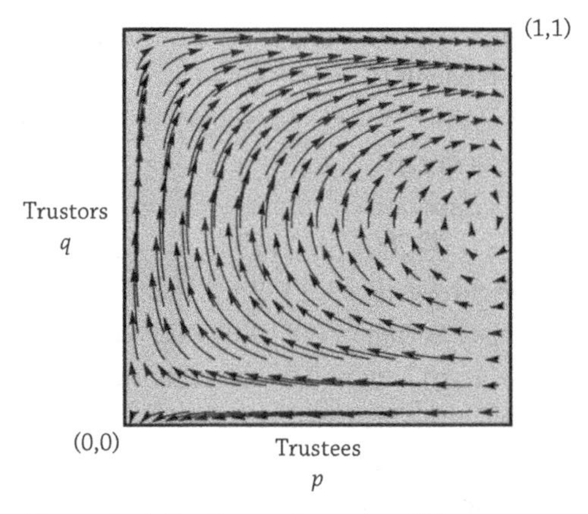

Figure 13.4 Replicator dynamics of the trust game with moral codes and parameters $e = .05$, $v = 2c = 1$; Self is referred to as Trusters and Other as Trustees.

we have

$$\Delta W^0 = q(2 + e(\frac{v}{c} - 3)) - 1. \qquad \textbf{(13.4)}$$

For reasonable parameter values, for example, $e = .05, c = 0.5, v = 2c = 1$, the loci $\Delta W^0 = 0$ and $\Delta W^S = 0$ intersect at an interior equilibrium point, approximately $(p^*, q^*) = (.945, .513)$ for the given parameter values. The trivial (zero cooperation, zero vengeance) equilibrium point $(p, q) = (0, 0)$ remains in this model as well.

Which equilibrium is more relevant? As always, we turn to dynamics to answer that question. A chapter end exercise invites you to work out replicator dynamics for this model, and to verify that the eigenvalues at $(p^*, q^*) = (.945, .513)$ are very close to pure imaginary, but with positive real part (approximately 0.001). The implication is that the interior equilibrium is locally unstable under replicator dynamics, and trajectories very near it tend to slowly spiral outwards, clockwise, as in Figure 13.4. The figure also shows that the trivial equilibrium has a rather small basin of attraction.

As usual, however, stability of a center equilibrium point depends on which monotone dynamics are imposed. The large asymptotic cycles under replicator dynamics seem implausible, but for group meme adaptation the inherent assumption of replicator dynamics, of proportional share adjustment, seems even less plausible. Some sort of noisy best response makes more sense, so we investigate logit dynamics. As you can verify in an exercise, the interior equilibrium point is now stable under reasonable parameter values, as shown in Figure 13.5.

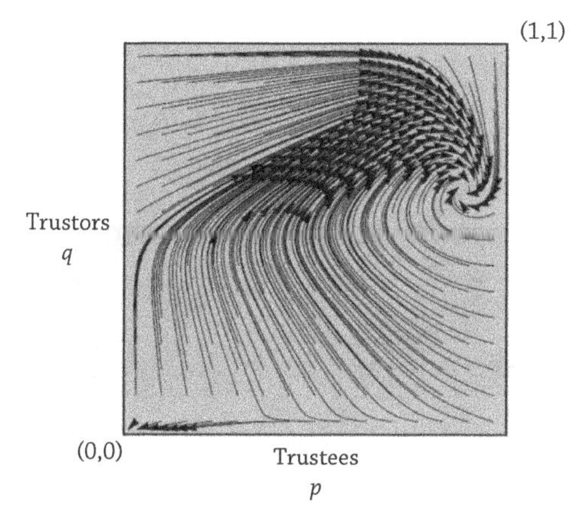

Figure 13.5 Logit dynamics with precision $\beta = 10$ for the same game.

This sequence of models illustrates the following points.

- A transparently vengeful disposition can rationalize efficient cooperation in a social dilemma, but its evolutionary viability is questionable. Cheap mimics have a cost advantage, and they render dispositions opaque. Once mimics become sufficiently common, efficient cooperation breaks down.
- A moral code can stabilize efficient cooperation, as code compliers share the fitness cost (and utility pleasure) of punishing miscreants. Code compliers must be able to recognize each other and exclude noncompliers. They share information, albeit imperfectly ($e > 0$).
- An interior equilibrium exists supporting a high degree of efficiency (p^* not far below 1) even though the share q^* of code compliers is far short of 100%.
- The model suggests cycles around equilibrium, where the share of compliers rises and falls and the fraction of defectors fluctuates. Under logit dynamics (arguably more plausible in this application than replicator dynamics), the cycle amplitude diminishes over time, and from a wide range of initial conditions the system converges to the interior equilibrium.

13.8 PREHISTORIC AND HISTORIC MORAL CODES

Although we can't directly observe our prehistoric ancestors' moral codes, we do have access to anthropologists' observations of human societies isolated from civilization, and we can make some inferences from DNA samples and

Figure 13.6 Two !Kung San tribesmen making fire. (From open-source Wikipedia article.)

from historical linguistics. Insights from behavioral ecology and evolutionary anthropology are also helpful.

Consider the !Kung San, the short-statured tribesmen of the Kalahari Desert. Their click language and genes point to a branching from other human lineages about 50,000 years ago. Their hunter-gatherer lifestyle persisted late into the 20th century and has been studied extensively (e.g., Lee 1979).

Like that of other known hunter-gatherers, the traditional moral code of the !Kung San was very egalitarian. They had essentially no leadership structure beyond the nuclear family. Most of the year they lived in nomadic groups of 2 to 6 nuclear families, and practiced a loose division of labor. Women gathered ground nuts and other edibles, men hunted game using poison arrows, and kids helped out.

Anthropologists sometimes called them "the gentle people," since their moral code involved relatively little violence. Code violators, such as hunters who didn't fully share the game they brought down, were dealt with by mainly by ridicule and snubs. Normally the worst punishment was to invite a culprit to leave, or to move the camp without telling him about the plan. However, violence was not unknown among the !Kung San. Extensive interviews by anthropologists document a few dozen homicides over a six-decade span, and in some cases the victims' relatives hunted down the murderers.

Every dry season, many groups of the !Kung San would gather at the best oases. Up to 200 people would swap stories and goods, initiate the youngsters, find marriage partners for the adolescents, and generally socialize. This group "fusion" phase would only last a few weeks because eventually the local game and plants would become depleted. Then frictions could easily arise, such as over who spotted the antelope first. This naturally led to "fission": small groups, usually no more than six nuclear families, would leave and resume nomadic wandering.

Most anthropologists believe that our Pleistocene ancestors had a somewhat similar hunter-gatherer lifestyle, with seasonal fusion and fission of groups speaking the same language and sharing essentially the same moral code. Of course, different specific codes would evolve in response to different local conditions, and not all elements of the code dealt with what we would consider moral issues. For example, the proper way to make arrows might be part of the shared understanding transmitted to the young by skilled older tribesmen, and older women would transmit their knowledge of the proper seasons for gathering particular root vegetables, and the proper way to cook them. Hunter-gatherer codes seamlessly cover what we regard as distinct realms, including technology, arts, religion, and politics.

We know precious few of the particular elements of our ancestors' codes over the hundreds of millennia before history began to be recorded. We do know, from their bones and the stone tools they left behind, that our ancestors evolved codes enabling flexible cooperation in niches of the African kaleidoscope. By 50,000 years ago, their evolving memes allowed our ancestors to spread out across Asia and Europe, adapting ways to cooperate on hunting and gathering to niches from mountains to the sea, from jungle to polar ice.

Beginning about 10 or 12 thousand years ago, the climate got a bit warmer and much steadier, as shown in Figure 13.7. Some of our ancestors exploited the stable new niches by becoming sedentary pastoralists or herdsmen. Both of these new lifestyles required major changes to moral codes, and minor changes in our genes.

Take herdsmen. Their codes describing proper behavior towards the herd animals evolved from stalking and killing them on sight, to keeping them nearby and driving off other predators, to actively managing them and their migrations. Their codes towards miscreants also evolved, and became much harsher. When

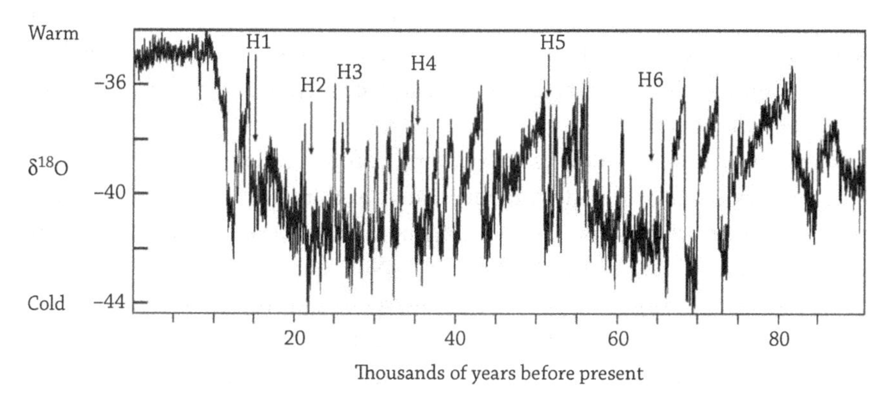

Figure 13.7 Temperature cycles in the Holocene and recent Pleistocene era. Vertical axis is a temperature proxy obtained from Greenland ice samples. Chart compiled from original sources by Richerson (2012), Figure 4.

your group's livelihood depends on deterring rustlers, you must maintain a fearsome reputation or your herd will soon disappear, and your group may not survive. Ideas of honor and valor begin to displace gentler ideas of just moving on when neighbors are troublesome. To a much lesser extent, their genes also evolved. For example, genes enabling adults to digest lactose made milk from the herd a more sustainable and valuable resource.

The transition to subsistence farming and settled villages required an even more drastic evolutionary transition. Agriculture requires techniques for storing harvests, and these require defense against raiders. Agriculture can support larger groups; indeed, populations in river valleys grew by a factor of 100 or more. However, egalitarian moral codes break down as the group size increases, and people don't know each other personally. Hierarchies are far more efficient at large scale in organizing defense and in maintaining irrigation systems and grain storage facilities.

Agriculture eventually leads to larger and more complex societies, with more complex moral codes. Hierarchical doctrines like the "divine right of kings" gain share. Genes evolve too, for example, for greater germ resistance, but these changes are dwarfed by changes in the moral code. Religion, politics, warcraft, law, arts and crafts, and commerce gradually develop separate spheres. Eventually life becomes civilized, and order is maintained by specialists (usually in warcraft, sometimes allied with specialists in religion or other spheres). The moral system, part of our human legacy, eventually becomes a trellis for law and politics. It is still needed, however, in each sphere. For example, political "legitimacy" means that people's moral sentiments are engaged. It was (and is) key to political success.

Friedman and McNeill (2013) argue that the human moral system experienced another major evolutionary transition over the past 200 years. After dominating the civilized world for about 5000 years, hierarchical empires have been displaced by a more dynamic and flexible market system. That transformation has required qualitative changes in the moral system—honor and loyalty memes have lost share to "bourgeois virtues" such as prudence, civility, patient investment, and acquisition of property.

Starting from small countries in Western Europe around 1800, the market system grew decade by decade and now spans the globe. Two hundred years ago most people were subsistence farmers, relying almost entirely on local resources and obedient to the local duke or emperor. Now most of us purchase all our necessities (and luxuries) on the market, and spend our workdays producing for the global market. This is cooperation on a scale that puts eusocial insects to shame—it commands the lives of billions of humans and much of our planet's resources. Friedman and McNeill argue that the challenge of our times is to evolve a suitable moral system, one that reconciles global market imperatives to a satisfying human society.

13.9 DISCUSSION

This chapter touches on many themes, but its main argument can be summarized briefly. Social creatures somehow must cope with the tension between self-interest and group interest. Two well-known solutions are kin selection and bilateral reciprocity. Human cooperation, however, relies on a third solution, which we refer to as the moral system. It builds on the previous two solutions, and is what makes us human.

The moral system co-evolved with our large brains, and special abilities such as face recognition and spoken language. Arguably the moral system was the driver. It was especially advantageous in the kaleidoscope world of Pleistocene Africa, where attractive niches and mortal dangers appeared and disappeared within decades. The moral code consists of shared understandings of proper behavior or norms, which evolve on a rapid timescale, and which allow groups of humans to cooperate flexibly in diverse and changing environments.

Our moral system enabled our ancestors to spread across the globe in Upper Paleolithic times. Since then it has undergone two major transformations – one at the dawn of civilization less than 10,000 years ago, and the other with the beginnings of the market system about 200 years ago. Still, the moral system remains at the core of our humanity.

A final thought. Scientific understanding of human cooperation continues to evolve rapidly, and our treatment is far from definitive. This chapter offers an idiosyncratic sample of recent thinking on deep questions, but new ideas are still bubbling up. We have no idea which particular models will dominate when a scientific consensus finally emerges, but we are confident that evolutionary game ideas, and perhaps even some of the material presented in this chapter, will play a central role.

EXERCISES

1. Recall the 2×2 payoff matrix in Section 13.7 for vengeance backed by a moral code. Verify that the pro-rata punishment cost is $\frac{ev(1-p)}{q}$ for all compliers, where v is the cost per punishment, e is the fraction of the $1-p$ share of defectors, and that the cost is spread over share q of code compliers. Use this and the Hadamard product with the given adjustment matrix A to confirm all entries of the payoff matrix.
2. Write out replicator dynamics for that model, using equations (13.3)–(13.4) and the usual binomial expressions. Confirm that, for parameter values $e = 0.05, v = 2c = 1$, there is an interior equilibrium point near $(p^*, q^*) = (.945, .513)$. Also compute the Jacobian matrix and show that the

characteristic function is approximately $0 = \lambda^2 - .003\lambda + .023$ with approximate roots (eigenvalues) $\lambda = .001 \pm .152\sqrt{-1}$.

3. Using your favorite alternative dynamics for meme transmission, assess the local stability of the interior equilibrium.

NOTES

The negative impact of a large number of prey on the predator's ability to target an individual prey is known as the *confusion effect* (Shaw 1978).

Social dilemmas

The four-quadrant diagram (Figure 13.1) and surrounding discussion assumes that payoffs (fitness or utility) are comparable across players. In particular, it assumes that Self's gains or losses can be added or averaged with Others' gains or losses, and so iso-efficiency lines have slope -1. Comparability of payoffs across players is not a problem when payoffs are biological fitness, and the logic can be extended to memetic fitness. Comparability is more controversial when payoffs are utility. In some standard economic applications (e.g., evolution of firms) the payoffs can be regarded as monetary, and hence comparable across players. In other applications, utility represents the subjective level of personal well-being, and economists often hesitate to compare it across individuals. In such cases, the utility functions for Self and Other use the the money metric (e.g., Varian 1992, p.108ff), which restores comparability for our purposes.

The Tragedy of the Commons game is nicely formalized as an infinite population game. Write $u(x_i, Ex) = x_i - h(Ex)$, where Ex is the population mean action and $h' > 1$. Here the personal harm is negligible since it is so widely dispersed (i.e., since $\frac{dEx}{dx_i} \approx 0$), but if as a group players increase their actions so that Ex increases, then everyone experiences increased harm h' that exceeds the average increased benefit of 1.

Kin selection

The basic algebra in Section 13.2 assumed that every individual $i = 1, ..., n$ received the same benefit b. If, as seems more realistic, they receive possibly different benefits b_i that sum to B, then Hamilton's rule still works if the average degree of relatedness is redefined as $r = B^{-1}\sum_{i=1}^{n} b_i r_i$ (i.e., as the benefit-weighted average). To deal also with spiteful behavior, where C and B can have their signs reversed, the biological literature (referring to fitness increments, not levels) writes $W^{inclusive} = W^{direct} + \sum_{i=1}^{n} W_i^{indirect} \times r_i$. Such expressions still assume that the externality is additive across individuals and linear, but generalized versions of r and W (or B and C) can accommodate Hamilton's rule to other situations, although the expressions become rather messy when traits are correlated.

In a provocative critique, Nowak et al. (2010) argue that inclusive fitness is not useful because it holds in its original form only in simple cases, and whenever it (in simple or messier form) is valid, it merely agrees with evolutionary dynamics defined using the standard definition of fitness. Many researchers responded to the critique, most of them defending inclusive fitness. The controversy clarified that, in our context, inclusive fitness $V = -C + \sum_{i=1}^{n} b_i r_i$ is the direct effect on Self plus (a) the expected indirect effect via others that share the gene. By contrast, ordinary fitness in our context is $W = -C + \sum_{j \neq Self} s_j b_j$, where the last term accounts for (b) the impact b_j on Self of Others' behavior. The main message of the controversy, it seems to us, is that equilibrium shortcuts, such as maximizing inclusive fitness (or other sorts of modified objective functions), are useful to the extent that they aid intuition, but their validity always rests on the fundamental construct, evolutionary dynamics.

Gardner et al. (2012) argue that haplodiploidy by itself does not really promote the evolution of eusociality. Although sisters are more closely related, females are less closely related to males, and this creates some evolutionary tensions. They argue that monogamy is more important in promoting eusociality in hymenoptera. Boomsma (2009) and others make a similar argument for monogamy (and against inbreeding) to explain eusociality in termites.

Reciprocal altruism

Game theorists in the 1960s greatly generalized the example in Section 13.3 of rational cooperation in the repeated prisoner's dilemma. They showed that essentially any payoff vector of the underlying game (including the most socially efficient but not including those that give any player less than she can achieve on her own) can be supported as a NE outcome of the repeated game, as long as the discount factor is sufficiently close to 1. This result is known as the Folk Theorem, because various versions of it occurred to many researchers about the same time, and it was widely known (and mentioned in textbooks) before definitive versions were written down, for example, by J. Friedman (1971) and Fudenberg and Maskin (1986). Note that the Folk Theorem applies only when the game repeats indefinitely into the future, and that efficient cooperation unravels when the end date is known.

Wilkinson's support for Trivers's theory has been questioned in recent work such as Hammerstein (2003). The underlying problem seems to be Wilkinson's small sample sizes, and the lack of other non-primate examples. In an October 2007 personal communication, Wilkinson writes "I don't think it is an accident that vampire bats have the largest brains for their body size of any other bat," and argues that these bats clearly have the capacities for reciprocity. He acknowledges that the data have limitations, but remains confident that they demonstrate true reciprocity. See Clutton-Brock (2009) for a recent review.

Social preferences

A large research literature on social preferences has appeared in recent years. One popular approach assumes that people care directly about the entire distribution of material payoffs across all players, not just their own material payoff (and not about how the payoff distribution was achieved). This approach is exemplified in the Fehr and Schmidt (1999) inequality aversion model, the Bolton and Ockenfels (2000) mean preferring model, and the Charness and Rabin (2002) social maximin model.

Other strands of literature model reciprocity, assuming that people care about how a payoff distribution was achieved. The psychological games approach captures reciprocity by postulating that my preferences regarding your payoff depend on my beliefs about your intentions. Building on the Geanakopolis et al. (1989) model, Rabin (1993) constructs reciprocity equilibria for two-player normal-form games, and Dufwenberg and Kirchsteiger (2004) and Falk and Fischbacher (2006) adapt the idea to extensive-form games. Levine (1998) provides a more standard game theoretic alternative by replacing beliefs about others' intentions with estimates of others' types. Cox, Friedman, and Sadiraj (2008) propose a general model where social preference components (like θ in Section 13.4) depend on Self's status and experience (positive or negative) with Other.

The Gold, Silver, and Brass rules are from Hirshleifer (1987), who uses the terms to refer specifically to the repeated prisoner's dilemma game strategies Always Cooperate, Tit for Tat, and Always Defect. Tit for Tat consists of choosing Cooperate in the first period and in subsequent periods imitating the opponent's action in the previous period; it is a more forgiving version of Grim Trigger.

Indirect evolution and vengeance

The terminology and perspective is due to Guth and Yaari (1992); precursors include Becker (1976), and Rubin and Paul (1979). Indirect evolution is featured in a number of recent papers such as Guth (1995), Dekel et al. (2007), Ely and Yilankaya (2001), Kockesen et al. (2000), Ok and Vega-Redondo (2001), Samuelson and Swinkels (2006), and Possajennikov (2002).

Spite, the desire to harm another person without regard to that person's behavior, should be distinguished from vengefulness (or, more briefly, vengeance), the desire to harm someone contingent on some sort of "bad" behavior by that someone. Besides articles already mentioned, the literature includes "altruistic punishments" articles by Fehr and Gächter (2002); Boyd et al. (2003); Dufwenberg et al. (2013). As emphasized in Henrich (2004), the analysis relies on structured populations and group selection.

Herold (2004) shows that positive as well as negative reciprocity (vengeance) preferences can survive in a "haystack" model, in which people interact in small groups that are occasionally remixed. This also ties to a large literature on group selection, mentioned in the Chapter 5 notes.

Friedman and Singh (2009) takes on the problem of how vengeance evolves in large anonymous groups. The analysis is static, and proposes the equal fitness principle, that (in a Bayesian Nash framework) in each population all types in the support of the distribution achieve equal and maximal expected fitness. It shows that this principle refines seven continuous families of perfect Bayesian equilibria down to two points: one is the usual inefficient equilibrium with no vengeance and no cooperation, and the other is a particular mix of vengefulness and nonvengefulness (and of trusting and avoiding interaction) that supports a highly efficient outcome of the Noisy Trust game. A follow-up paper, Rabanal and Friedman (2014) performs a dynamic analysis and shows that the more efficient equilibrium involves cycles around a center point. The specific model in Section 13.7 is analyzed in Rabanal and Friedman (2015).

BIBLIOGRAPHY

Bartz, Stephen H. "Evolution of eusociality in termites." *Proceedings of the National Academy of Sciences* 76, no. 111 (1979): 5764–5768.

Becker, Gary S. *The economic approach to human behavior*. University of Chicago Press, 1976.

Bolton, Gary E., and Axel Ockenfels. "ERC: A theory of equity, reciprocity, and competition." *American Economic Review* 90, no. 1 (2000): 166–193.

Boomsma, Jacobus J. "Lifetime monogamy and the evolution of eusociality." *Philosophical Transactions of the Royal Society B: Biological Sciences* 364, no. 1533 (2009): 3191–3207.

Boyd, Robert, Herbert Gintis, Samuel Bowles, and Peter J. Richerson. "The evolution of altruistic punishment." *Proceedings of the National Academy of Sciences* 100, no. 6 (2003): 3531–3535.

Charness, Gary, and Matthew Rabin. "Understanding social preferences with simple tests." *Quarterly Journal of Economics*, 117, no. 3 (2002): 817–869.

Clutton-Brock, Tim. "Cooperation between non-kin in animal societies." *Nature* 462, no. 7269 (2009): 51–57.

Cox, James C., Daniel Friedman, and Vjollca Sadiraj. "Revealed altruism." *Econometrica* 76, no. 1 (2008): 31–69.

Darwin Charles, R. "On the origin of species by means of natural selection, or the preservation of favoured races in the struggle for life" (1859).

Dawkins, Richard. *The selfish gene*. Oxford University Press, 1976.

Dekel, Eddie, Jeffrey C. Ely, and Okan Yilankaya. "Evolution of preferences." *The Review of Economic Studies* 74, no. 3 (2007): 685–704.
Dufwenberg, Martin, Alec Smith, and Matt Van Essen. "Hold-up: With a vengeance." *Economic Inquiry* 51, no. 1 (2013): 896–908.
Dufwenberg, Martin, and Georg Kirchsteiger. "A theory of sequential reciprocity." *Games and Economic Behavior* 47, no. 2 (2004): 268–298.
Ely, Jeffrey C., and Okan Yilankaya. "Nash equilibrium and the evolution of preferences." *Journal of Economic Theory* 97, no. 2 (2001): 255–272.
Falk, Armin, and Urs Fischbacher. "A theory of reciprocity." *Games and Economic Behavior* 54, no. 2 (2006): 293–315.
Fehr, Ernst, and Klaus M. Schmidt. "A theory of fairness, competition, and cooperation." *Quarterly Journal of Economics* (1999): 817–868.
Fehr, Ernst, and Simon Gächter. "Altruistic punishment in humans." *Nature* 415, no. 6868 (2002): 137–140.
Friedman, Daniel, and Daniel McNeill. *Morals and markets: The dangerous balance.* Palgrave Macmillan, 2013.
Friedman, Daniel, and Nirvikar Singh. "Equilibrium vengeance." *Games and Economic Behavior* 66, no. 2 (2009): 813–829.
Friedman, James W. "A non-cooperative equilibrium for supergames." *The Review of Economic Studies* 38, no. 1 (1971): 1–12.
Fudenberg, Drew, and Eric S. Maskin. "Folk theorem for repeated games with discounting or with incomplete information." *Econometrica* 54, no. 3 (1986): 533–554.
Gardner, Andy, Joao Alpedrinha, and Stuart A. West. "Haplodiploidy and the evolution of eusociality: Split sex ratios." *The American Naturalist* 179, no. 2 (2012): 240–256.
Geanakoplos, John, David Pearce, and Ennio Stacchetti. "Psychological games and sequential rationality." *Games and Economic Behavior* 1, no. 1 (1989): 60–79.
Grafen, Alan. "Natural selection, kin selection and group selection." In J. Krebs, and N. Davies (eds), *Behavioural Ecology: An evolutionary approach,* pp. 62–84. Blackwell, 1984.
Güth, Werner. "An evolutionary approach to explaining cooperative behavior by reciprocal incentives." *International Journal of Game Theory* 24, no. 4 (1995): 323–344.
Güth, Werner, and Menahem Yaari. "An evolutionary approach to explain reciprocal behavior in a simple strategic game." In U. Witt (ed.), *Explaining process and change: Approaches to evolutionary economics,* pp. 23–34. Ann Arbor, 1992.
Haldane, John BS. "Population genetics." *New Biology* 18 (1955): 34–51.
Hamilton, William D. "The evolution of altruistic behavior." *The American Naturalist* 97, no. 896 (1963): 354–356.
Hamilton, William D. "The genetical evolution of social behaviour." *Journal of Theoretical Biology* 7, no. 1 (1964): 1–16.
Hammerstein, Peter, ed. *Genetic and cultural evolution of cooperation.* MIT Press, 2003.
Henrich, Joseph. "Cultural group selection, coevolutionary processes and large-scale cooperation." *Journal of Economic Behavior and Organization* 53, no. 1 (2004): 3–35.
Herold, Florian. "Carrot or stick? The evolution of reciprocal preferences in a haystack model." *American Economic Review* 102, no. 2 (2012): 914–940.
Hirshleifer, Jack. *On the emotions as guarantors of threats and promises.* MIT Press, 1987.

Kelly, Raymond C. *The Nuer conquest: The structure and development of an expansionist system*. University of Michigan Press, 1985.

Kockesen, Levent, Efe A. Ok, and Rajiv Sethi. "The strategic advantage of negatively interdependent preferences." *Journal of Economic Theory* 92, no. 2 (2000): 274–299.

Lee, Richard B. *The !Kung San: Men, women, and work in a foraging society*. Cambridge University Press, 1979.

Levine, David K. "Modeling altruism and spitefulness in experiments." *Review of Economic Dynamics* 1, no. 3 (1998): 593–622.

Mekel-Bobrov, Nitzan, Sandra L. Gilbert, Patrick D. Evans, Eric J. Vallender, Jeffrey R. Anderson, Richard R. Hudson, Sarah A. Tishkoff, and Bruce T. Lahn. "Ongoing adaptive evolution of ASPM, a brain size determinant in *Homo sapiens*." *Science* 309, no. 5741 (2005): 1720–1722.

Nowak, Martin A., Corina E. Tarnita, and Edward O. Wilson. "The evolution of eusociality." *Nature* 466, no. 7310 (2010): 1057–1062.

Ok, Efe A., and Fernando Vega-Redondo. "On the evolution of individualistic preferences: An incomplete information scenario." *Journal of Economic Theory* 97, no. 2 (2001): 231–254.

Possajennikov, Alex. "Cooperative prisoners and aggressive chickens: Evolution of strategies and preferences in 2×2 games." No. 02-04. Sonderforschungsbereich 504, Universität Mannheim and Sonderforschungsbereich 504, University of Mannheim, 2002.

Rabanal, Jean Paul, and Daniel Friedman. "Incomplete information, dynamic stability and the evolution of preferences: Two examples." *Dynamic Games and Applications* 4, no. 4 (2014): 448–467.

Rabanal, Jean Paul, and Daniel Friedman. "How moral codes evolve in a trust game." *Games* 6 (2015): 150–160.

Rabin, Matthew. "Incorporating fairness into game theory and economics." *The American Economic Review* 83, no. 5 (1993): 1281–1302.

Richerson, Peter J, and Robert Boyd. "Rethinking paleoanthropology: A world queerer than we supposed." in Hatfield, Gary, and Holly Pittman, (eds), *Evolution of Mind, Brain, and Culture*, Vol. 5. University of Pennsylvania Press, 2013.

Rilling, James K., David A. Gutman, Thorsten R. Zeh, Giuseppe Pagnoni, Gregory S. Berns, and Clinton D. Kilts. "A neural basis for social cooperation." *Neuron* 35, no. 2 (2002): 395–405.

Robson, Arthur, and Larry Samuelson. "The evolution of decision and experienced utilities." *Theoretical Economics* 6, no. 3 (2011): 311–339.

Rubin, Paul H., and Chris W. Paul. "An evolutionary model of taste for risk." *Economic Inquiry* 17, no. 4 (1979): 585–96.

Samuelson, Larry. "Introduction to the evolution of preferences." *Journal of Economic Theory* 97, no. 2 (2001): 225–230.

Samuelson, Larry, and Jeroen M. Swinkels. "Information, evolution and utility." *Theoretical Economics* 1, no. 1 (2006): 119–142.

Shaw, Evelyn. "Schooling fishes: The school, a truly egalitarian form of organization in which all members of the group are alike in influence, offers substantial benefits to its participants." *American Scientist* (1978): 166–175.

Smith, Eric Alden. "Inuit foraging groups: Some simple models incorporating conflicts of interest, relatedness, and central-place sharing." *Ethology and Sociobiology* 6, no. 1 (1985): 27–47.

Trivers, Robert L. "The evolution of reciprocal altruism." *Quarterly Review of Biology* (1971): 35–57.

Varian, Hal R. *Microeconomic analysis*. Norton, 1992.

White, Tim D., Berhane Asfaw, Yonas Beyene, Yohannes Haile-Selassie, C. Owen Lovejoy, Gen Suwa, and Giday WoldeGabriel. "*Ardipithecus ramidus* and the paleobiology of early hominids." *Science* 326, no. 5949 (2009): 64–86.

Wilkinson, Gerald S. "Reciprocal food sharing in the vampire bat." *Nature* 308, no. 5955 (1984): 181–184.

14

Speciation

Consequently, I cannot doubt that in the course of many thousands of generations, the most distinct varieties of any one species [...] would always have the best chance of succeeding and of increasing in numbers, and thus of supplanting the less distinct varieties; and varieties, when rendered very distinct from each other, take the rank of species. (Darwin 1859, p. 155)

Darwin's great work was motivated by a simple but deep question: how do new species begin? He argued that a common ancestor gave rise to the bewildering variety of life forms, plant, animal, and microbial—but how, exactly? Darwin gave some partial answers, and much has been learned since then, but we still lack a full account of how new biological species come into existence.

In this chapter, we'll present some recent insights based on evolutionary games. These are not confined to the biological world, but also tell us something about social and virtual worlds. How did humans acquire so many distinct languages and cultures? More prosaically, why do modern economies have such a huge variety of products, services, and industries? Can the biological models suggest models of product branching and hybridization? Take computer operating systems, for example. When desktop computers became feasible, three separate operating systems emerged and dominated the industry: Windows, Mac OS, and Unix. Yet the latter two fused in OS X in 1999, and since then all operating systems continue to co-evolve with the hardware they run on (now especially mobile devices) and with the applications that run on them.

The chapter is rooted in evolutionary biology. Darwin's main answer, roughly speaking, was that species emerge very gradually, as geographic isolation allows groups to follow separate adaptive trajectories (or just separate drifts) until interbreeding is no longer viable. To use modern jargon, *allopatry* is the established dogma, as outlined by Ernst Mayr (1985): geographically subdivided populations build up genetic differences.

Darwin also hinted at a more endogenous or ecological process of speciation but was not able to pursue the idea. Some game theoretic treatments in recent decades (see notes for more specifics) argue for *sympatry,* in which multiple species diverge from a common ancestor while remaining in geographic proximity. Two processes are required for sympatric speciation: (a) disruptive frequency-dependent selection generates distinct phenotypes and (b) assortative mating mechanisms create reproductive isolation and swamp ongoing hybridization.

After a preliminary discussion of long-run versus short-run evolution, we present a leading version of process (a) called adaptive dynamics. Next we will build on ideas presented in earlier chapters to suggest a new alternative account in which mate choice favoring currently rare morphs is pitted against favoring own morph to maintain coadapted social genes.

Of course, both approaches are incomplete, in that they do not specify process (b). A later section begins to model that separation process, using cellular automata. To round out the chapter, the final sections discuss how these ideas can be applied to cultural evolution, and new considerations that arise in the social and virtual worlds.

14.1 LONG-RUN EVOLUTION

Most models discussed so far in this book assume a finite list of prespecified alternative strategies, for example, in the iconic one-population game, the three strategies R, P, and S are specified at the beginning. The analysis is short run, or perhaps medium run: it focuses on how the shares of those prespecified strategies might adjust and eventually settle down.

A different and longer-run perspective will help us think about speciation. The emergence and stabilization of a new species should allow for the possibility of *new* strategies, not present at the outset, yet somehow related to the original strategies. That is, we may need to deal with a *range of mutations,* which might individually be quite small but cumulatively could be substantial. Examples are everywhere. In biology examples include foot speed in terrestrial animals, and beak size or migration dates of birds. In human social institutions they include road speed limits, the age of adult privileges like voting or drinking, and college admission requirements.

Thus we should consider spaces of potential strategies that include entire intervals such as $[0,1]$ or $(0,\infty)$, rather than just a handful of specific alternative strategies such as $\{R,P,S\}$.

From the long run evolutionary perspective, we should focus on the ultimate response to new mutant strategies: which can invade, and what is their cumulative effect? Details of short-to-medium run share adjustment among a given set of strategies matter only to the extent that they change answers to those questions.

This distinction between short and long run evolution, and the idea of continuous strategy spaces, go back at least to Sewell Wright (1932). He introduced the metaphor that the graph of a fitness function defined over a continuous space can be thought of as a landscape, with peaks that represent local fitness maxima separated by valleys of lower fitness. From our perspective, the main shortcoming is that Wright's landscape is static. Strategic interdependence implies that an individual's fitness, and thus the fitness landscape, moves and reshapes itself as the population trait distribution changes over time. The chapter notes point the interested reader to recent attempts to make the landscape metaphor more dynamic and mathematically precise.

14.2 ADAPTIVE DYNAMICS

A long-run approach to frequency dependent selection, dubbed adaptive dynamics, gained prominence in the 1990s. Like much of this book, it suppresses the complications of diploid genetics, and just models the evolution of phenotypes, but in a continuous trait (or pure strategy) space such as $A = (0, \infty)$.

To avoid messy mixes of short-run and long-run analysis, adaptive dynamics assumes that the relevant population is mostly monomorphic: at almost all times t there is a single ("resident") strategy $r(t) \in A$ used by 100% of the relevant population. The analysis focuses on whether an infinitesimal fraction of a particular nearby mutant strategy, $m \in A$, can gain share under replicator dynamics. If so, basic adaptive dynamics assumes that it soon becomes the new resident. This assumption greatly simplifies the analysis of how, on a longer time scale, the resident strategy evolves as successive new mutants take over from previous residents. For example, suppose that increased foot speed increases fitness for a species of antelope over some range $[0, \hat{r}]$, but beyond $\hat{r}$ the metabolic cost (e.g., special dietary requirements) lowers overall fitness. Here adaptive dynamics would model how the antelope's foot speed approaches $\hat{r}$ in (long run) evolutionary time, and would analyze the dynamic stability of $\hat{r}$.

After writing out the basic mathematics, culminating in the so-called canonical equation, we'll show how adaptive dynamics explains evolutionary branching, yielding two proto-species from a single ancestral species. The math assumes a twice continuously differentiable fitness function $f(m,r)$ for mutant trait $m \in A$ when the share of the resident trait $r \in A$ is 1.0. It is natural to normalize so that $f(r,r) = 0$ for all $r \in A$; that is, fitness is measured as the long-run expected growth rate of the trait's share, which is of course zero when the share remains at 1.0.

The analysis focuses on slight mutants (i.e., those for which $|m-r|$ is small). In this case, Taylor's theorem tells us that

$$f(m,r) \approx f(r,r) + (m-r)\frac{\partial f(m,r)}{\partial m}|_{m=r} = 0 + (m-r)f_1(r,r) = (m-r)g(r), \tag{14.1}$$

where $g(r) \equiv f_1(r,r) \equiv \frac{\partial f(m,r)}{\partial m}|_{m=r}$ is called the selection gradient. If $g(r)$ is positive, then mutants m with trait value slightly higher than the resident can invade, and if the selection gradient is negative, then mutants with slightly lower trait values can invade since, in either case, $f(m,r) \approx (m-r)g(r) > 0 = f(r,r)$. Long run interior steady states, where neither sort of nearby mutant can invade, can only occur at critical points of the selection gradient (i.e., at roots r^* of g).

The standard second derivative test of an interior critical point r^* says that it is dynamically stable if the second derivative is negative at that point, so that it is a local fitness maximum. Biologists refer to such critical points as evolutionarily stable (or, stretching the term a bit, as ESS), although a more consistent term (referring to short-run dynamics) would be ecologically stable. In any case, we shall say that the critical point r^* is ES if $0 > f_{11}(r^*,r^*) \equiv \frac{\partial^2 f(m,r)}{\partial m^2}|_{m=r=r^*}$. The idea is that an ES critical point is dynamically stable in the short-run sense that it is immune to invasion from nearby mutants because it is a local fitness maximum.

Figure 14.1 illustrates. The critical point r^* is at the intersection of the diagonal (the locus $m=r$) with the locus $g=0$. Panel A is for an ES, where $0 > f_{11}$. It follows that $g = f_1(m,r) < 0$ at points above the locus $g = f_1(m,r) = 0$, and $g > 0$ at points below the locus. Thus mutants have positive fitness and can invade in the shaded regions where $(m-r)g(r) > 0$. The critical point r^* is immune to mutant invasion in the short run, as can be seen from the fact that all points on the vertical line through it (except r^* itself) have negative fitness.

Stability analysis takes a different twist when we consider the long-run adaptive process. Although immune to local invasion, there is no guarantee that long-run evolutionary dynamics would actually take us to an evolutionarily stable r^* if we started from a nearby monomorphism $r \neq r^*$. We follow the adaptive dynamics literature in defining these long-run dynamics via the so-called canonical equation

$$\dot{r} = kg(r). \tag{14.2}$$

That is, we assume that the resident population evolves according to the selection gradient at a rate $k > 0$ which, it can be argued, is proportional to the mutation rate, the variance of mutations and the population size.

Steady states for (14.2) are the critical points of the selection gradient g, and we know from earlier chapters that such points are dynamically stable at downcrossings. For historical reasons, biological theorists refer to this property

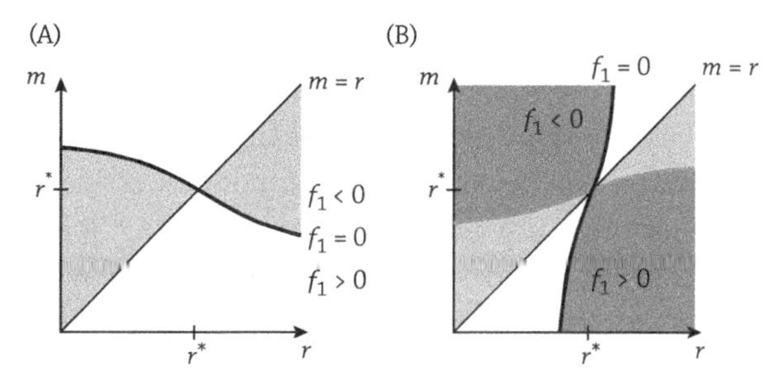

Figure 14.1 Pairwise Invasion Plots for two CS critical points r^*. In Panel A, mutations $m > r^*$ have positive fitness only when below the diagonal (and mutations $m < r^*$ have positive fitness only when above the diagonal) so mutations succeed only when they lie between r and r^*. It is also true in Panel B that a mutation is successful only when it is closer than r to r^*, although there it may lie on the opposite side of r^*. As explained in the text, r^* is also ES (hence CSS) in Panel A, while r^* in Panel B is not ES (and hence is a branch point).

as convergence stability. More formally, a monomorphism $r^* \in A$ is said to be *convergence stable* (CS) if it is a downcrossing of the selection gradient, that is, if

$$0 > g'(r^*) = \frac{df_1}{dr}|_{m=r=r^*} = \left[\frac{\partial^2 f(m,r)}{\partial m^2} + \frac{\partial^2 f(m,r)}{\partial r \partial m}\right]_{m=r=r^*}$$
$$\equiv f_{11}(r^*,r^*) + f_{12}(r^*,r^*). \quad \textbf{(14.3)}$$

Thus the CS condition is that the total derivative of the selection gradient is negative, while ES refers to a partial derivative. The condition implies that, starting at some r near the critical point r^*, small mutations can invade only if they are closer than r to r^*.

What is the relation between evolutionary stability and convergence stability? At first it might seem that one should imply the other, but not so. Chapter-end exercises 2 and 3 show that (ignoring nongeneric cases) an interior critical point r^* is CS iff $f_{11}(r^*,r^*) < f_{22}(r^*,r^*)$. We have already seen that it is ES iff $f_{11}(r^*,r^*) < 0$. The exercises show that you can graph the pure second partials of f at the critical point and see that, of the eight subquadrants, three are neither ES nor CS, one is CS but not ES, one is ES but not CS, and three are both ES and CS. Biologists after Eshel (1981) refer to the last case, illustrated in Figure 14.1A, as continuously stable strategy (CSS).

The most interesting case for us is a critical point r^* that is CS but not ES, i.e., where $0 < f_{11}(r^*,r^*) < f_{22}(r^*,r^*)$. The adaptive dynamics literature refers to such critical points, illustrated in Figure 14.1B, as *branch points*. It argues that the monomorphism assumed elsewhere is disrupted at branch points, and suggests that the population then separates into two distinct morphs, which then can become separate species. More specifically, mutations in the shaded region of Figure 14.1B satisfy $f(m,r) > 0$ so trait level m can invade resident level r, while in the darker shaded region (obtained by reflecting the $f_1 = 0$ locus across the diagonal) it is also true that $f(r,m) > 0$ so trait level r can invade when the resident level is m. Since r^* is a local minimum (not maximum) of the vertical line through it, the figure suggests that successive waves of mutations may split the resident level.

An algebraic example may help fix ideas. Consider the fitness function $f(m,r) = 2m^2 + 3r^2 - m^4 + r^4 - 5mr$ for trait values $r, s \in (-\infty, \infty)$. The function is already normalized so that $f(r,r) = 0$ for all r. The selection gradient is $g(r) = f_1(r,r) = \frac{\partial}{\partial m}[2m^2 + 3r^2 - m^4 + r^4 - 5mr]|_{m=r} = 4m - 4m^3 - 5r|_{m=r} = -r - 4r^3$. Clearly we have a unique critical point at $r^* = 0$, and it is a downcrossing and therefore convergence stable (CS). To confirm, note that

$$f_{11}(0,0) = 4 - 12m^2|_{m=r=0} = 4 > 0, \tag{14.4}$$

$$f_{12}(m,r) = \frac{\partial}{\partial r}[4m - 4m^3 - 5r] = -5, \tag{14.5}$$

$$f_{22}(0,0) = 6 + 12r^2|_{m=r=0} = 6 > 0, \tag{14.6}$$

$$g'(0) = f_{11}(0,0) + f_{12}(0,0) = 4 + (-5) = -1 < 0. \tag{14.7}$$

But inequality (14.4) shows that the critical monomorphism is evolutionarily unstable (not ES), and so is a branch point.

The adaptive dynamics literature typically lacks technical detail on exactly how a monomorphic population separates into two distinct subpopulations at a branch point, but Cressman and Hofbauer (2005) show how to apply replicator dynamics to the space of probability measures. They conclude that in the example above, an initial population distribution clustered near the critical point $r^* = 0$ (but with full support) will converge towards a bimodal distribution with all population mass at the two points $r = \pm 1$. (If the 4th degree terms were dropped, the population mass evidently would diverge to $\pm\infty$.)

To gain intuition, suppose that a remote island initially has a single species of seed-eating birds and several species of seed-bearing plants. The birds' beak size is well-adapted to eating mid-range seeds but not especially efficient for the largest and toughest seeds or the smallest. Now imagine that, due to climate change, the plants with mid-range seeds become scarce. Then, although the intermediate size beak remains CS, it is no longer ES. Selective pressure encourages subpopulations of large-beaked and of small-beaked birds.

As noted in the introduction, disruptive selective pressure is not sufficient to cause the birds to split into two species. Point (b) is that speciation also requires a process that somehow causes the gene pools to separate. Otherwise, interbreeding will reverse the phenotypic analysis just presented. Geographical isolation could do the job: in the story just told, we'd eventually have true speciation if one of the subpopulations but not the other gained a foothold in a neighboring island.

There are other processes that can separate gene pools that don't rely on special geographical situations. Behavior, especially mating behavior, can also do the job. This seems a good juncture, therefore, to take another look at assortative mating and sexual selection.

14.3 MORPH LOSS IN RPS

Earlier chapters have had much to say about male rock-paper-scissors games, but we will now see that female mate preferences are the key to understanding the stability of RPS and its ability to spin off new species. We build upon the effects of the assortative mating for RPS alleles first introduced in Chapter 7. To see the basic tensions, consider a true RPS game such as the version in Chapter 2 of the male side-blotched lizard game. We will see that there are selective advantages for females to prefer males with genes like her own, and there are also advantages to prefer males whose sons will have enhanced fitness. Following up on empirical observations by Corl et al. (2010), we will show that the first sort of preference can lead to loss of one or two morphs in a given population. With physical or other separation from other populations that continue to play the full RPS, that given population can become a new species. Conversely, in a species with only one or two morphs, decreasing such preference parameters, or increasing preference for males whose sons will have enhanced fitness, can help a new mutant invade and that ultimately can produce a new trimorphic RPS species.

Before introducing models to illustrate female preference-induced speciation, there is an important background issue to discuss. Suppose that a new morph appears (e.g., a yellow sneaker male) in a game with two existing morphs (e.g., orange and blue in a stable HD contest). Even if that new morph arises from a mutation at a single locus, the genome will continue to evolve. Other mutations at other loci, called "modifiers," will arise from time to time, and due to correlational selection they will stick around if they boost the fitness of the new morph (Sinervo et al. 2008). For example, yellow sneaker males benefit from genes that promote crypsis, while orange aggressive males benefit from genes that make their disposition more conspicuous.

These sort of modifiers promote female preference for males with genes like her own. Why? A female carrying modifiers beneficial to yellow that mates with orange or blue males would have progeny that, due to recombination, would be missing pieces of the co-adpated gene complexes of alleles that enhance yellow, and so the progeny (e.g., conspicuous yellow sons) would have lower fitness. By contrast, mating with her own type would preserve all the beneficial modifiers in her progeny, increasing their fitness. Indeed, she would benefit from a preference for well-integrated male morphs who carry all the useful modifiers. For example, the yellow female should prefer cryptic yellow males over conspicuous yellow males. Likewise orange females should prefer conspicuous orange males, a proposition supported by field and lab evidence (Bleay and Sinervo 2007; Lancaster et al. 2009, 2010). Similarly, there is evidence for blue females preferring well-integrated blue males (Sinervo et al. 2006). We will refer to such female preferences for well-integrated males of her own type as *self-preference* for coadapted genes.

A different type of female preference might be advantageous when males play an RPS cyclic game. Females increase their own fitness if they produce sons that best respond to the common morph (Alonzo and Sinervo 2001). For example, suppose that R is currently common (and is likely to remain common when offspring sired today mature). Then P sons will fare especially well, so females might evolve a conditional preference for P males, as well as a conditional preference for S males when P is common and for R males when S is common. For short, we will refer to this as a preference for *rare winner sons*.

Clearly female preferences for rare winner sons are in tension with self-preference. How should we model the evolution of such preferences? Most of this book uses simple replicator dynamics on phenotypes. We will do that here by simply adjusting relevant entries of the 3×3 fitness matrix to represent extra progeny for the male morphs that females prefer. But that is a shortcut whose validity is seldom explored, and a major goal of this section is to see how well the shortcut works for speciation via female preferences in the context of RPS male mating strategies.

A full evolutionary analysis of mating preferences requires the more complicated techniques introduced in Section 4.5. As in Chapter 4, we assume haploid sexual dynamics. Adapting those techniques to deal with a single trinary locus instead of two binary loci, the analysis proceeds as follows. Begin with the morph shares of the young in some initial period, and assume that the shares are the same in males and females (although of course the phenotypic expression differs). Then determine the unnormalized shares of potential moms and dads using payoff matrices representing selection, that is, survival to sexual maturity. Then combine matrix operations for attraction (i.e., for departures from random pairings due to mating preferences, in this case by females) with genetic transmission from the two parents. Finally, renormalize to obtain the morph frequencies in the next generation.

Genetic transmission is a key ingredient, and is represented here by a trimatrix since we are dealing with a single locus with three alternative genes, labeled R, P, and S (or 1, 2, 3 in matrix operations). For now we assume that each parent fully imparts its own morph type to half its progeny, as captured in the trimatrix $[Q^1, Q^2, Q^3] =$

		Males:	
Females:	**1. R**	**2. P**	**3. S**
1. R	$[1,0,0]$	$[\frac{1}{2},\frac{1}{2},0]$	$[\frac{1}{2},0,\frac{1}{2}]$
2. P	$[\frac{1}{2},\frac{1}{2},0]$	$[0,1,0]$	$[0,\frac{1}{2},\frac{1}{2}]$
3. S	$[\frac{1}{2},0,\frac{1}{2}]$	$[0,\frac{1}{2},\frac{1}{2}]$	$[0,0,1]$

For example, the upper-right entry $[Q^1_{13}, Q^2_{13}, Q^3_{13}] = [\frac{1}{2}, 0, \frac{1}{2}]$ indicates that a R-female mating a S-male will produce progeny of which 50% carry R and 50% carry S on the locus in question.

Female self-preference is captured in an attraction matrix A, which specifies departures from random mating. When all females prefer own-type with intensity $a > 1$, the attraction matrix is $A = \begin{pmatrix} a & 1 & 1 \\ 1 & a & 1 \\ 1 & 1 & a \end{pmatrix}$ for a given attraction parameter $a > 1$.

We now combine the ingredients to obtain the morph distribution $\mathbf{s}'$ of progeny, given the current morph distribution $\mathbf{s} = (s_1, s_2, s_3)$ for R, P, and S respectively, in the recruit population (live progeny of the previous generation). The first step is to apply replicator dynamics to $\mathbf{s}$, using the (survival only) payoff matrices M for males and F for females to obtain the shares of potential dads and moms. In this application, there is no differential survival rate for females carrying genes for the alternative male strategies, so F is the identity matrix, and the distribution of potential moms is simply $s^F = s$.

As further explained in the notes, the unnormalized shares of potential dads are the components of the vector $\mathbf{s}^M = (s_1\mathbf{e}^1 M\mathbf{s}, s_2\mathbf{e}^2 M\mathbf{s}, s_3\mathbf{e}^3 M\mathbf{s})$. The unnormalized fraction z_i of progeny of morph i is obtained from Q^i, the ith part of the genetic transmission trimatrix, adjusted for mate preferences A. That is, we treat the Hadamard product $H^i = A \circ Q^i$ (see Chapter 4 for a review) as a quadratic form, and set $z_i = \mathbf{s}^F \cdot H^i \mathbf{s}^M$. Finally, to get the desired progeny shares, we just normalize by dividing by $z_T = z_1 + z_2 + z_3$, so that $\mathbf{s}' = \mathbf{z}/z_T$.

For example, suppose that $\mathbf{s} = (.5, .3, .2)$, and let M be a true RPS matrix, for example $M = \begin{pmatrix} 1 & 0.4 & 1.4 \\ 1.5 & 1 & 0.6 \\ .5 & 1.6 & 1 \end{pmatrix}$. Let $A = \begin{pmatrix} 2 & 1 & 1 \\ 1 & 3 & 1 \\ 1 & 1 & 1 \end{pmatrix}$. Then shares of potential moms are still $\mathbf{s} = (.5, .3, .2)$. To get the unnormalized shares s^M of potential dads,

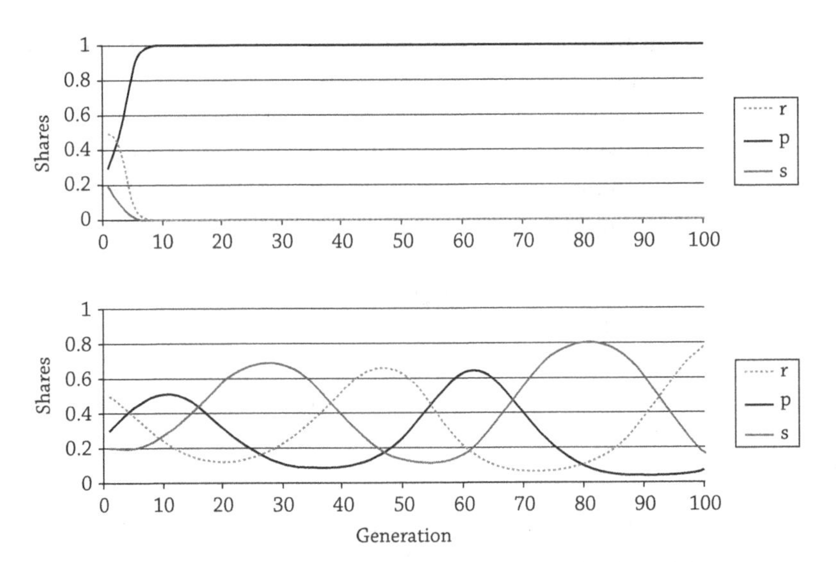

Figure 14.2 Upper panel shows share evolution under haploid sexual dynamics given an attraction (adjustment) matrix A with large positive diagonal entries, as in text. Lower panel shows dynamics for the same RPS game except that A matrix entries are all 1.0.

first note that $Ms = (0.9, 1.17, 0.93)^T$, so $s^M = (0.450, 0.351, 0.186)$. The quadratic form for $i = 1$ (or rock) is $H^1 = \begin{pmatrix} 2 & 0.5 & 0.5 \\ 0.5 & 0 & 0 \\ 0.5 & 0 & 0 \end{pmatrix}$, so $z_1 = \mathbf{s} \cdot H^1 s^M = 0.69675$. Similar calculations give $z_2 = \mathbf{s} \cdot H^2 s^M = 0.53415$ and $z_3 = \mathbf{s} \cdot H^3 s^M = 0.1917$. Thus $z_T = 1.4226$ and $\mathbf{s}' \approx (.490, .375, .135)$, which are the shares in the next generation. In general, above critical levels the large values along the diagonal of A destabilize the game dynamics and the system will lose morphs; see Figure 14.2.

How about the impact of preferences for rare winner sons? In the all-male shortcut, we could capture such preferences by a matrix $B = \begin{pmatrix} 1 & 1 & b \\ b & 1 & 1 \\ 1 & b & 1 \end{pmatrix}$ For example, when R is prevalent, females are relatively (at rate $b > 1$) attracted to P males, as this pairing is more likely to produce P sons, who will be advantaged to the extent that R remains prevalent when those sons mature. It is less obvious how to capture such preferences in a two parent model using a transmission multimatrix, but see the chapter-end exercises for related material. Figure 14.3, adapted from Alonzo and Sinervo (2001), shows the anticipated impact of b on game dynamics. Allowing females the choice of sires that maximize a son's fitness will accelerate the RPS dynamics as females bring up the invading strategy from low frequency to high frequency at a faster rate than mating at random among males and females.

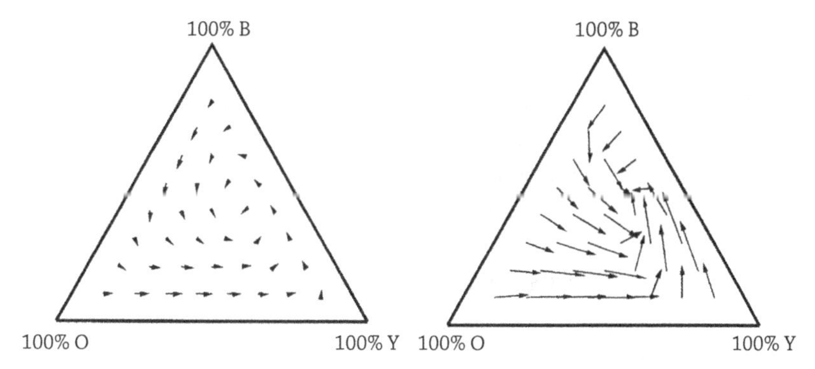

Figure 14.3 Phase portrait on the left shows haploid sexual dynamics schematically when b = 1. Portrait on the right is the same except that $b > 1$, indicating female preference for rare winner sons. (modified from Alonzo and Sinervo 2001).

The analysis so far takes the attraction matrices as given, and ignores the iterative updating required for *B* preferences. But where do the parameters $a, b > 1$ come from? As in Chapter 4, a more complete analysis would consider which values of those parameters can become fixed in a population where initially some (or most) females have no mating preferences $(a, b = 1)$. The evolution of given level of self-preference could be captured in 2-locus model with genotypes [Ra, RA, Pa, PA, Sa, SA], where the upper-case A refers to females with a particular value $a > 1$ of the self-preference parameter, while the lowercase are those with no (i.e., random) preference, $a = 1$.

To consider the simultaneous evolution of rare-winner-son preference as well as self-preference, we'd need a three-locus model with genotypes [Rab, RAb, RaB, RAB, Pab, PAb, PaB, PAB, Sab, SAb, SaB, SAB], with a similar convention for a particular value $b > 1$ of the relevant preference parameter. A simpler formulation of joint evolution assumes that alleles for self-preference segregate at the same locus as alleles for rare mating preference, and assume away alleles for no preference. In this two-locus model the genotypes are [RB, RA, PB, PA, SB, SA], thus pitting assortative self-preference directly against rare-winner-son preference.

In all these multi-locus models, recombination becomes an issue, and in the chapter-end exercises, we invite interested readers to work out the consequences. Of course, free recombination will limit speciation. The intuition here is that female preference loci hitchhike a fitness ride off of the RPS locus: all of their fitness is derived from the males' future prospects. If the loci are linked and the recombination rate r is very small, then the female choice (A, B) will be very close to the RPS locus and can hitch a ride to high fitness. Because all females have a preference for rare winners, the context-dependent strategy will be a winning strategy for all females and should invade and take over the population.

Like all existing literature of which we are aware, we do not really come to grips with a more fundamental issue: how do the attraction parameters a, b themselves evolve in $(0, \infty)$? One could try the adaptive dynamics approach, but that approach is silent at the point of speciation where the issue is especially acute. A simple approach is to assume the ancestral strategy is random choice at each of the a and b loci, but we will not develop this idea here.

Returning to the main line of argument, assume that a given population of a species engaged in an RPS mating game has particular values $a, b > 1$ of the female preference parameters at a given moment of evolutionary time. Over a longer time scale (decades or perhaps milennia) the values change, driven perhaps by environmental change or perhaps via internal dynamics. At some point, qualitative behavior changes, and we have a bifurcation. This typically involves morph loss, which in turn puts evolutionary pressure on modifiers of the remaining morphs. Genomes change, and absent recurrent interactions with other populations still playing full RPS, we can get speciation. Corl et al. (2012) provide data showing that the preference for self (described by Bleay et al. 2007) can be converted into a preference for one's own population in as little as 20,000 years of evolution, in a side-blotched lizard population that has lost the yellow allele, relative to the ancestral RPS mating system surrounding it, generating the conditions for speciation.

14.4 EMERGENT BOUNDARY LAYERS IN CELLULAR AUTOMATA

To illustrate the role of spatial structure in separating gene pools, we examine a cellular automaton model by Hochberg, Sinervo, and Brown (2003). While this model was developed for the four alternative social strategies of altruism, competition, mutualism and spite, the approach can also be applied to social morphs in general and the side-blotched lizard system in particular. The viscosity of the systems that favor altruistic behaviors can, in turn, promote assortative mating (by spatial proximity) between like phenotypes, with prezygotic reproductive isolation as a possible end product (Lande 1981; Kirkpatrick 2000).

Hochberg, Sinervo, and Brown (2003) simulated an $N \times N$ cellular automaton with Moore neighborhoods on the torus. Each cell is either empty or else occupied by a player characterized by three evolving haploid characters: a phenotypic signaling marker M (with alleles M1 and M2), and a tolerance T for social interactions with, and a mating preference P for, its eight immediate neighbors according to their markers.

During a generation the following sequence of events occurs. Each player has a constant probability of mortality p. Each surviving player then has two social interactions with each of the up to eight adjacent players: once as potential donor and once as a potential recipient. The player's fitness is the sum of total payoffs (positive or negative) received from or paid to each of its neighbors.

The donation choice is a simple function of the player's T and each neighbor's M: if a player has the T_{all} allele (Tolerate all), then it always performs its social act to each of its neighbors, whereas if it has the T_{self} (Tolerate self), $T_{nonself}$ (Tolerate nonself), or T_{none} (Tolerate none) allele, then it respectively donates to neighbors with the same marker only, the alternative tag only, or to no players at all.

Vacant cells are filled as follows. A neighboring player with the highest fitness chooses a preferred mate from the remaining neighboring cells. If that player has the P_{all} allele (promiscuous), then it always mates with a randomly chosen individual, whereas if it has the P_{self}, or $P_{nonself}$ allele, then it respectively restricts its random choice to neighboring cells with the same marker or the alternative marker. If there are no preferred mates available, then the cell remains vacant.

The model will generate assortative mating if the self preference allele, P_{self}, becomes coupled to the M locus alleles, M1 and M2. Speciation can then occur, as in Figure 14.4. The simulation tends towards groups of alternatively tagged players (M1 and M2), separated by a boundary layer. The probability of convergence to such a configuration is related to the relative strength of selection due to social acts, which can be considered donation levels in this current example. At low donation levels, defectors that tolerate no other players (T_{none}) dominate,

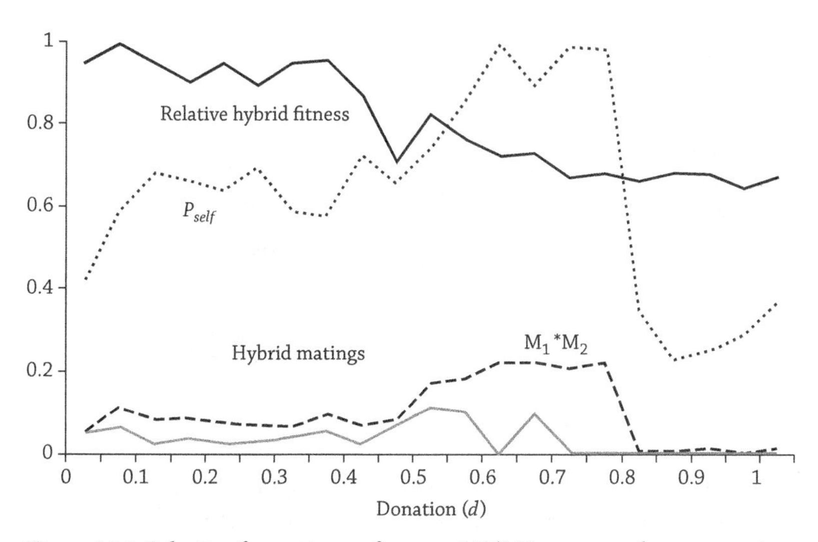

Figure 14.4 Selection for mating preferences. M1*M2 measures the evenness in the frequencies of the two tags and reflects the long-term stability of tag coexistence (from Hochberg et al. 2003). The P_{self} allele sweeps across the CA and drives the evolution of low hybrid fitness.

and there is little fitness difference between hybrid and pure-bred progeny, arising respectively from matings between parents with opposite markers and parents with the same markers. The two markers coexist but do not separate spatially or form species. For intermediate donations ($d = 0.4 - 0.6$), the T_{self} discriminating allele invades (due to benefits of donations to similarly tagged neighbors) and hybrid offspring have somewhat lower fitness. At species borders this, in turn, selects for the P_{self} allele, implying strict assortative mating by type in the new evolving species.

Once this spatial pattern of the T_{self} and M alleles begins to crystallize, hybrid unfitness at speciation border selects for the P_{self} allele at the mating preference locus, and this mating preference allele rapidly spreads to species centers due to assortative mating (Figure 14.5). Hochberg, Sinervo, and Brown (2003) also found that diverse social behaviors could generate the speciation effect, and these correspond to the strategies exhibited by RPS lizards (cooperation and competition).

Intriguingly, the positive/negative effect of selfish behavior is reminiscent of interspecific competition, and we suspect that this type of act could be at the heart of certain speciation events attributed to generic competition (Schluter 2000). The idea is that the two sub-populations evolve interactions that are neutral within each group, but negative across group boundaries. Individuals at the species boundary hybridize and produce inappropriate negative effects on self types and neutral effects on non-self, which has lower fitness than pure neutral

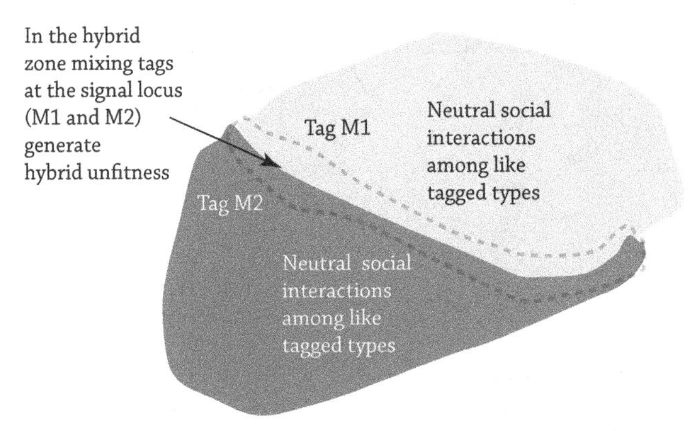

Figure 14.5 In socially mediated speciation, social acts evolve to be neutral in the center of the species range because all types share the same tag. However, at the border between species, where different tagged types might hybridize, there is constant strong selection against hybrids, and the P_{self} allele is fixed across the CA in both species, but for two alternative tags (M1 and M2).

to self and negative to non-self. At this point, the two subpopulations have gene pools that no longer mingle, and they become separate species.

14.5 SPECIATION IN SOCIAL AND VIRTUAL WORLDS

What does it mean to have a new social or virtual species? The biological definition—that crossbreeding is impossible or at least yields no viable offspring—doesn't seem helpful. In this section, we simply say that social or virtual speciation occurs if now there are multiple distinct entities where previously there was only one (or none). This section will not probe for deeper definitions. Instead, using a series of simple models and examples, it will explore processes that produce such multiplicity.

We begin with a simple mathematical model known as the supercritical pitchfork bifurcation. Let $y \in [-\infty, \infty]$ represent a cultural trait, and let $b \in [-a, a]$ represent an environmental parameter. Assume that adaptation of the trait is described by

$$\dot{y} = g(y) = by - y^3. \tag{14.8}$$

The gradient g vanishes at (i.e., its roots are) $y = 0, \pm\sqrt{b}$ if $b > 0$, while only $y = 0$ is a root if $b \leq 0$. The root $y = 0$ is stable iff $g'(0) = b < 0$, that is, it is stable when unique except at the bifurcation point $b = 0$, and unstable when $b > 0$. In that case the other roots $\pm\sqrt{b}$ exist and they are both stable, since $0 > g'(\pm\sqrt{b}) = b - 3b = -2b$ when $b > 0$. Thus we have the situation illustrated in Figure 14.6.

Here is a down-to-earth example. The trait value y is the location of the annual tribal gathering along a long river. Historical examples include late summer/early

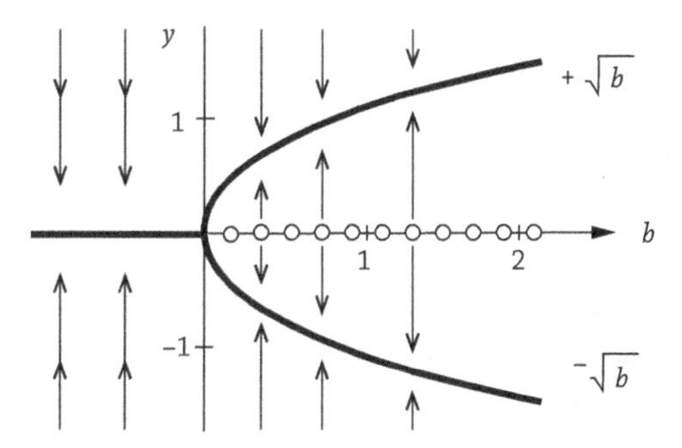

Figure 14.6 Supercritical pitchfork bifurcation. For each value of the bifurcation parameter b, the stable equilibria of equation (14.8) are shown as solid circles (forming a thick line as b varies) and the unstable equilibria as open circles. Arrows show the evolution of the trait value y.

fall gatherings of the Shoshone and fission-fusion tribes of North America (e.g., Murphy and Murphy 1986), and the herdsmen's fall gathering on the Oxus River featured in Michener's 1963 best-seller, *Caravans*. The bifurcation parameter b is the tribe's population relative to the meeting site's carrying capacity.

To get to the tidy canonical equation (14.8), we have to normalize the variables. For example, y might be the distance (measured, say, in horseback travel days) upstream from the traditional meeting site $y = 0$, and b might be [population minus carrying capacity] divided by carrying capacity.

The mathematics in previous paragraphs then has a simple interpretation. The unified tribal structure for negative b becomes less stable as b increases, perhaps because hungry tribesmen increasingly deplete the local game and vegetation. At the critical value $b = 0$, the tribe begins to split. As b increases further, the two branches of the tribe increasingly separate, a precondition for cultural speciation. The pitchfork model doesn't exactly pin down the time, but speciation occurs at the point when members of each branch regard members of the other, now distant, branch as "them," and no longer as "us."

Early human habitation of the Hawaiian Islands is a more complex but instructive example. Polynesian navigators and colonists first arrived as early as 300 AD, but (due to the hazards of the long voyage and lack of profitable exports) contact with the ancestral culture was tenuous. The colonists lived in small, egalitarian groups along the coast, gathering seafood supplemented by pigs, coconuts, taro, and other crops they had brought with them. Abundant resources allowed the population to double every century. As in the pitchfork model, villages branched and moved into vacant spots along the coast, occasionally jumping to new islands.

The islands' carrying capacity for that lifestyle is only about 15,000 people, and the population reached that level by 1400 AD. A more dramatic form of cultural speciation then began. It featured large inland settlements supported by labor-intensive agriculture, including irrigation systems and pond fields. Within a century or two, the homogeneous society began to separate into social classes. The ruling class, the ali'i, and the priestly and professional class, the kahuna, were on top, supported by the field managers, called konohiki. Of course, most people were commoners, subordinate to the kohohiki and their superiors.

Population growth resumed, and that led to territorial conflict among the ali'i. Local rulers with a larger population, supported by more efficient farming and organization, did better because they could afford more warriors and better gifts to cement alliances. The most successful ali'i expanded their domains, some of them eventually encompassing an entire island. Contact with the outside world, starting in 1778 with Captain James Cook and the crew of *HMS Resolution*, interrupted this internal dynamic.

The birth of a new cultural species distant from predecessors can be modeled by the fold bifurcation, also known as the saddle-node or blue sky bifurcation.

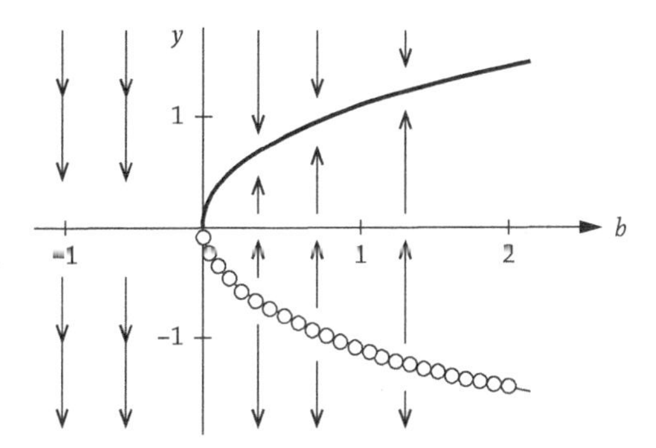

Figure 14.7 Fold bifurcation. For each value of the bifurcation parameter b, the stable equilibria of equation (14.8) are shown as solid circles (forming a thick line as b varies) and the unstable equilibria as open circles. Arrows show the evolution of the trait value y.

The normal form is

$$\dot{y} = g(y) = b - y^2. \tag{14.9}$$

The gradient g vanishes at $\pm\sqrt{b}$ iff $b \geq 0$. The root $y = +\sqrt{b} \geq 0$ is stable since $g'(+\sqrt{b}) = -2\sqrt{b} < 0$, while the other root is unstable since $g'(-\sqrt{b}) = -2(-\sqrt{b}) > 0$. For $b < 0$ the gradient is always negative, and so there is no equilibrium point at all; hence the stable equilibrium at $y = +\sqrt{b}$ appears "out of the blue" when the bifurcation parameter first becomes positive. (Of course, the trajectories slow down considerably near $y = 0$ when b is just below zero.) Thus we have the situation illustrated in Figure 14.7.

A possible Hawaiian interpretation is that y represents the intensivity of inland agriculture while b is again population in excess of carrying capacity, here of the coastal lifestyle. Although some inland forays doubtless occurred before 1300 or 1400, they didn't really take off until the more desirable coastal niches reached capacity. Thereafter, inland intensivity increased and absorbed an increasing population share.

A few remarks about boundary layers may be in order. The class boundaries that emerged in Hawaii 600 years ago were roughly similar to those that arrived more than 5000 years earlier in the great river valleys of the Mideast and Asia. Boundary layers also are key to the emergence of new languages; and conversely the disappearance of languages in the modern world can be attributed to its corrosion of traditional boundary layers. Consider for example two versions of the French language, français québécois in Canada, versus français in France. Arguably speciation was on its way until 100 years ago, but then better transportation and communication and political intervention stabilized both variants. Now français québécois is just a dialect or family of dialects.

Some sorts of social and virtual speciation are not well described by the preceding models. In particular, consider a trait space already inhabited by several distinct varieties when increasing resources (or technical improvements) allow additional varieties to appear and to thrive. The new varieties alter the niches of incumbent varieties, which must then adapt to restore equilibrium.

New commercial products are perhaps the most obvious illustrations of that sort of speciation process. More subtle illustrations include national identities emerging or re-emerging within large empires—recently Scottish nationalism, but the classic examples are all the nations that emerged from the Ottoman and Austro-Hungarian empires in the aftermath of World War I. An appealing illustration (at least to professors like us) is a new academic discipline. Computer science, genomics, and critical studies emerged as separate disciplines in recent decades, while in previous centuries natural philosophy speciated into physics, chemistry and biology, while moral philosophy spun off psychology, economics, and political science.

Although the bifurcation models already introduced can describe particular instances, for example, a pitchfork bifurcation could describe how chemistry acquired a distinct identity from physics, these local models offer limited insight into the overall process. To gain a more global perspective, we introduce the last model of this book, a variant of the Hotelling (1929) model of spatial competition.

For simplicity, consider a 1-dimensional trait space (of product characteristics or attributes) with no boundary. With little further loss of generality, the trait space is the unit circle $\Re$ mod 1, as in Figure 14.8. (The mod notation means that we conflate all real numbers that have the same residual after subtracting their integer part, for example, 3.02 and 17.02 are both plotted on the circle at 0.02.) In keeping with Hotelling's original interpretation, we refer to each existing species $i = 1, \ldots, n$ as a product produced by its own firm, all with the same fixed cost $k > 0$ and the same constant marginal cost $c \geq 0$. Firm i is located at $x_i \in [0,1)$ and chooses price $p_i \geq c$. Consumers, of total mass $M > 0$ uniformly distributed over the circle, each inelastically demand a single unit of the product. The consumer at location $x \in [0,1)$ purchases the variety with lowest delivered price $p(x) = \min_i[p_i + t|x - x_i|]$, where $t > 0$ is the marginal cost of transporting (or transforming) a unit of the product around the circle. Figure 14.9 illustrates, after cutting the circle at the top point and spreading it out to show locations on the horizontal axis and price on the vertical axis.

What prices will $n \geq 2$ firms choose, given equally spaced locations? A chapter-end exercise invites you to show that the symmetric strategy profile with

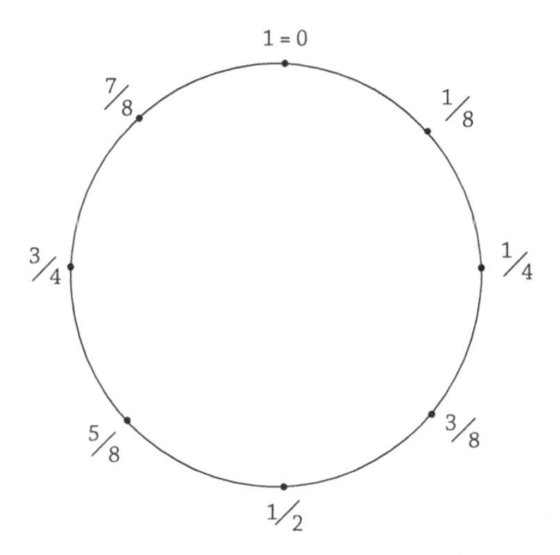

Figure 14.8 Hotelling location model. Equilibrium with eight varieties is shown.

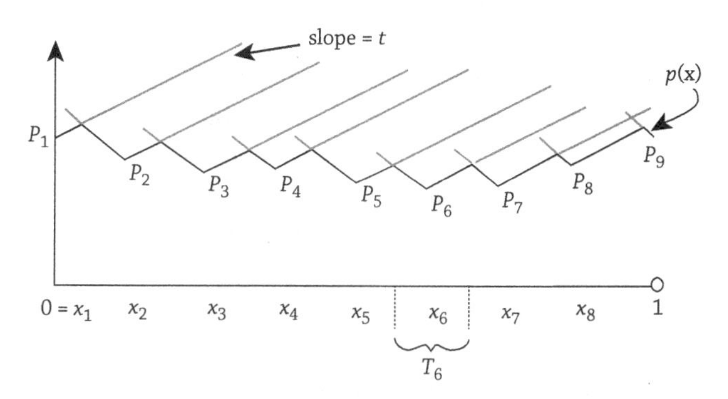

Figure 14.9 Pricing and territories in the Hotelling model. Delivered price is the dark zig-zag line, and the territory served by firm 6 is marked as T_6.

$p_i = c + \frac{t}{n}$ and payoffs $\pi_i = tMn^{-2} - k$ is a NE of this game, and it points to literature proving that, under mild technical conditions, this NE is unique.

Now consider a version of Hotelling's model in which all firms can freely adjust location $x_i \in [0,1)$ but automatically pick the same price $p_i = p = c + m$ with fixed profit margin $m > 0$. It is easy to see that the best response of firm i^* to a profile x_{-i^*} of other firms' locations is to move into the widest interval that they leave, that is, to choose $x_{-i^*} \in (x_j, x_{j+1})$, where $j = i$ if $i < i^*$ and $j = i+1$ if $i > i^*$. It follows that equal spacing, with territorial length $T_i = 1/n$, is the only NE of this location game. This well-known result is sometimes called the "principle of maximum differentiation." At NE, each firm's profit is $mMT_i - k = mM/n - k$.

How many firms can survive? If $n > B \equiv \frac{mM}{k}$, then all firms have negative profit and some will go extinct (or will exit, to use economists' blander jargon). If $n + 1 < B$ then there is the possibility of a viable new species (or entry, as economists would say). How might this come about?

Hotelling's original paper (and its successors) say nothing formal about the speciation process, but the intuition seems consistent with the following description. Incumbent firms move their locations randomly a bit (think of a stream of slight mutations) but on average remain at equilibrium locations. There are always potential entrants lurking around, probing random entrance points. They get a rough estimate of the profitability, and enter if it appears that it will be positive so that they can cover their fixed cost. Myopic firms will enter only when they probe an interval of length at least $2/B$, but farsighted firms that can tolerate losses until the incumbents reposition themselves will enter intervals of length at least $1/(B-1)$. The upshot is that entry will occur (i.e., a new species can find a viable niche among n existing species) when $B = \frac{mM}{k}$ is sufficiently large. How large? In the best case we need $B \geq n+1$ but even in the most pessimistic case $B \geq 2n$ suffices.

In this spatial competition model, the composite parameter $B = \frac{mM}{k}$ is the key. It predicts that speciation will occur as the resource base M or extraction efficiency m increases, or as the base maintenance cost k decreases. To illustrate, consider tablet computers. Apple Inc. introduced the Newton in 1993, but it was not viable—the market size M was too small given its level of performance and its marginal and fixed cost. However, over the next two decades the market became much larger while the fixed cost k of developing a new product shrank as Apple's supplier network and development teams matured. By 2010 Apple CEO Steve Jobs believed that there was a viable niche in the product spectrum between smartphones and laptop computers, and so he introduced the iPad. He was right; unit profits m from iPad sales quickly covered the fixed cost k. As M continued to increase and k to decrease while keeping a nice profit margin m, Apple found it profitable to introduce the iPad mini in 2012, filling a niche between the iPad and iPhone. More new products from Apple and its competitors will become viable as long as $B = \frac{mM}{k}$ continues to increase.

The Hotelling location model (price competition suppressed) is also a standby in political science, where it is known as the Downs model. There the trait space is interpreted as political position in the Left-Right spectrum, rather than in a product spectrum. We leave it to interested readers to re-interpret k, m, M in terms of emerging (or submerging) national identities, and to consider further applications.

14.6 DISCUSSION

This chapter touches on many themes, but its main argument can be summarized briefly. New biological species emerge and radiate in the presence of disruptive selection, and frequency dependence (or strategic interaction) is a major source of disruption. But disruption alone is not enough; there must also be separation of the gene pools, and here assortative mating can play a leading role.

Insights from biology inspire more general approaches that apply to human society and to virtual computer-based ecosystems. General patterns of splitting and merging can be drawn from mathematics and historical experience, key parameters can be identified, and these models can be taken to the data to provide sharper insights and useful predictions. We hope that we have inspired our readers to do so in ways beyond our own imagining.

EXERCISES

1. Analyze the fitness function $f(m,r) = r^2 - m^2 + m - r$. Does it have the usual normalization? Compute its selection gradient, find all critical point(s), and analyze their convergence stability and evolutionary stability. Does this fitness function have a CSS? A branch point?
2. Let $f(m,r)$ be an arbitrary fitness function, assumed smooth and normalized. Take the total derivative (with respect to r) twice of the normalization identity $f(r,r) = 0$ to verify that $f_{11}(r,r) + 2f_{12}(r,r) + f_{22}(r,r) = 0$ holds for all r. Use this expression to eliminate the cross-partial derivative in the CS inequality $f_{11}(r^*,r^*) + f_{12}(r^*,r^*) < 0$ for a critical point r^*, and conclude that an equivalent expression for CS is $f_{11}(r^*,r^*) < f_{22}(r^*,r^*)$. Now draw a diagram with $x = f_{11}(r^*,r^*)$ shown on the horizontal axis and $y = f_{22}(r^*,r^*)$ on the vertical axis. Bisect the four quadrants (using the lines $x+y=0$ and $x-y=0$) to obtain eight subquadrants. Finally, label each subquadrant according to its stability properties for critical points. Hint: For example, there are three subquadrants to the left of the vertical axis and above the main diagonal $x - y = 0$. The critical point of any fitness function whose pure second partial derivatives fall into any of these three subquadrants is a CSS, because it satisfies ES by virtue of being left of vertical axis and satisfies CS by lying above the diagonal.
3. Each subquadrant in the previous exercise has its own pairwise invasibility plot (PIP). Which two of these were shown in Figure 14.1? Choose one of the remaining six subquadrants and draw its PIP as follows. Draw a square and its main diagonal; the horizontal axis is r and the vertical axis is m, so

the main diagonal is the locus $m = r$, and near the middle of the diagonal is the critical point r^* in question. Draw another line to represent the locus $f_1(m,r) = 0$; use standard calculus (and the cross partial identity on the diagonal mentioned in the previous exercise) to convince yourself that this line goes through (r^*, r^*) and has slope there of $\frac{2f_{11}}{f_{11}+f_{22}}$, which has limited range for each subquadrant. Use a bit more calculus to convince yourself that this locus and the diagonal break up the square into four regions, in each of which either $f(m,r) > 0$ (in which case rare mutants can invade) or $f(m,r) < 0$ (in which case they can't invade). Shade and label the four regions respectively with a plus or minus, and voila, you have a PIP!

4. Do genetic modifiers have a memic analogue? Think of some major innovation, such as electric power generation or representative democracy, and consider what sort of minor innovations have gained fitness by improving the fitness of the major innovation.
5. This exercise includes details for the simple two-locus model for self-attraction mentioned in Section 14.3 with genotypes [Ra, RA, Pa, PA, Sa, SA], in a sexual haploid, where the upper case A refers to females with a particular value $a > 1$ of the self-preference parameter, while the lower-case are those with no (i.e., random) preference, $a = 1$. With six genotypes, genetic transmission is governed by a hexmatrix. Verify that, in the general case with a free parameter $r \in [0,1]$ for recombination, that hexmatrix is as shown below.

		Males:				
Fem:	**Ra**	**RA**	**Pa**	**PA**	**Sa**	**SA**
Ra	[1,0,0,0,0,0]	$[\frac{1}{2},\frac{1}{2},0,0,0,0]$	$[\frac{1}{2},0,\frac{1}{2},0,0,0]$	$[\frac{1-r}{2},\frac{r}{2},\frac{r}{2},\frac{1-r}{2},0,0]$	$[\frac{1}{2},0,0,0,\frac{1}{2},0]$	$[\frac{1-r}{2},0,0,\frac{r}{2},\frac{r}{2},\frac{1-r}{2}]$
RA	$[\frac{1}{2},\frac{1}{2},0,0,0,0]$	[0,1,0,0,0,0]	[„„,]	[„„,]	[„„,]	[„„,]
Pa	$[\frac{1}{2},0,\frac{1}{2},0,0,0]$	$[\frac{r}{2},\frac{1-r}{2},\frac{1-r}{2},\frac{r}{2},0,0]$	[„„,]	[„„,]	[„„,]	[„„,]
PA	$[\frac{1-r}{2},\frac{r}{2},\frac{r}{2},\frac{1-r}{2},0,0]$	$[0,\frac{1}{2},0,\frac{1}{2},0,0]$	[„„,]	[„„,]	[„„,]	[„„,]
Sa	$[\frac{1}{2},0,0,0,\frac{1}{2},0]$	$[\frac{1-r}{2},\frac{r}{2},0,0,\frac{r}{2},\frac{1-r}{2}]$	[„„,]	[„„,]	[„„,]	[„„,]
SA	$[\frac{1-r}{2},0,0,\frac{r}{2},\frac{r}{2},\frac{1-r}{2}]$	$[\frac{1}{2},0,0,0,0,\frac{1}{2}]$	[„„,]	[„„,]	[„„,]	[„„,]

Choose a value of $a > 1$, perhaps $a = 1.5$, and an initial recruit distribution of genotypes, perhaps (.25, .25, .13, .17, .1, .1). Assume that female preference parameters in males' genomes have no impact on male survival rates and that females' self-preference applies to RPS but not to Aa. Recycle the basic male selection matrix M used in the text to get the appropriate 6x6 selection matrix, and similarly write out the 6x6 attraction matrix. Then compute the unnormalized vectors of moms and dads, and use the hexmatrix (and normalization) to compute the progeny distribution assuming (a) $r = 0$, (b) $r = \frac{1}{2}$, and (c) $r = 1$.

6. If you enjoyed the previous exercise and are looking for more, repeat the computations for the simplified two-locus model with self-preference competing against rare-winner-son preference. Here the hexmatrix is as shown below.

		Males.				
Fem:	**RA**	**RB**	**PA**	**PB**	**SA**	**SB**
RA	[1,0,0,0,0,0]	$[\frac{1}{2},\frac{1}{2},0,0,0,0]$	$[\frac{1}{2},0,\frac{1}{2},0,0,0]$	$[\frac{1-r}{2},\frac{r}{2},\frac{r}{2},\frac{1-r}{2},0,0]$	$[\frac{1}{2},0,0,0,\frac{1}{2},0]$	$[\frac{1-r}{2},0,0,\frac{r}{2},\frac{r}{2},\frac{1-r}{2}]$
RB	$[\frac{1}{2},\frac{1}{2},0,0,0,0]$	[0,1,0,0,0,0]	[„„,]	[„„,]	[„„,]	[„„,]
PA	$[\frac{1}{2},0,\frac{1}{2},0,0,0]$	$[\frac{r}{2},\frac{1-r}{2},\frac{1-r}{2},\frac{r}{2},0,0]$	[„„,]	[„„,]	[„„,]	[„„,]
PB	$[\frac{1-r}{2},\frac{r}{2},\frac{r}{2},\frac{1-r}{2},0,0]$	$[0,\frac{1}{2},0,\frac{1}{2},0,0]$	[„„,]	[„„,]	[„„,]	[„„,]
SA	$[\frac{1}{2},0,0,0,\frac{1}{2},0]$	$[\frac{1-r}{2},\frac{r}{2},0,0,\frac{r}{2},\frac{1-r}{2}]$	[„„,]	[„„,]	[„„,]	[„„,]
SB	$[\frac{1-r}{2},0,0,\frac{r}{2},\frac{r}{2},\frac{1-r}{2}]$	$[\frac{1}{2},0,0,0,0,\frac{1}{2}]$	[„„,]	[„„,]	[„„,]	[„„,]

7. Recall the Hotelling circle model introduced in Section 14.5. Prove that the strategy profile $p_i^* = c + \frac{t}{n}$, $i = 1,...,n$ is indeed a NE of the game with exogenous equally spaced locations. Hint: Assume that prices are regular, that is, that $p_{j+1} - p_j < t$ for all j, so that territories are sharply defined. Show that, given neighbors' prices p_{i-1} and p_{i+1}, player i's territory has length $T_i(p) = \frac{1}{n} + \frac{0.5(p_{i-1}+p_{i+1})-p}{t}$ and that his payoff is $\pi_i = (p-c)M$ $T_i(p)$. Use this condition to compute the best-response function, and verify that p_i^* is indeed a mutual best response. Comment: The proof that this NE is unique is a bit tedious; a complete version can be found in Selten and Ostmann (2001).

NOTES

Sympatric speciation and polymorphism

Building on theory by Levene (1953), Maynard Smith (1966) showed that polymorphism might be a first step in speciation. These models of polymorphism were based on simple assumptions for ecological performance (two niche types) and mate choice (two preference alleles). In a modern version of sympatric speciation, frequency-dependent competition between similar ecological types can lead to disruptive selection and diversification. Rosenzweig (1978) framed the idea in terms of concepts of competitive speciation. Seger (1985) presented the first mathematical model showing that frequency-dependent competition for occupation of a niche continuum can induce sympatric speciation under certain conditions. West-Eberhard reviewed cases of morphs and speciation, and first pointed out (Ritchie 2007) that speciation may not merely arise from ecology per se, but might arise from social competition (West-Eberhard 1983). Later, Hochberg, Sinervo, and Brown (2003) developed a model explicitly couched in terms of social interactions, not ecological interactions per se, and showed that Hamiltonian interactions of altruism and selfishness can generate new species, reframing the requirement of ecological competition in terms of social competition.

While previous authors have covered the topic of competitive speciation driven by ecological niches, these previous efforts to explain speciation are not couched in a general framework of the role of frequency dependent strategies. West-Eberhard (1983) first suggested that social competition, not merely ecology per se, might generate speciation. However, a key step involved in the speciation process remains to be solved, the build-up of assortative mating for social phenotypes, which will ultimately generate hybrid unfitness between two new incipient species.

Simulation results from cellular automata speciation models (Hochberg, Sinervo, and Brown 2003) hint that this requires a key tipping point that can be expressed in terms of the relative contribution of competition among social strategies (playing a game within a species) to the overall strength of selection arising from mating within one's own type in a contact zone between species where unfit hybrids are common. When the former force predominates, the game of speciation can co-opt the loci for assortative mating to be used in the speciation process, which may even have been originally a component of the within-species social game. If the later force predominates, the species remains locked in the within-population game and does not speciate. Nevertheless a tension between these two forces can generate a balance (Lancaster et al. 2014), awaiting some other event that might generate speciation (e.g., geographic isolation followed by loss of a strategy type in one of the isolated populations).

Long run evolution and landscapes

Sewall Wright thought of the landscape as static, that is, he ignored strategic interdependence. Stuart Kauffman (1993) includes some strategic interdependence, albeit co-evolution across species (especially predator and prey) rather than general frequency dependence.

Friedman and Ostrov (2010, 2013) provide a mathematical development of the idea that the fitness landscape shifts as the relevant populations evolve. They work with continuous action spaces, as in adaptive dynamics, but with non-trivial distributions that evolve according to the landscape gradient. The mathematics involves partial differential equations beyond the scope of the present book.

Adaptive dynamics

Two minor technical qualifications to the statements in Section 14.2 may be worth mentioning. First, as usual, critical points are not quite the only possible candidates for steady states. If the range of traits has finite upper or lower endpoints, then these can also be long-run steady states. The adaptive dynamics literature usually neglects this possibility, and focuses on interior steady states. Second, the inequality in equation (14.3) can be weak and still have a downcrossing given appropriate higher order terms of the selection gradient. Following the adaptive dynamics literature, we also ignore this non-generic case.

The algebraic example is adapted from Cressman and Hofbauer (2005, p. 57). For an extensive bibliography on adaptive dynamics, comprising hundreds of articles and many recent applications, see http://mathstat.helsinki.fi/~kisdi/addyn.htm

Trimatrix computation

Some comments on the trimatrix computation in Section 14.3 may be helpful. Here we treat vectors as column vectors in matrix operations, but write them out as row vectors to conserve space. Recall that $\mathbf{e}^i$ is the ith basis vector, for example, $\mathbf{e}^2 = (0,1,0)$ so $\mathbf{e}^2 M\mathbf{s} = M_2$ is the fitness of strategy 2. Thus dividing $\mathbf{s}^M$ by the weighted mean (or sum) $\mathbf{s} \cdot M\mathbf{s}$ gives the normalized shares under discrete replicator dynamics. But there is no reason to normalize at this stage of the calculations since later steps would still require further normalizations; it seems cleaner to just normalize once, in the final step.

Material on Hawaii

is drawn from Friedman (2008), which in turn draws on Boone (1992, pp. 301–337). The former reference draws on numerous other sources for examples of cultural evolution.

Hotelling model

For simplicity we assumed that the profit margin m is independent of the number of firms n, but readers charmed by the earlier result can assume instead that $m = \frac{t}{n}$ for some fixed $t > 0$, and can check that the main qualitative conclusions below remain unchanged after appropriately altering expressions, for example, setting $B = \sqrt{\frac{m}{k}}$ instead of $\frac{mM}{k}$.

BIBLIOGRAPHY

Alonzo, Suzanne H., and Barry Sinervo. "Mate choice games, context-dependent good genes, and genetic cycles in the side-blotched lizard, *Uta stansburiana*." *Behavioral Ecology and Sociobiology* 49, no. 2–3 (2001): 176–186.

Bleay, Colin, and Barry Sinervo. "Discrete genetic variation in mate choice and a condition-dependent preference function in the side-blotched lizard: Implications for the formation and maintenance of coadapted gene complexes." *Behavioral Ecology* 18, no. 2 (2007): 304–310.

Boone, James L. "Competition, conflict, and the development of social hierarchies." In *Evolutionary ecology and human behavior*, E. A. Smith and B. Winterhalder eds., Aldine Transaction (1992).

Corl, Ammon, Alison R. Davis, Shawn R. Kuchta, and Barry Sinervo. "Selective loss of polymorphic mating types is associated with rapid phenotypic evolution during morphic speciation." *Proceedings of the National Academy of Sciences* 107, no. 9 (2010): 4254–4259.

Corl, Ammon, Lesley T. Lancaster, and Barry Sinervo. "Rapid formation of reproductive isolation between two populations of side-blotched lizards, *Uta stansburiana*." *Copeia* 2012, no. 4 (2012): 593–602.

Cressman, Ross, and Josef Hofbauer. "Measure dynamics on a one-dimensional continuous trait space: Theoretical foundations for adaptive dynamics." *Theoretical Population Biology* 67, no. 1 (2005): 47–59.

Darwin Charles, R. "On the origin of species by means of natural selection, or the preservation of favoured races in the struggle for life." (1859).

Eshel, Ilan. "Evolutionary and continuous stability." *Journal of Theoretical Biology* 103, no. 1 (1983): 99–111.

Friedman, Daniel. *Morals and markets: An evolutionary account of the modern world.* Macmillan, 2008.

Friedman, Daniel, and Daniel N. Ostrov. "Gradient dynamics in population games: Some basic results." *Journal of Mathematical Economics* 46, no. 5 (2010): 691–707.

Friedman, Daniel, and Daniel N. Ostrov. "Evolutionary dynamics over continuous action spaces for population games that arise from symmetric two-player games." *Journal of Economic Theory* 148, no. 2 (2013): 743–777.

Hochberg, Michael E., Barry Sinervo, and Sam P. Brown. "Socially mediated speciation." *Evolution* 57, no. 1 (2003): 154–158.

Hotelling, Harold. "Stability in competition." *The Economic Journal* 39, no. 153 (1929): 41–57.

Kauffman, Stuart A. *The origins of order: Self-organization and selection in evolution.* Oxford University Press, 1993.

Kirkpatrick, Mark. "Reinforcement and divergence under assortative mating." *Proceedings of the Royal Society of London. Series B: Biological Sciences* 267, no. 1453 (2000): 1649–1655.

Lancaster, Lesley T., Andrew G. McAdam, and Barry Sinervo. "Maternal adjustment of egg size organizes alternative escape behaviors, promoting adaptive phenotypic integration." *Evolution* 64, no. 6 (2010): 1607–1621.

Lancaster, Lesley T., Andrew G. McAdam, Christy A. Hipsley, and Barry R. Sinervo. "Frequency-dependent and correlational selection pressures have conflicting consequences for assortative mating in a color-polymorphic lizard, *Uta stansburiana*." *The American Naturalist* 184, no. 2 (2014): 188–197.

Lancaster, Lesley T., Christy A. Hipsley, and Barry Sinervo. "Female choice for optimal combinations of multiple male display traits increases offspring survival." *Behavioral Ecology* 20, no. 5 (2009): 993–999.

Lande, Russell. "Models of speciation by sexual selection on polygenic traits." *Proceedings of the National Academy of Sciences* 78, no. 6 (1981): 3721–3725.

Levene, Howard. "Genetic equilibrium when more than one ecological niche is available." *American Naturalist* 87 (1953): 331–333.
Maynard Smith, John. "Sympatric speciation." *American Naturalist* 100, no. 916 (1966): 637–650.
Mayr, Ernst. "How biology differs from the physical sciences." In *Evolution at a crossroads: The new biology and the new philosophy of science* (1985): 43–63.
Murphy, Robert F., and Yolanda Murphy. "Northern Shoshone and Bannock." *Handbook of North American Indians* 11 (1986): 284–307.
Ritchie, Michael G. "Sexual selection and speciation." *Annual Review of Ecology, Evolution, and Systematics* 38 (2007): 79–102.
Rosenzweig, Michael L. "Competitive speciation." *Biological Journal of the Linnean Society* 10, no. 3 (1978): 275–289.
Seger, J. M. "Interspecific resource competition as an ecological cause of speciation." *Evolution: Essays in honor of John Maynard Smith, P. J. Greenwood, P. H. Harvey and M. Slatkin*. CUP Archive (1985): 43–53.
Schluter, Dolph. *The ecology of adaptive radiation*. Oxford University Press, 2000.
Selten, Reinhard, and Axel Ostmann. "Imitation equilibrium." *Homo Oeconomicus* 43 (2001): 111–149.
Sinervo, Barry, Alexis Chaine, Jean Clobert, Ryan Calsbeek, Lisa Hazard, Lesley Lancaster, Andrew G. McAdam, Suzanne Alonzo, Gwynne Corrigan, and Michael E. Hochberg. "Self-recognition, color signals, and cycles of greenbeard mutualism and altruism." *Proceedings of the National Academy of Sciences* 103, no. 19 (2006): 7372–7377.
Sinervo, B., J. Clobert, D. B. Miles, A. McAdam, and L. T. Lancaster. "The role of pleiotropy vs signaller-receiver gene epistasis in life history trade-offs: Dissecting the genomic architecture of organismal design in social systems." *Heredity* 101, no. 3 (2008): 197–211.
West-Eberhard, Mary Jane. "Sexual selection, social competition, and speciation." *Quarterly Review of Biology* 58, no. 2 (1983): 155–183.
Wright, Sewall. "The roles of mutation, inbreeding, crossbreeding, and selection in evolution." *Proceedings of the Sixth International Congress on Genetics* (1932).

GLOSSARY

Section numbers (in parentheses) indicate locations of more complete explanations. For example, (14.1) refers the reader to Section 1 of Chapter 14.

allele	one alternative copy of a gene in either haploids or diploids (1.10)
allopatry	populations or species separated in space (14.1)
altruism	behavior that decreases own payoff (or fitness) in order to increase others' payoff (5.6, 7.4)
anti-apostatic selection	positive frequency dependence (7.1)
aposematic	a signaling strategy in which prey develop a toxin or other adaptation that makes predation undesirable, and a conspicuous signal of undesirability (7.5)
apostatic selection	negative frequency dependence (7.1)
basin of attraction	of a stable steady state is the open set of initial conditions from which the trajectory converges to that steady state (1.9, 3.9)
Batesian mimic	a false signaling strategy in which prey save the cost of carrying the toxin but still display the conspicuous signal of undesirability (7.5)
belief learning	an adaptive process in which an individual updates beliefs as she accumulates experience, and updates actions accordingly (8.1)
best response	or best reply is a feasible action that maximizes payoff (or fitness) given the current population state (4.3)
bet hedging	employing a randomized (mixed) strategy that includes best (or at least good) responses to all likely states (9.0)

bilateral reciprocity	conditional cooperation between two individuals: each behaves altruistically towards the other as long as the other also behaves altruistically (13.3)
branch point	in adaptive dynamics, an equilibrium that is ecologically stable (ES) but not convergence stable (CS), and therefore is associated with speciation (14.2)
centroid	center of mass in n-simplex, for example, $(\frac{1}{3}, \frac{1}{3}, \frac{1}{3})$ when $n = 3$ (7.2, 7.8)
choke-price	highest price with positive demand (12.2)
chromosomes	very long sequences of many genes coded by DNA with special parts for replication (centromere) and end caps (telomeres) (5.3, 7.3)
comparative statics	sensitivity analysis of equilibrium points to parameter values (4.1)
continuous replicator dynamics	ordinary differential equations in which each strategy share's growth rate is proportional to its fitness relative to the population average (1.8)
continuously stable strategy (CSS)	in adaptive dynamics, an equilibrium that is both ecologically stable (ES) and convergence stable (CS) (14.2)
convergence stable (CS)	downcrossing of the selection gradient (14.2)
cost margin	choke-price minus marginal cost (12.3)
crossbreeding	or interbreeding: when parents belong to different races or species (14.5)
crypsis	prey strategy of being difficult to detect, for example, via camouflage (7.3, 14.2)
demography	specifies a population's age and sex structure and resultant population growth rate (9.3)
despotic	aggressive strategy that uses violence or threat of violence to acquire resources (7.1)
deterministic	dynamics in which current state and environment determine future states; in contrast to stochastic dynamics (1.6, 3.2, 3.9)
differential inclusions	generalize ordinary differential equations by specifying that the tangent vector belongs to a given set rather than that it is equal to a given expression (4.7)
diploid transmission	process involving two copies of genes from each of two parents (1.10)
discrete replicator dynamics	difference equations in which each strategy share's growth rate is proportional to its fitness normalized by its population average (1.8, 4.7)

dominant strategy	strategy that is a best response to every state (2.3)
downcrossing	root of an equation where the derivative is negative; often implies dynamic stability (2.3, 14.2)
ecologically stable (ES)	in adaptive dynamics for a continuous trait, refers to a critical point of the fitness function where the second-order condition holds indicating short-run dynamic stability (14.2)
eigenvalues	roots of the characteristic function for a matrix (4.7)
eigenvectors	vectors associated with eigenvalues (4.7)
eusociality	a caste system where queens do all the reproduction and workers collect all food, take care of young and defend the colony (13.2)
ESS	evolutionarily stable state: equilibrium point satisfying certain inequalities that often imply dynamic stability (2.1, 4.1, 14.2)
externality	costs borne or benefits received by bystanders (11.4, 12.1)
fecundity	reproduction rate of adult organisms (3.1, 3.3, 7.3)
free rider	player who enjoys a public good but doesn't contribute to its production (13.4)
frequency contingent	a plastic strategy that selects a specific action based on cues involving the current action frequencies (9.1)
frequency dependent	selection fitnesses are a function of the current strategy shares (14.2)
genotype	of an individual specifies which allele it carries in each relevant locus (1.10, 3.9)
globally stable	steady state whose basin of attraction is the entire interior of the state space (1.9)
grim trigger	repeated game strategy in which the player cooperates until another player defects, and thereafter the given player always defects (13.3)
haplodiploid	species (including ants and bees) in which unfertilized eggs develop into haploid males and fertilized eggs develop into diploid females (13.2)
haploid asexual	transmission process in which reproduction occurs by direct replication from a single parent (1.10)
haploid sexual	transmission process in which two haploid parents (one copy of each gene) each contribute an allele to produce diploid progeny, which generates haploid daughters by meiosis (1.10)
homozygous state	condition in which both alleles in a diploid organism are the identical (13.2)

inclusive fitness	includes fitness impact on genetically related other individuals (13.2, 13.11)
inheritance term	in Price equation, captures "all else" besides selection (5.4)
isopods	marine crustaceans (7.3)
juvenile plumage	a strategy in which an adult male morphs mimics juveniles (7.3)
kin selection	evolution when behavior affects genetically related other individuals (13.2, 13.9)
laissez-faire	a symmetric game with a dominant strategy that maximizes mean fitness (2.3)
locally stable	equilibrium has a non-empty basin of attraction (1.9, 4.7)
locus	the physical location of a gene on the chromosome (3.4)
Lyapunov stable	equilibrium whose nearby trajectories remain nearby; includes neutrally stable equilibria whose basin of attraction is empty (1.9, 4.7)
macroevolution	major evolutionary transitions such as those that produced multicellular organisms (13.5)
Markov property	or stationarity: update depends on the current state only, not on previous history (6.1)
MLE	maximum likelihood estimation: statistical procedure that obtains parameters which maximize the likelihood of the observed data (3.2)
mean matching	in population games, a protocol that averages each player's payoff over all possible matches (8.2)
mechanism	or format or institution: rules that determine players' payoffs (e.g., in markets), how transactions are determined from bids (8.1)
memes	behaviors whose shares evolve via imitation or other forms of cultural (non-genetic) transmission and selection (1.5)
microevolution	share dynamics with fixed set of alternative genes or memes (13.5)
mimic	a disguise strategy, intended to resemble another ("model") strategy (7.3–7.5)
monogamy	long-term exclusive match between male and female sexual partners (13.2)
monomorphic	at any given time a single strategy is used by 100% of the relevant population (14.2)
monotone dynamic	evolutionary process in which changes (or growth rates) of shares have the same rank order as their fitnesses (3.9, 4.1, 4.6)

Moore neighborhood	in a cellular array, the cells that share a corner or edge with a given cell (6.1)
moral code	shared understanding of proper behavior (13.6)
morphs	alternative phenotypes (i.e., visibly or behaviorally different alternatives) (2.4)
multiplicative updates	machine learning analogue of discrete replicator dynamic (10.1)
mutualism	behavior in which a player helps others at the same time that she helps herself (5.6, 13.1)
naïve	predator strategy of ignoring prey's conspicuous apsematic signal (7.5)
Nash equilibrium	state at which no player can gain by changing her strategy (4.1, 4.2)
negative definite	a matrix M such that the quadratic form $vM \cdot v < 0$ for all $v \neq 0$ (4.3)
negative frequency dependence	biological term for decreasing returns; payoff decreases in own share (2.1)
non-atomic	a distribution with no clumps (i.e., its cumulative function is continuous) (11.1)
norms	elements of a moral code (13.6)
n-simplex	the set of n-vectors whose components are non-negative and sum to 1 (1.4)
PLE	perfect lifetime equilibrium: each player always maximizes the sum of current and expected discounted future payoff (9.5)
phenotype	in contrast to genotype, specifies only observable behavior or appearance (3.2, 7.1, 7.3)
plastic	strategy responsive to cues about current circumstances, can switch between two or more actions (9.0, 9.5)
pollution haven	hypothesis that the nation with the lower tax rate on pollution produces the polluting good (12.5)
polygynous	male reproductive strategy to seek out multiple female partners (13.2)
positive frequency dependence	biological term for increasing returns; payoff increases in own share (2.1, 2.3)
preferences	characterize (usually via a utility function) flexible behavior responsive to varying costs and benefits (13.5, 13.7)
price of anarchy	the ratio of social optimum welfare to Nash equilibrium welfare (11.2, 12.2)

prisoner's dilemma	symmetric game with a dominant strategy that minimizes mean fitness (2.3)
random pairwise	in contrast to mean matching, the protocol assigns the payoff a player receives from the match with one randomly chosen opponent (8.6)
reciprocal altruism	synonym for bilateral reciprocity (13.3, 13.11)
recombination	a genetic process that can transmit alleles from different parents at two loci on the same autosomal (non-sex) chromosome (5.3)
reinforcement learning	in contrast to belief learning, players directly increase the future probability of using the current strategy to the extent that it yields a high payoff (8.8)
resident	for Ruff shorebirds, a morph with dark plumage that defends territory with vigor (7.3); in adaptive dynamics, the current monomorphic strategy (14.2)
responsive	in contrast to naïve, the predator strategy of avoiding prey with conspicuous signal of toxicity (7.5)
rigid strategy	in contrast to plastic, a strategy not contingent on cues (9.0)
role contingent	a plastic strategy responsive to cues regarding current role, for example, resource owner or intruder (9.1)
saddle	a steady state to which some trajectories converge but others diverge (1.9)
satellite	light-colored morph that defends territory, but lurks close to Residents (7.3)
segregation	in an organism with multiple chromosomes, the random process by which chromosomes derived from each parent are distributed to a given progeny (1.10)
selection gradient	the derivative of the fitness function with respect to the prevailing trait (14.2)
self-preference	tendency of females to mate with genetically similar males, which are not necessarily related (14.2)
sensitivity analysis	synonym for comparative statics (4.1)
sink	a dynamically stable steady state (1.9)
smooth	continuously differentiable, for example, no jumps or kinks (1.6)
social dilemma	a game with tension between individual rationality and social efficiency (2.3)
socially efficient	a state that maximizes the mean payoff in every population (13.1)

source	a steady state from which all trajectories diverge (1.9)
stable limit cycle	a closed loop towards which nearby trajectories spiral (3.7)
state	the vector of shares in each population, describing the evolving system at a moment in time (1.4)
state contingent	plastic strategy responsive to the current state as a cue (9.1)
state space	the set of all possible states (2.6, 3.5)
steady state	a state that remains constant under evolutionary dynamics (1.9)
stochastic	random, as opposed to deterministic (3.2, 3.9, 9.8)
sympatry	multiple species diverge from a common ancestor while remaining in geographic proximity (14.0)
Thévenin equivalent	a simple two-port traffic network with the same resistance and voltage drop as the more complex given network (11.6)
trace	the sum of the diagonal elements of a matrix (4.7)
traffic assignment	the state of a traffic game; specifies a feasible flow of traffic through a network (11.1)
torus	a doughnut-shaped surface generated from a stretchable rectangle by gluing together the top and bottom edges, and also gluing together the left and right edges (6.1)
upcrossing	root of an equation where the derivative is positive; often implies dynamic instability (2.3, 8.1, 14.2)
Von Neumann	neighborhood in cellular array consisting of cells that share an edge (but not just a corner) with the given cell (6.1)
Wardrop equilibrium	Nash equilibrium in a non-atomic traffic game: no player can improve his payoff by switching strategies (routes), and that the payoff (delay) of a route is unchanged by a single player changing strategies (11.1)
zero-flow zero-price	equilibrium satisfies the intuitive restriction that if the flow through a link is zero, its price must be zero (11.5)

MATHEMATICAL NOTATION

iff	if and only if
$\dot{D}$	the time derivative dD/dt of quantity D
Σ	summation
E	expectation operator, e.g., $E_{\mathbf{p}}\mathbf{x} = \sum x_i p_i$ if outcomes $\mathbf{x}$ have probability $\mathbf{p}$.
$Pr[e]$	probability of event e
$\mathbf{m}^{\mathbf{T}}$	transpose of matrix $\mathbf{m}$. For example, if $\mathbf{m} = \begin{pmatrix} \mathrm{a} & \mathrm{c} \\ \mathrm{b} & \mathrm{d} \end{pmatrix}$, then $\mathbf{m}^{\mathbf{T}} = \begin{pmatrix} \mathrm{a} & \mathrm{b} \\ \mathrm{c} & \mathrm{d} \end{pmatrix}$
$\mathbf{v} \cdot \mathbf{u} = \mathbf{v}\mathbf{u}^{\mathbf{T}}$	dot product $\sum_{i=1}^{n} v_i u_i$ of row vectors $\mathbf{v}$, $\mathbf{u}$
λ	eigenvalue: solution of the characteristic equation $\lvert\mathbf{J} - \lambda\mathbf{I}\rvert = 0$
$\mathbf{B} \circ \mathbf{A}$	Hadamard product of matrices, defined by entry-by-entry multiplication
$s_i = \frac{N_i}{N}$	fraction (or share) of given population using strategy (or trait) i
$\mathbf{s}$	the state vector $(s_1, \ldots, s_n)$
$\mathcal{S}$	the state space, consisting of row vectors whose components s_i are non-negative and sum to 1
N_i'	the number of progeny of i-strategists
w_i, W_i	fitness: ability of a strategy (or trait) to increase its share; typically $W_i = \frac{N_i'}{N_i}$. Upper (lower) case indicates discrete- (continuous-) time applications. In two population application u_i and v_i are also used to denote fitness.
w_{ij}	fitness impact of strategy j on strategy i.
$w_i = \sum_{j=1}^{n} w_{ij} s_j$	fitness of strategy i given payoff matrix $\mathbf{w} = ((w_{ij}))$
$\overline{w} = \mathbf{s}\mathbf{w} \cdot \mathbf{s}$	is mean fitness $\sum_{i=1}^{n} \sum_{j=1}^{n} w_{ij} s_i s_j$ in the population

$\frac{W_i}{\overline{W}}$	relative fitness = discrete time growth factor under replicator dynamics
Δw_{i-j}	fitness advantage or fitness difference $w_i(\mathbf{s}) - w_j(\mathbf{s})$ of strategy i over strategy j
$\bullet, \odot, \otimes$	respectively indicate sinks (stable), neutrally stable, and saddle equilibria
$v, c > 0$	in Hawk-Dove game, the value of the resource and cost of losing a battle
$A(\mathbf{s})$	the active (non-extinct) strategies at state $\mathbf{s}$, so $A(\mathbf{s}) = \{i : s_i > 0\}$
$a(x)$	Bergstrom assortativity index
δ	discount factor

Note: Vectors and matrices are shown in bold font in Chapters 1–3; later chapters use bold only when it helps avoid confusion.

INDEX